Zinc in Biology

Molecular Structures, Cellular Processes and Living Systems

Coordination Chemistry Perspectives

For a list of titles in this series see: rsc.li/ccp

How to obtain future titles on publication:
A standing order plan is available for this series. A standing order will bring delivery of each new volume immediately on publication.

For further information please contact:
Book Sales Department, Royal Society of Chemistry, Thomas Graham House, Science Park, Milton Road, Cambridge, CB4 0WF, UK
Telephone: +44 (0)1223 420066, Fax: +44 (0)1223 420247
Email: booksales@rsc.org
Visit our website at books.rsc.org

Zinc in Biology
Molecular Structures, Cellular Processes and Living Systems

Edited by

Hiroyuki Yasui
Kyoto Pharmaceutical University, Japan
Email: yasui@mb.kyoto-phu.ac.jp

and

Taiho Kambe
Kyoto University, Japan
Email: kambe.taiho.7z@kyoto-u.ac.jp

ROYAL SOCIETY
OF **CHEMISTRY**

Coordination Chemistry Perspectives No. 2

Hardback ISBN: 978-1-83767-456-5
EPUB ISBN: 978-1-83767-659-0
PDF ISBN: 978-1-83767-658-3
Print ISSN: 3029-0589
Electronic ISSN: 3029-0597

A catalogue record for this book is available from the British Library

The Royal Society of Chemistry is a charity, registered in England and Wales, Number 207890, and a company incorporated in England by Royal Charter (Registered No. RC000524), registered office: Burlington House, Piccadilly, London W1J 0BA, UK, Telephone: +44 (0)20 7437 8656.

For further information see our website at www.rsc.org

For general enquiries, please contact books@rsc.org

For EU product safety enquiries, please email books@rsc.org or contact Royal Society of Chemistry Worldwide (Germany) GmbH, Römischer Hof, Unter den Linden 10, 10117 Berlin.

Printed in the United Kingdom by CPI Group (UK) Ltd, Croydon, CR0 4YY, UK

Preface

In the second volume of *Coordination Chemistry Perspectives* titled **Zinc in Biology: Molecular Structures, Cellular Processes and Living Systems**, particular emphasis is placed on the multifaceted role of zinc (Zn) within biological systems. Zn, the second most abundant trace metal in the human body, is indispensable to all living organisms due to its critical involvement in a multitude of physiological and biochemical processes. Its versatility stems from its ability to act as a catalytic agent in diverse enzymatic reactions, provide structural support to an array of proteins such as Zn finger proteins, and function as a signalling molecule in complex cellular signal transduction pathways. Zn deficiency poses significant health risks, manifesting as growth retardation, dermatitis, alopecia, testicular disorders, impaired gonadal function, and a compromised immune response. Furthermore, disruptions in Zn homeostasis have been increasingly linked to the pathogenesis of various chronic and degenerative diseases, including cancer, neurodegenerative disorders, and metabolic syndromes such as diabetes. These findings highlight the importance of maintaining tightly regulated systemic and cellular Zn homeostasis. Processes such as absorption, sensing, storage, transport, and utilisation collectively ensure that Zn concentrations are optimised for vital cellular functions, including cell proliferation, differentiation, gene expression, and intricate regulation of signal transduction pathways.

This volume begins by discussing Zn-containing small molecules, including those developed for delivering Zn into cells, imaging, and monitoring of cellular Zn dynamics. This discussion bridges the disciplines of biological inorganic chemistry, biochemistry, and cell biology and offers a comprehensive perspective on the molecular roles of Zn. The text then shifts to the significant advancements in understanding of the cellular functions of Zn, emphasising its regulatory influence on gene expression, protein homeostasis, cell differentiation, and enzyme activation. Expanding beyond

Coordination Chemistry Perspectives No. 2
Zinc in Biology: Molecular Structures, Cellular Processes and Living Systems
Edited by Hiroyuki Yasui and Taiho Kambe

Published by the Royal Society of Chemistry, www.rsc.org

the cellular context, this volume explores the systemic implications of Zn biology. Recent discoveries regarding Zn-related diseases such as diabetes, cancer, neurodegenerative diseases, and inflammatory conditions, including inflammatory bowel disease (IBD), have been meticulously reviewed. These sections offer insights into the molecular mechanisms underlying the contribution of Zn to potential therapeutic strategies. By highlighting the growing intersection of Zn research and clinical medicine, this volume aims to foster a deeper understanding of Zn-related disorders and stimulate interest in the development of innovative Zn-based diagnostics and therapeutics.

The editors express their sincere hope that this work will inspire interest in Zn biology among readers, encouraging them to pursue further exploration and contribute to the expanding body of knowledge in this essential field of study. By bringing together cutting-edge research and diverse perspectives, this volume aspires to inspire the next generation of scientists to unlock the full potential of Zn in improving human health.

Taiho Kambe, PhD
Hiroyuki Yasui, PhD

Contents

Section I: Recent Progress in Zinc Complexes and Zinc Probes for Imaging

Coordination Chemistry Perspectives No. 2
Zinc in Biology: Molecular Structures, Cellular Processes and Living Systems
Edited by Hiroyuki Yasui and Taiho Kambe

Published by the Royal Society of Chemistry, www.rsc.org

Section II: Regulation of Systemic and Cellular Zinc Homeostasis and Its Importance in Cellular Functions

Section I: Recent Progress in Zinc Complexes and Zinc Probes for Imaging

CHAPTER 1

Quantitative Imaging of Labile Zinc in Intracellular Organelles Using Fluorescent Probes

RONG LIU, TOSHIYUKI KOWADA AND SHIN MIZUKAMI*

Institute of Multidisciplinary Research for Advanced Materials, Tohoku University, 2-1-1 Katahira, Aoba-ku, Sendai, Miyagi 980-8577, Japan
*Email: shin.mizukami@tohoku.ac.jp

1.1 Introduction

Zn^{2+} is the second most abundant transition metal ion in the human body and is generally considered to play three major physiological roles in living cells: as an enzyme cofactor, a modulator of protein structure, and a mediator of intracellular signaling.[1,2] The first two roles have long been the focus of biochemical studies on metalloproteins. Most of the total intracellular Zn^{2+} concentration, estimated to be 200–300 μM, is believed to be bound to proteins and to fulfil these roles. For instance, approximately 3000 proteins, accounting for about 10% of the human proteome, are known to contain zinc finger domains. In contrast, "labile" or "free" Zn^{2+}, which loosely binds to proteins or other biomolecules, has garnered significant attention as a mediator of intracellular signaling, particularly in immune cells and breast cancer cells.[3–5] These labile Zn^{2+} ions are characterized by their ready accessibility to chelating agents such as *N*,*N*,*N*′,*N*′-tetrakis(2-pyridylmethyl)-ethylenediamine (TPEN).[6]

The concentration of labile Zn^{2+} ($[Zn^{2+}]$) within cells is regulated by metallothioneins and two families of Zn^{2+} transporters: ZIPs (Zrt- and

Coordination Chemistry Perspectives No. 2
Zinc in Biology: Molecular Structures, Cellular Processes and Living Systems
Edited by Hiroyuki Yasui and Taiho Kambe

Published by the Royal Society of Chemistry, www.rsc.org

Irt-related proteins, SLC39s) and ZnTs (Zinc Transporters, SLC30s).[7] To date, more than 20 types of Zn^{2+} transporters have been identified in mammals, with their expression locations varying by subtype. The distribution and homeostasis of Zn^{2+} through zinc transporters and buffer proteins suggest various physiological roles for Zn^{2+} at the organellar level. Understanding the properties and functions of Zn^{2+} transporters, Zn^{2+}-binding proteins, and Zn^{2+} itself requires determining $[Zn^{2+}]$ at the organellar level, both at steady state and during induced fluxes.

A suite of analytical techniques capable of detecting metal ions with high sensitivity and sub-cellular spatial resolution—such as X-ray fluorescence microscopy (XRFM) and electron spectroscopic imaging (ESI)—can quantify the total amount of metal ions but cannot distinguish between 'labile' and 'protein-bound' metal ions.[8] In contrast, fluorescence imaging is one of the most promising methods for studying labile Zn^{2+} in living cells due to its high sensitivity, high spatiotemporal resolution, and low invasiveness.[2] Considerable effort has been devoted to developing fluorescent probes for studying labile Zn^{2+} in living cells, especially within subcellular compartments. Therefore, this chapter first provides an overview of fluorescent probes developed for labile Zn^{2+} detection—small-molecule (SM) probes and fluorescent protein-based (FP) probes—and then introduces a quantitative imaging technique based on small-molecule–protein hybrid probes developed by the authors.

1.2 Important Properties of Fluorescent Probes for Quantitatively Imaging Labile Zn^{2+} in Living Cellular Compartments

Fluorescent probes must possess several important properties to quantitatively image $[Zn^{2+}]$ in living cellular compartments. These include the ability to target specific organelles, a large dynamic range (DR) of signals, suitable affinity for Zn^{2+}, high selectivity under physiological conditions, and low sensitivity to various subcellular environments such as pH and oxidative stress. Here, we focus primarily on two of the most crucial factors for accurately measuring $[Zn^{2+}]$: affinity and DR.

Affinity is represented by the dissociation constant (K_d), which equals the ion concentration when 50% of the probe is bound. The K_d is determined by fitting a curve to experimental data obtained from titration experiments, typically by measuring the probe's fluorescence at different ion concentrations. Similar to pH titration, the fluorescence intensity of a probe changes most significantly when the ion concentration is near the K_d (Figure 1.1). Therefore, it is widely accepted that a probe is sensitive to the target ion within a specific range, typically from $0.1 \times K_d$ to $10 \times K_d$. Outside this range ($< 0.1 \times K_d$ and $> 10 \times K_d$) there is a higher likelihood of increased noise, necessitating the use of a probe with suitable affinity for accurate quantitative imaging.

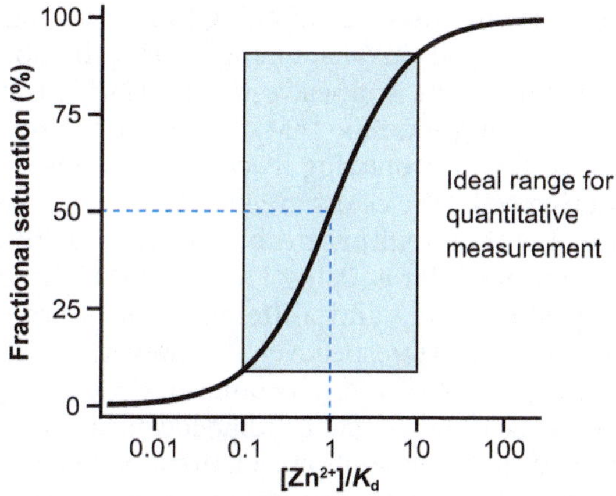

Figure 1.1 Relationship between [Zn^{2+}] and the fractional saturation value of the probe. When [Zn^{2+}] is equal to K_d, 50% of the population of the probe is bound to Zn^{2+}.

On the other hand, DR is defined as F_{max}/F_{min} or R_{max}/R_{min}, where F represents the fluorescence intensity and R represents the fluorescence ratio. The suffixes 'max' and 'min' indicate the maximum value under Zn^{2+}-saturated conditions and the minimum value under Zn^{2+}-depleted conditions, respectively. A larger DR indicates that the difference in fluorescence signals caused by the target ions becomes more pronounced relative to noise, thereby enhancing the sensitivity of the probe. The DR of a probe, especially within cells, does not necessarily remain consistent across different measurement systems and is susceptible to influences from both the microscopy system and the subcellular environment. Nevertheless, the DRs of ratiometric FP probes are generally smaller than those of SM probes.

1.3 Quantitative Imaging Using Fluorescent Probes

Fluorescence imaging plays a crucial role in studying zinc biology. However, relatively few research groups have focused on the quantitative imaging of intracellular labile Zn^{2+}. Although numerous fluorescent probes for Zn^{2+} have been reported, those with the necessary properties to accurately determine [Zn^{2+}] and its fluctuations at the organellar level remain limited. In this chapter, we highlight recent advances in developing fluorescent probes for the quantitative imaging of intracellular Zn^{2+} and discuss the characteristics of these probes used in quantification studies.

1.3.1 Small-molecule (SM) Probes

Development of SM probes for Zn^{2+} detection began in the 1980s, and they have been applied to various biological Zn^{2+} studies due to their ease of use and the fine tunability of their photophysical and coordination properties.

While SM probes are often used for imaging $[Zn^{2+}]$ throughout entire cells, some have been employed in visualizing $[Zn^{2+}]$ in specific subcellular compartments. In numerous studies, cytosolic $[Zn^{2+}]$ ($[Zn^{2+}]_{cyto}$) imaging has been performed using FluoZin-3AM. This probe is designed to possess membrane permeability by protecting its carboxy groups with acetoxymethyl esters. Upon entering cells, the esters are hydrolyzed by esterases, converting the probe to its membrane-impermeable form, FluoZin-3 ($K_d = 15$ nM), which is retained intracellularly. Using FluoZin-3AM, the $[Zn^{2+}]$ in the cells during resting, proliferation, and differentiation were estimated to be 614 pM, 784 pM, and 1.25 nM, respectively.[9] However, some studies employed FluoZin-3AM for imaging $[Zn^{2+}]$ in lysosomes and Golgi apparatus,[10,11] indicating uncertainty regarding the subcellular localization of such SM probes even if they exhibited high Pearson's correlation coefficient (PCC) values with organelle marker dyes. Their diffusion or accumulation in specific organelles could lead to false signal changes in quantitative precision.

To visualize $[Zn^{2+}]$ within specific subcellular compartments with greater precision, integrating targeting groups into SM probes is an effective strategy. Figure 1.2 illustrates examples of various organelle-targeting SM Zn^{2+} probes. Mitochondria are characterized by the two lipid bilayers and the crucial membrane potential that indicates their activity. Lipophilic cations, such as triphenylphosphonium (TPP), have been used to target SM probes to mitochondria. For example, Cho *et al.* reported a series of two-photon probes containing a TPP moiety. One example is SZn-Mito,[12] which exhibits a two-photon absorption cross-section of 75 GM at 760 nm and emission at 493 nm ($\Phi = 0.92$)

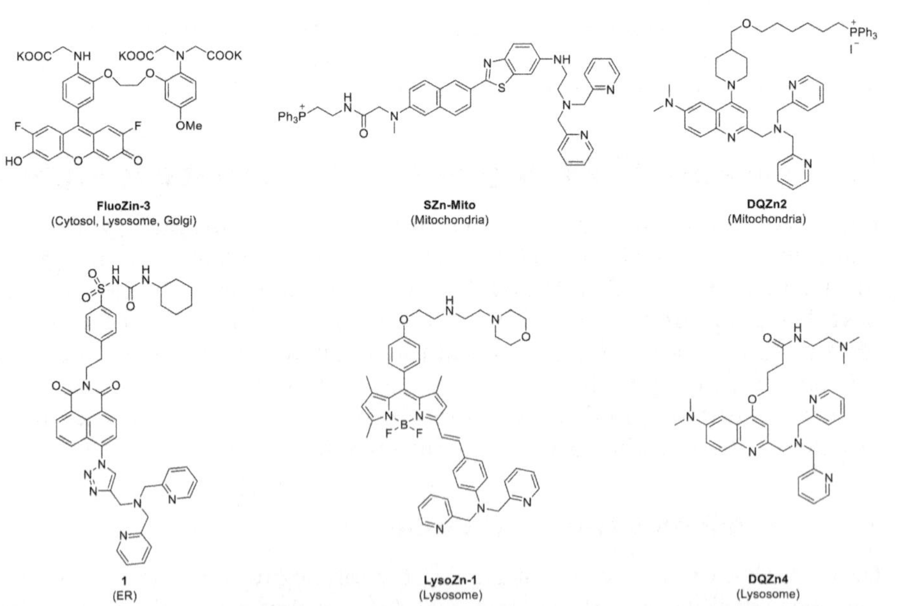

Figure 1.2 Structures of organelle-targetable SM probes.

in the presence of Zn^{2+} at pH 7.2. The K_d of SZn-Mito was determined to be 3.1 nM for both one-photon and two-photon excitation. Two-photon imaging revealed that SZn-Mito localizes well to mitochondria with a high PCC (0.89) with MitoTracker Red FM. The intracellular fluorescence signal increased upon the addition of Zn^{2+} pyrithione (ZPT) and decreased with the addition of TPEN. However, quantification of mitochondrial $[Zn^{2+}]$ ($[Zn^{2+}]_{mito}$) was not performed.

Jiang *et al.* developed a ratiometric sensor, DQZn2, which incorporates TPP as a mitochondrial targeting moiety, and used it to quantify $[Zn^{2+}]_{mito}$ at steady state.[13] Under neutral or weakly basic conditions, the quinoline nitrogen is protonated cooperatively with the Zn^{2+} ligand, *N*,*N*-di(2-pyridylmethyl)amine (DPA), forming a diaminoquinoline resonance structure with effective intramolecular charge transfer (ICT) from the electron-donating amino group at the 4-position of quinoline. Upon Zn^{2+} binding, the probe exhibited a shift in the emission peak from 550 nm ($\Phi = 0.11$) to 504 nm ($\Phi = 0.22$). The fluorescence intensity ratio at 504 nm and 550 nm (F_{504}/F_{550}) changed from 0.39 to 1.94, and the K_d was determined to be 0.45 nM at pH 7.4 and 0.15 nM at pH 8.0. The fluorescence ratio values remained nearly constant over the pH range from 5 to 8 in the presence or absence of Zn^{2+}. The mitochondrial localization of DQZn2 was confirmed in NIH3T3 cells (PCC = 0.86 with MitoTracker Deep Red). By acquiring fluorescence microscopy images at 430–490 nm (green) and 520–580 nm (red) upon excitation at 405 nm, the fluorescence ratio (F_{Green}/F_{Red}) of DQZn2 in the steady-state cells was calculated to be 1.17. This value changed to 1.59 upon the addition of $ZnSO_4$ (25 μM) and pyrithione (10 μM), and to 0.79 upon TPEN (10 μM) addition. The steady-state $[Zn^{2+}]_{mito}$ of the NIH3T3 cells was estimated to be 72 pM using the K_d (77 pM), which was determined in cells.

To target the endoplasmic reticulum (ER), sulfonylurea and sulfonamide moieties, which can be recognized by ATP-dependent K^+ channels on the ER membrane, have often been incorporated into SM probes.[14] Watkinson *et al.* utilized two sequential azide–alkyne click reactions to introduce glibenclamide, an ER-resident ATP-sensitive K^+ channel inhibitor, and a Zn^{2+} chelator into a naphthalimide fluorophore (compound **1** in Figure 1.2).[15] The fluorescence of the naphthalimide was quenched by photoinduced electron transfer (PeT) from the Zn^{2+} chelator and exhibited approximately a 16-fold fluorescence enhancement upon Zn^{2+} binding, with a maximum emission at 414 nm. The K_d of compound **1** was estimated to be 3.5 nM. The pH sensitivity of compound **1** was also tested, yielding $pK_{a1} = 0.81$, $pK_{a2} = 4.3$, $pK_{a3} = 6.8$, and $pK_{a4} = 11.6$, indicating that this probe is pH-sensitive within the physiological pH range. The probe demonstrated good targetability to the ER in various cell lines, including HeLa, MCF-7, and HepG2. The addition of ZPT significantly increased the fluorescence intensity of the probe, while the addition of TPEN almost completely quenched the fluorescence.

For lysosome targeting, probes with moieties such as morpholine and dimethylaminoethyl groups are used. Once these probes permeate into the lysosomes, the protonation of the amino group increases the hydrophilicity of the probes, promoting the accumulation inside the lysosomes. Using a dimethylaminoethyl group, Jiang *et al.* developed a lysosome-targeting ratiometric probe, DQZn4 ($K_d = 16$ nM at pH 5.2).[16] Similar to DQZn2, DQZn4 exhibited an emission shift from 542 nm ($\Phi = 0.11$) to 495 nm ($\Phi = 0.17$) upon Zn^{2+} binding. The probe could be easily localized in the lysosomes, and the fluorescence ratio (F_{Green}/F_{Red}) increased after the addition of $ZnSO_4$. Although the ratio decreased after the addition of TPEN, it was still larger than the baseline level. Thus, a calibration method for quantifying [Zn^{2+}] in lysosomes has not been sufficiently established.

When performing quantitative Zn^{2+} imaging using SM probes targeting organelles, several issues need to be considered. For example, the addition of ZPT or TPEN during the calibration process may cause the probe diffusion. In addition, if the probe exhibits significant changes in its fluorescence properties or dissociation constants for Zn^{2+} in response to pH changes within the physiological pH range, the quantitative results should be carefully interpreted.

1.3.2 Fluorescent Protein (FP)-based Probes

FP-based metal ion probes have achieved great success in intracellular Ca^{2+} imaging. One major advantage of FP probes is their ability to be targeted to various intracellular locations by fusing them with signal peptides. FP probes often consist of one or more FP fused to natural or artificially designed metal-binding domains. To quantify intracellular [Zn^{2+}], Förster resonance energy transfer (FRET)[17]-based probes, which have a structure with a metal-sensing domain between two different FPs, are frequently used. Upon metal ion binding, the structural change alters the FRET efficiency between the FPs, resulting in a change in the emission spectrum. By taking the fluorescence ratio of the two FPs, ratiometric measurements independent of the probe concentration can be achieved. Four families of FP probes have been developed: ZapCY, eCALWY, eZinCh, and ZnT72R (Figure 1.3, Table 1.1). For each family, several probes have been developed with different affinities for Zn^{2+} and different signal peptides to target various organelles.

The ZapCY family consists of CFP and YFP linked with a zinc-binding domain from the yeast transcription factor Zap1.[18,19] Palmer *et al.* developed ZapCY1 ($K_d = 2.5$ pM) and created ZapCY2 ($K_d = 811$ pM) by mutating two cysteines in the binding domain to histidines to reduce the affinity. By fusing different signal peptides, ZapCY1 was used to quantify [Zn^{2+}] in the cytosol, ER, and Golgi apparatus, revealing that the [Zn^{2+}] in the ER (0.9 pM) and Golgi (0.6 pM) were 100 times lower than in the cytosol (80 pM). ZapCY1 showed a DR of 4.15 in the cytosol but smaller values in the ER (DR = 2.16) and Golgi (DR = 2.09). [Zn^{2+}]$_{mito}$ was also estimated to be 0.14 pM in HeLa cells using the K_d (1.6 pM) at pH 8.0. The researchers also investigated [Zn^{2+}]$_{mito}$ in different cell lines and reported that it varied among different cell lines and under different experimental conditions.

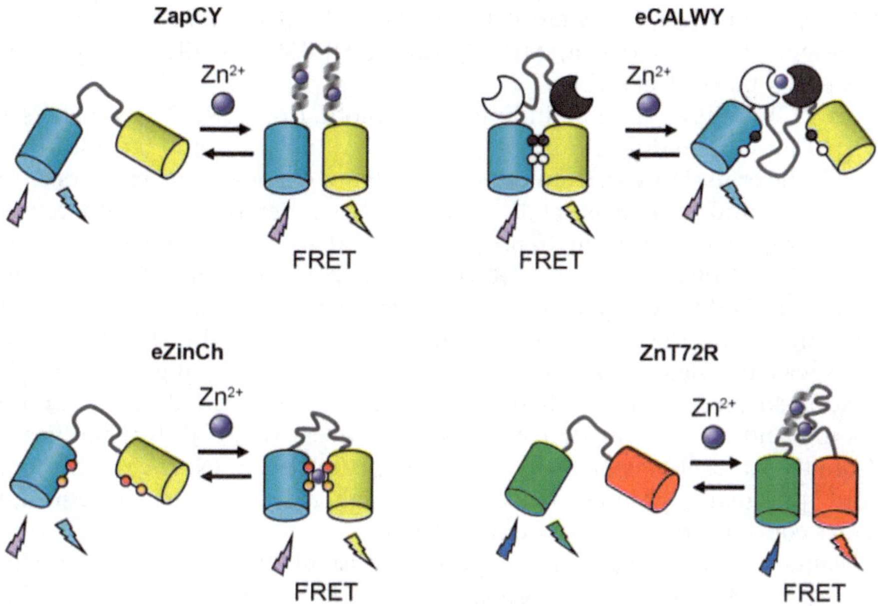

Figure 1.3 Illustration of FP probes for Zn^{2+} quantification.

The eCALWY family connects two fluorescent proteins, Cerulean and Citrine, with two Zn^{2+}-binding domains, Atox1 and WD4, *via* a long flexible linker.[20,21] Upon Zn^{2+} binding, the emission ratio ($F_{\text{Citrine}}/F_{\text{Cerulean}}$) of eCALWY-1 decreases by 2.4-fold, with a K_d of 2 pM at pH 7.1. Mutation of a cysteine (C416S) in the WD4 domain yielded eCALWY-4, which exhibits a 300-fold increase in K_d. Varying the linker length led to the development of eCALWY1–6, which have Zn^{2+} affinities ranging from picomolar to nanomolar levels. Using this sensor toolbox, the $[Zn^{2+}]_{cyto}$ in pancreatic cells was estimated to be 0.4 nM.[20] The eCALWY family probes were further expanded to localize to the ER and mitochondrial matrix by fusion with KDEL and COX8 peptides, respectively. eCALWY-4 showed appropriate affinity and was used for detecting Zn^{2+} in the ER and mitochondria. The K_d of this probe was determined to be 60 pM at pH 7.8, which is one-tenth of the K_d (630 pM) at pH 7.4. The eCALWY probes determined $[Zn^{2+}]_{ER}$ (>5 nM) and $[Zn^{2+}]_{mito}$ (0.23 nM) in HeLa cells, which were much higher than those determined by the ZapCY1 probes.[21]

In the eZinCh family, the Zn^{2+} coordination site was engineered directly on the surface between the β-barrels of the Cerulean and Citrine proteins, and the two FPs were connected by a flexible linker.[22] eZinCh-2 showed similar Zn^{2+} affinity ($K_d = 1.0$ nM at pH 7.1) to eCALWY-4, but exhibited a greater change in DR. eZinCh-2 demonstrated a significant difference (100-fold) in K_d at different pH values: 1.0 nM at pH 7.1 and 10 pM at pH 7.8. The probe was used to quantify $[Zn^{2+}]$ in the cytosol, ER, mitochondria, and secretory vesicles. The quantification results showed that the ranges of

$[Zn^{2+}]_{cyto}$ and $[Zn^{2+}]_{ER}$ were 0.4–0.9 nM and 0.5–0.8 nM, respectively, in different cell lines (HeLa, MCF-7, TamR, HEK293), while $[Zn^{2+}]_{mito}$ was about 3.3 pM in HeLa cells.

ZnT72R utilizes a zinc-binding RING motif from TRIM72, which is inserted into the middle of mNeonGreen and mRuby2.[23] Upon the addition of Zn^{2+}, the emission spectrum excited at 480 nm changes, showing a decrease at 517 nm and an increase at 584 nm. The K_d was determined in different pH solutions *in vitro*, resulting in 6.8 μM at pH 6.0, 6.1 μM at pH 6.5, and 7.8 μM at pH 7.4. However, the K_d determined at pH 6.0 in HEK293T cells was significantly different (56 μM) from the *in vitro* result, suggesting that there is room for improvement in the calibration method. The DR in HEK293T cells was also determined to be 1.4. Due to its relatively high K_d, $[Zn^{2+}]$ in insulin-sorting vesicles of diabetic and healthy mice were determined to be 28 μM and 41 μM, respectively. However, these values differ significantly from the value (120 nM) determined using eZinCh-2 in INS-1 cells. Although $[Zn^{2+}]$ in insulin granules is likely to be higher than that in the cytosol, its exact concentration remains under debate.

Palmer *et al.* compared the performance of the SM probe FluoZin-3 ($K_d = 15$ nM) and the FP probe ZapCY2 ($K_d = 811$ pM) for quantifying $[Zn^{2+}]_{cyto}$.[10] FluoZin-3 tended to accumulate in subcellular regions, presumed to be the Golgi apparatus or lysosomes, and exhibited a relatively slow response and dependence on probe concentration for quantifying $[Zn^{2+}]$ values. In contrast, ZapCY2 showed excellent subcellular targeting by fusion with a signal peptide. Such direct comparison studies between SM and FP probes demonstrated the superiority of FP probes in quantifying intracellular $[Zn^{2+}]$, especially regarding the precise probe targeting. However, FP probes have issues with pH sensitivity affecting their emission properties and K_d values, suggesting that inaccurate pH estimation leads to significant errors in the calculated $[Zn^{2+}]$ values. Additionally, Palmer *et al.* reported that FP probes with cysteines in the coordination site can undergo diverse disulfide formations, particularly under oxidative conditions such as those in the ER.[24] Overall, the $[Zn^{2+}]$ values determined by FP probes still require further validation using other methods, including optimization of the intracellular calibration method.

1.3.3 Small Molecule–Protein Hybrid Fluorescent Probes

The third type of fluorescent probes, known as hybrid probes, are designed to combine the organelle-targeting ability provided by protein engineering techniques with the spectral diversity, pH stability, and high DR characteristics of SM probes. By fusing a signal peptide or a specific protein, a tag protein, such as HaloTag or SNAP-tag, can be precisely targeted to subcellular regions of interest. This section summarizes the reported hybrid Zn^{2+} probes, including their photophysical properties and cellular applications, as an introduction to the quantitative imaging technique developed by the authors, which will be detailed in the following section. Figure 1.4 summarizes all the reported hybrid Zn^{2+} probes.

Table 1.1 Photophysical and coordination chemical properties of genetically-encoded Zn^{2+} sensors.

FP probes	Donor	Acceptor	ε^f (cm^{-1} mM^{-1})	$\lambda_{ex}{}^f$ (nm)	$\lambda_{em}{}^f$ (nm)	Φ^f	K_d pH 7.1	pH 7.4	pH 7.8	pH 8.0	DR
ZapCY1[a]	ECFP	Citrine	32.5/77	434/516	477/529	0.4/0.76	—	2.5 pM	—	1.6 pM	4.15
ZapCY2[a]							—	811 pM	—	—	1.4
eCALWY-1[b]	Cerulean	Citrine	43/77	433/516	475/529	0.62/0.76	2 pM	—	—	—	1.6
eCALWY-4[c]							—	630 pM	60 pM	—	1.83
eZinCh-2[d]							1.0 nM	—	10 pM	—	2.06
ZnT72R[e]	mNeonGreen	mRuby2	116/113	506/559	517/600	0.8/0.38	—	7.8 µM	—	—	1.4

[a] Data were obtained from ref. 18.
[b] Data were obtained from ref. 20.
[c] Data were obtained from ref. 21.
[d] Data were obtained from ref. 22.
[e] Data were obtained from ref. 23.
[f] Values are presented for donor/acceptor.

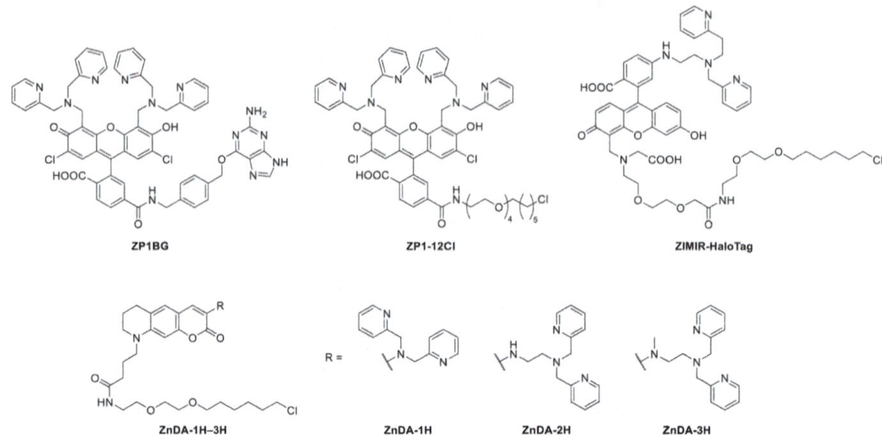

Figure 1.4 The structures of Zn^{2+} probes with a tag protein ligand.

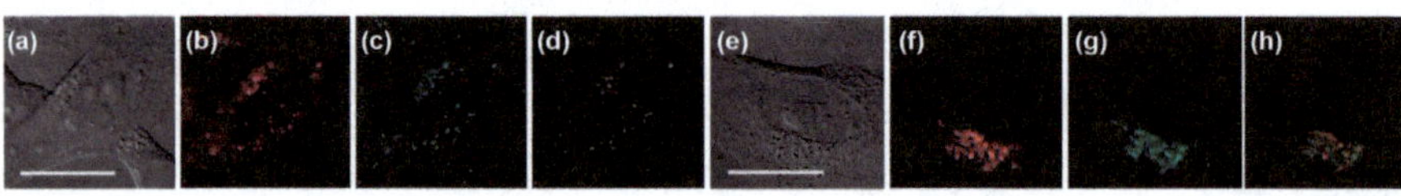

Figure 1.5 HeLa cells expressing AGT in the mitochondria (a)–(d) and in the Golgi apparatus (e)–(h) were labeled with ZP1BG (10 μM). The organelles were marked in red by staining with Mitotracker Red (b) or by expression of mCherry in the Golgi apparatus (f). For each series, DIC images (a) and (e) and overlays of the red and green fluorescence images (d) and (h) are shown. Scale bars: 25 μm. Reproduced from ref. 25 with permission from American Chemical Society, Copyright 2008.

Lippard *et al.* developed ZP1BG by conjugating a Zn^{2+} probe, Zinpyr-1, with a benzylguanine structure,[25] which is a covalent ligand for a mutant of human O^6-alkylguanine transferase (AGT), also known as SNAP-tag[26] (Figure 1.4). Free ZP1BG showed little increase in fluorescence intensity upon Zn^{2+} addition, attributed to fluorescence quenching by the guanine moiety. After labeling with AGT, the ZP1-AGT conjugate exhibited Zn^{2+}-dependent fluorescence enhancement (approximately 2.7-fold) without changes in the absorption and emission wavelengths. GalT and COX8 were used as localization motifs for specific expression of AGT in the Golgi apparatus and mitochondria, respectively. In HeLa cells expressing these fusion proteins, ZP1BG was localized to the target organelles and detected local Zn^{2+} (Figure 1.5). Lippard *et al.* also developed another Zinpyr-1 derivative for HaloTag labeling technology,[27] called ZP1-12Cl[28] (Figure 1.4). After binding to HaloTag, ZP1-12Cl showed similar photophysical properties to the free probe, such as approximately 2.7-fold fluorescence enhancement upon Zn^{2+} binding and a K_d of 0.11 nM. Additionally, ratiometric Zn^{2+} imaging was achieved by utilizing HaloTag fused to mCherry. However, due to its small DR and high pH sensitivity, the quantification of subcellular $[Zn^{2+}]$ has not been achieved.

Li *et al.* conjugated a HaloTag ligand to a green fluorescent Zn^{2+} probe, a ZIMIR derivative (Figure 1.4), and covalently labeled it on the outer leaflet of the plasma membrane of pancreatic islet β cells.[29] An ethylene glycol linker was incorporated between the ZIMIR derivative and the HaloTag ligand to ensure efficient labeling. This probe exhibited a 15-fold increase in fluorescence upon Zn^{2+} addition, with a K_d determined to be 0.13 μM at pH 7.5. The probe was then targeted to a fusion protein, HaloTag/R-GECO1.2, which was constructed with a HaloTag and a red FP Ca^{2+} probe, R-GECO1.2, positioned at the extracellular and intracellular ends of the transmembrane domain of pDisplay, respectively. Interestingly, this hybrid system was used for simultaneous monitoring of intracellular Ca^{2+} influx and insulin Zn^{2+} release upon KCl addition and channelrhodopsin activation (Figure 1.6). However, quantitative Zn^{2+} imaging has not been performed using this hybrid probe.

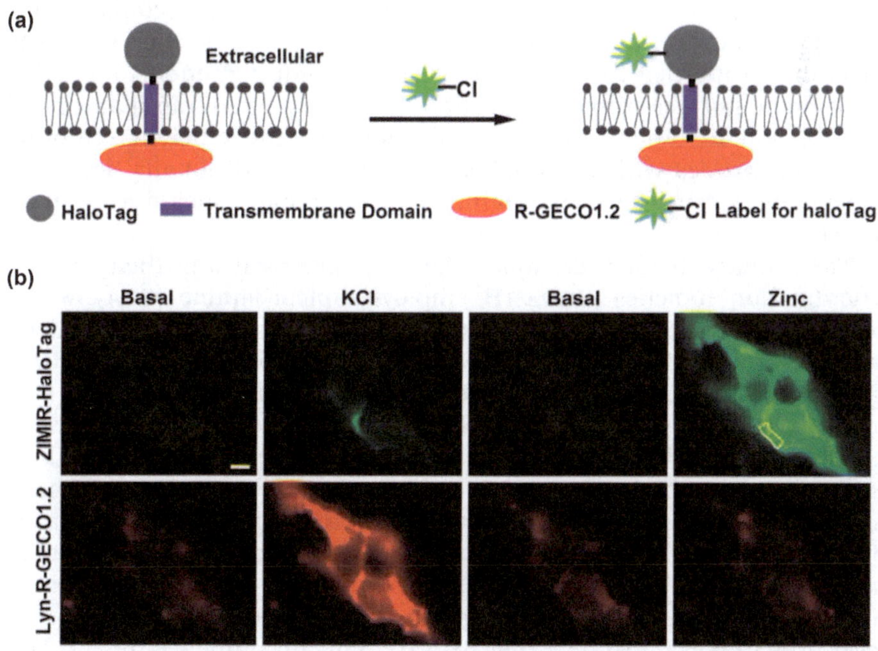

Figure 1.6 (a) Schematic illustration of a fusion protein (HaloTag/R-GECO1.2) containing HaloTag (extracellular) and R-GECO1.2 (intracellular) linked by a transmembrane peptide for simultaneously monitoring Zn^{2+} release from cells and Ca^{2+} influx. (b) Live-cell imaging of Zn^{2+}/insulin release with ZIMIR-HaloTag. Images of MIN6 beta cells co-expressing ZIMIR-labeled HaloTag (pDisplay) and Lyn-R-GECO1.2 and sequentially incubated in the basal medium with 10 μM EDTA and 2 μM dipicolylamine *N*-ethylsulfonate, 40 mM KCl, basal medium, and a high Zn^{2+} buffer (1 μM). Scale bar: 5 μm. Adapted from ref. 29 with permission from American Chemical Society, Copyright 2015.

1.4 Quantitative Imaging of Intracellular Zn^{2+} Using Hybrid Fluorescent Probes

In this section, we provide a detailed description of our recently developed aminocoumarin-based fluorescent probes, ZnDAs (Zn^{2+}-detectable aminocoumarin-based fluorescent probes), organized into several subsections. These subsections include discussions on the molecular design, physical properties, subcellular targeting, quantification principles, and quantitative results concerning intracellular labile Zn^{2+} concentrations.

1.4.1 Design and Properties of ZnDA Probes

We have developed three probes, ZnDA-1H, 2H, and 3H (Figure 1.4), with different affinities for Zn^{2+}. All of these probes incorporate a HaloTag ligand, enabling covalent binding to HaloTag proteins expressed in specific subcellular locations.[30,31] Utilizing these SM–protein hybrid systems, Zn^{2+} probes can be localized to various subcellular compartments. The parent aminocoumarin fluorophore was selected because of its small size and properties to meet several critical requirements. Its good cell membrane permeability and low lipophilicity facilitate rapid and precise localization to target organelles, allowing for the efficient removal of unbound probes with minimal washing steps. Additionally, its low pH sensitivity is crucial for maintaining stable signal values during quantitative imaging.

The primary distinction among the three probes lies in their metal ion coordination moieties. ZnDA-1H employs dipicolylamine (DPA), whereas ZnDA-2H and ZnDA-3H feature Zn^{2+} binding motifs with higher coordination numbers. The fluorescence of the ZnDA probes is quenched in the absence of Zn^{2+} through the twisted intramolecular charge transfer (TICT) mechanism. This quenching mechanism contrasts with that of conventional SM Zn^{2+} probes, which exhibit the fluorescence quenching in the unbound state *via* PeT. Upon Zn^{2+} coordination, the fluorescence of ZnDA probes is restored due to the suppression of C–N bond rotation at the 3-position. This fluorescence activation mechanism ensures that the probe fluorescence remains stable across a pH range of 5–8, maintaining consistent fluorescence intensity during imaging despite pH fluctuations within this range. However, it is important to note that pH variations do influence the K_d. Therefore, to accurately quantify $[Zn^{2+}]$ in various organelles, K_d values measured under specific organellar pH conditions should be used.

The photophysical and coordination properties of the ZnDA probes were evaluated both as free SM probes and as HaloTag conjugates (Halo–ZnDAs) (Table 1.2). With increasing free Zn^{2+} concentration in the buffer, the absorption maximum wavelength of the probes underwent a red shift, accompanied by the appearance of an isosbestic point and a slight increase in absorbance. The stoichiometry of Zn^{2+} binding for all three probes was confirmed to be 1 : 1. In the absence of Zn^{2+}, both ZnDA-1H and ZnDA-3H

Table 1.2 Photophysical and coordination chemical parameters of ZnDA probes and the HaloTag-conjugated probes.[a]

	Free probe[b]				Zn^{2+} complex[b]				K_d (nM)				
Compound	λ_{abs} (nm)	λ_{em} (nm)	ε[d] (cm^{-1} mM^{-1})	Φ	λ_{abs} (nm)	λ_{em} (nm)	ε[d] (cm^{-1} mM^{-1})	Φ	pH 6.5	pH 7.0	pH 7.4	pH 8.0	DR
ZnDA-1H	409	518	21.7	0.028[c]	440	506	30.9	0.52[c]	3.2×10^2	2.9×10^2	2.4×10^2	—	53
Halo–ZnDA-1H	403	516	21.7	0.11[c]	438	507	27.6	0.91[c]	5.4×10^2	3.5×10^2	3.0×10^2	—	30
ZnDA-2H	399	520	19.5	0.22[c]	417	506	24.0	0.56[c]	—	—	5.0	—	5.6
Halo–ZnDA-2H	394	514	19.6	0.16	414	495	23.4	0.60	2.8	2.5	2.7	2.2	11
ZnDA-3H	407	508	19.2	0.029[c]	433	506	26.8	0.57[c]	—	—	0.16	—	40
Halo–ZnDA-3H	402	504	20.6	0.11	431	503	27.6	0.77	0.37	0.083	0.082	0.11	20

[a] Data were obtained from ref. 30 and 31.
[b] Measured in HEPES buffer (pH 7.4).
[c] Measured in HEPES buffer (pH 7.5).
[d] Measured at λ_{abs}.

showed minimal fluorescence (quantum yields, $\Phi_{ZnDA\text{-}1H} = 0.028$ and $\Phi_{ZnDA\text{-}3H} = 0.029$ at pH 7.5), whereas ZnDA-2H displayed weak fluorescence ($\Phi_{ZnDA\text{-}2H} = 0.22$ at pH 7.5). The DR of the fluorescence intensity was 53-fold for ZnDA-1H and 40-fold for ZnDA-3H, while it was 5.6-fold for ZnDA-2H. The probes demonstrated high selectivity for Zn^{2+} over other physiological metal ions, whereas similar to other probes with a chelator composed of amino groups and/or heterocyclic amines, Cu^{2+} and Cu^{+} showed stronger affinity than Zn^{2+}, resulting in fluorescence quenching. This characteristic must be carefully considered when using these probes under conditions where copper ions are present.

To accurately estimate the [Zn^{2+}] in various organelles, it is essential to determine the K_d under the pH conditions specific to each organelle. The three probes were prepared as HaloTag conjugates, which function within cells, and their K_d values were evaluated at several different pH values in the range of 6.5–8.0. Among the three ZnDA probes, Halo–ZnDA-1H exhibited the lowest affinity, with K_d values ranging from 0.30 to 0.54 μM, and a 1.8-fold difference between pH 7.4 and pH 6.5. Halo–ZnDA-2H demonstrated moderate affinity, with K_d values gradually varying from 2.2 nM at pH 8.0 to a maximum of 2.8 nM at pH 6.5. Halo–ZnDA-3H showed the highest affinity, with K_d values slightly changing from 0.083 to 0.11 nM at neutral pH (pH 7.0–8.0) but increasing approximately 4-fold to 0.37 nM at pH 6.5. In comparison, FP probes such as ZapCY1 and eZinCh-2 exhibit significantly higher pH sensitivity, with a 10-fold change in Zn^{2+} affinity at pH 7.4–8.0 and a hundred-fold change at pH 7.1–7.8, respectively.[21,22] The lower pH sensitivity of ZnDA probes is advantageous for accurately detecting intracellular [Zn^{2+}] and monitoring Zn^{2+} dynamics without significant interference from pH fluctuations.

1.4.2 Subcellular Targeting of ZnDA Probes and Labile Zn^{2+} Quantification Using Colocalization Ratiometry

HaloTag proteins without any signal peptide are expressed in both the nucleus and cytosol in mammalian cells. When fused with specific signal peptides, they can be expressed in various organelles, including the nucleus, ER, mitochondria, Golgi apparatus, and lysosomes. Due to their excellent membrane permeability and low pH sensitivity, the three ZnDA probes specifically localized to the target subcellular compartments, as demonstrated by the high PCCs upon co-incubation with the target organelle markers. The covalent targeting offers additional practical benefits: it suppresses probe dissociation from target regions despite fluctuations in intracellular environments and enables quantitative ratiometric imaging by colocalizing standard fluorophores within the same regions.

In our developed *in situ* standard quantification method (Figure 1.7),[32] cells expressing HaloTag are treated with ZnDA probes along with HTL-TMR, a Zn^{2+}-insensitive red fluorophore, tetramethylrhodamine (TMR), linked to a HaloTag ligand. While one molecule of the HaloTag proteins can covalently bind to only one molecule of either the ZnDA probes or TMR, many

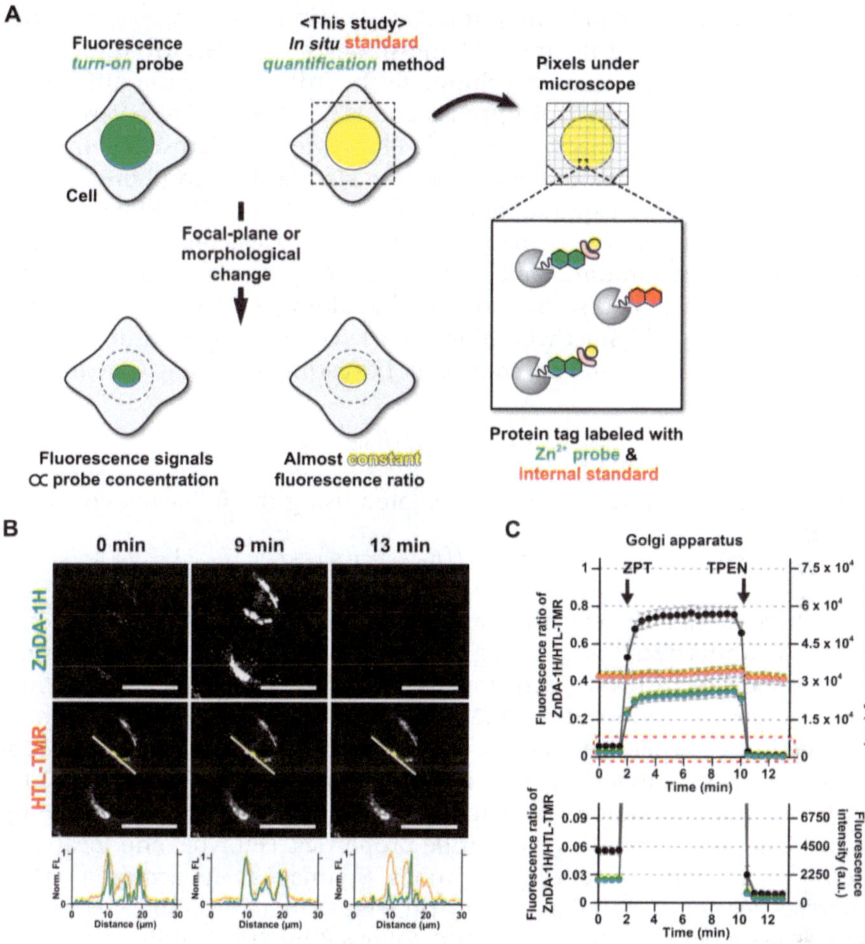

Figure 1.7 (A) Schematic illustration of an *in situ* standard quantification method for monitoring of [Zn^{2+}] changes using ZnDA-1H and HTL-TMR. (B) Confocal microscopy images of ZnDA-1H and HTL-TMR targeted into the Golgi apparatus in the steady state (0 min) and under conditions of labile Zn^{2+} saturation (9 min) and depletion (13 min). (C) Time-course of the fluorescence ratios (black) and intensities of ZnDA-1H (green) and HTL-TMR (orange) in response to Zn^{2+}. The region enclosed by the red dotted line in the upper plot is magnified in the lower panel. Adapted from ref. 30 with permission from Elsevier, Copyright 2020.

molecules of both the fluorophores co-localize within one pixel at a certain ratio according to their concentrations, membrane permeability, and so on. Conventional intensiometric fluorescent probes yield errors on the estimation of the target ion concentrations from the probe fluorescence intensity because the number of probes per pixel are varied by cell morphology changes during time-lapse imaging. On the other hand, by measuring the fluorescence signal ratio from two fluorophores ($R = F_{ZnDA}/F_{TMR}$) within one pixel, $[Zn^{2+}]$ changes can be monitored while minimizing the external effects such as focal plane drift and cell morphology changes. However, a calibration process is necessary to eliminate the influence by the difference in the labeling efficiency of ZnDA probes and HTL-TMR, and also variations in HaloTag expression levels among cells. Fluorescence microscopic images of ZnDA probes (green fluorescence) and the standard fluorophore TMR (red fluorescence) can be acquired, and steady-state R value (R_{SS}) images can be obtained through image calculation. Subsequently, calculated images of the maximum and minimum R values (R_{max} and R_{min}) are obtained through an *in situ* calibration process by sequentially adding ZPT and Zn^{2+} chelators (TPEN or tris(2-pyridylmethyl)amine (TPA)). The fractional saturation (θ) of ZnDA probes at the steady state can be estimated using the following equation:

$$\theta_{ss} = (R_{ss} - R_{min})/(R_{max} - R_{min}).$$

$[Zn^{2+}]$ at the steady state can be calculated using the following equation:

$$[Zn^{2+}] = K_d \cdot (R_{ss} - R_{min})/(R_{max} - R_{ss}) = K_d \cdot \theta_{ss}/(1 - \theta_{ss}).$$

We here explain the experimental procedure for quantifying $[Zn^{2+}]$ in the Golgi apparatus ($[Zn^{2+}]_{Golgi}$) using ZnDA-1H. To localize ZnDA-1H within the Golgi lumen, ManII-Halo, a fusion protein of HaloTag and the N-terminal 117 residues of mannosidase II (ManII), a *cis-/medial*-Golgi marker, was expressed in HeLa cells. The cells were incubated with 250 nM of ZnDA-1H and 5 nM of HTL-TMR to achieve co-labeling of both fluorophores to ManII-Halo proteins. The incubation concentrations of each compound must be optimized based on factors such as probe properties, cell type, and localization site. After removing the unreacted probes, R_{ss} images were acquired once the fluorescence signal stabilized during time-lapse imaging. Subsequently, R_{max} images were obtained by adding ZPT, followed by TPEN addition to acquire R_{min} images. This calibration protocol may vary depending on the sample type to minimize cytotoxicity; for instance, TPEN was added before ZPT in the subsequent report.[31] The variability of R_{max}, R_{min}, and R_{ss} among cells requires calculating the average θ_{ss} from the average R values in each cell. Using a θ_{ss} of 4% in the Golgi at the steady state and a K_d of 540 nM at pH 6.5, the typical $[Zn^{2+}]_{Golgi}$ was estimated to be 25 nM.

1.4.3 Organelle-level Labile Zn^{2+} Mapping Using ZnDA Probes

We attempted to detect $[Zn^{2+}]$ in various subcellular compartments using ZnDA-1H. However, no sufficient probe signals were observed in the cytosol, nucleus, mitochondria, and ER. Given that ZnDA-1H has a moderate binding affinity for Zn^{2+} and its detection limit is near the $[Zn^{2+}]_{Golgi}$, probes with higher affinity, such as ZnDA-2H and ZnDA-3H, would be more suitable for quantifying $[Zn^{2+}]$ in most subcellular compartments.

To address this hypothesis, ZnDA-2H and ZnDA-3H were co-labeled with HTL-TMR to HaloTag expressed at specific intracellular target sites. Because both ZnDA probes showed sufficient fluorescence signals from the cytosol and nucleus, we applied them to quantify $[Zn^{2+}]$ in these compartments ($[Zn^{2+}]_{cyto}$ and $[Zn^{2+}]_{nuc}$). ZnDA-2H showed a θ_{ss} of approximately 6.0% in the cytosol and 5.7% in the nucleus. Using the K_d of about 2.7 nM for Zn^{2+} at pH 7.4, this θ_{ss} value allowed estimation of $[Zn^{2+}]_{cyto}$ and $[Zn^{2+}]_{nuc}$ to be 0.22 nM and 0.17 nM, respectively. Since the nuclear membrane possesses nuclear pores, through which small substances like metal ions can permeate, the difference in $[Zn^{2+}]_{cyto}$ and $[Zn^{2+}]_{nuc}$ should be an insignificant deviation. However, significant enhancement of the ZnDA-2H signal was not detected in the ER and mitochondria, suggesting $[Zn^{2+}]$ levels in these organelles are even lower.

ZnDA-3H, identified as the most promising probe due to its highest affinity for Zn^{2+}, effectively detects very low $[Zn^{2+}]$ in various organelles. This probe demonstrated detectable signals across all examined subcellular compartments. When colocalized with HTL-TMR similarly to other ZnDA probes, stable signal ratio changes were observed upon the addition of TPEN or TPA followed by ZPT (Figure 1.8). Utilizing the K_d values of Halo–ZnDA-3H, 0.082 nM at pH 7.4 and 0.11 nM at pH 8.0, we estimated subcellular $[Zn^{2+}]$ in HeLa cells as $[Zn^{2+}]_{cyto} = 0.13$ nM ($\theta_{ss} = 60\%$), $[Zn^{2+}]_{nuc} = 0.11$ nM ($\theta_{ss} = 58\%$), $[Zn^{2+}]_{ER} = 14$ pM ($\theta_{ss} = 15\%$), and $[Zn^{2+}]_{mito} = 60$ pM ($\theta_{ss} = 29\%$) (Figure 1.9). Considering that the high affinity of ZnDA-3H might disrupt the homeostasis of labile Zn^{2+} in the examined organelles, we quantified $[Zn^{2+}]_{ER}$ using various concentrations of ZnDA-3H ranging from 50 nM to

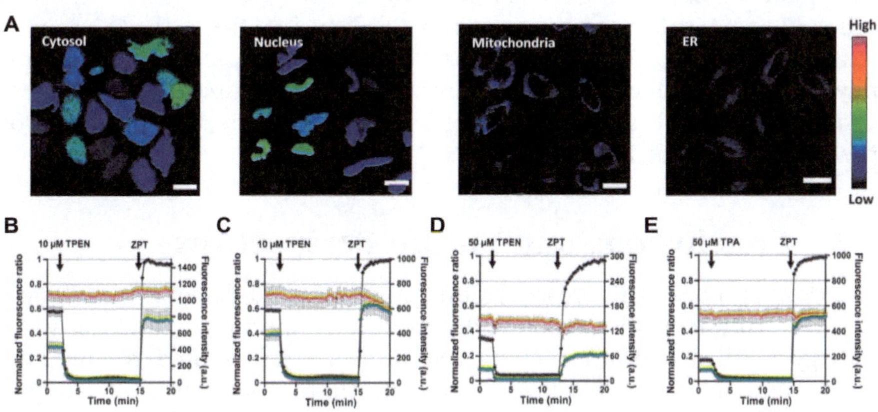

Figure 1.8 (A) Pseudocolor fluorescence ratio images of ZnDA-3H/HTL-TMR at the steady-state in each organelle. Scale bars: 30 μm. (B)–(E) Time-course of *in situ* calibration of normalized fluorescence ratio (black) and fluorescence intensity of ZnDA-3H (green) and HTL-TMR (orange) for $[Zn^{2+}]$ quantification in the cytosol (B), nucleus (C), mitochondria (D), and ER (E). Error bars represent the s.e.m. Reproduced from ref. 31 with permission from American Chemical Society, Copyright 2022.

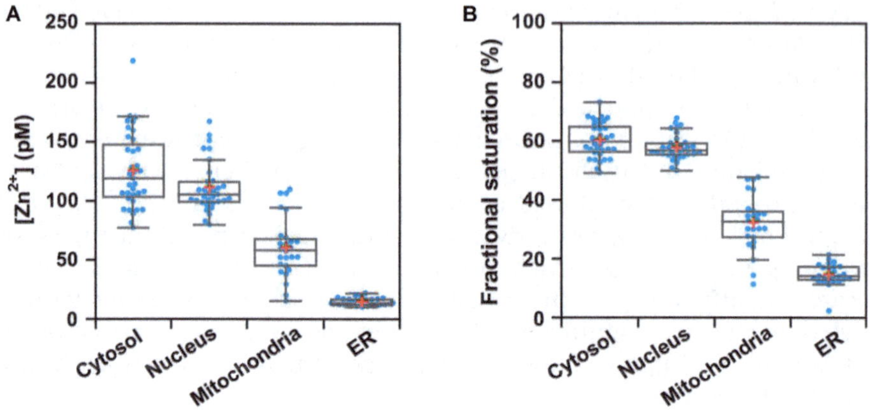

Figure 1.9 Quantitative mapping of organellar $[Zn^{2+}]$ by using ZnDA-3H (A) and fractional saturation of ZnDA-3H (B) at the steady state in various organelles. Red plus symbols represent the mean values. Reproduced from ref. 31 with permission from American Chemical Society, Copyright 2022.

1.0 μM. However, no significant differences were observed in the quantified values, suggesting that our estimated $[Zn^{2+}]_{ER}$ values reflect physiological levels at the steady state. These results indicate that even low-level labile endogenous Zn^{2+} in some organelles is sufficiently buffered by endogenous Zn^{2+} pools, likely stored in proteins or other coordinating molecules.

In summary, by developing three HaloTag-conjugated ZnDA probes, we quantified the organelle-level labile Zn^{2+} distribution in HeLa cells. The resulting concentration order in the steady state is $[Zn^{2+}]_{ER} < [Zn^{2+}]_{mito} < [Zn^{2+}]_{cyto} \approx [Zn^{2+}]_{nuc} < [Zn^{2+}]_{Golgi}$. Understanding the physiological mechanisms and significance of this subcellular labile Zn^{2+} distribution, which spans several orders of magnitude, is crucial for comprehending cellular homeostasis.

1.5 Conclusions and Future Perspectives

The development of fluorescence imaging probes to detect intracellular labile Zn^{2+} began in the 1980s and played an important role in early Zn^{2+} biology studies. However, many SM probes were limited to qualitative measurements of whole-cell zinc concentrations. Some research groups attempted the organelle-targeting of the probes using specific targeting moieties, but few pursued quantification. The probe retention at target sites during time-lapse imaging under various experimental conditions has also not been thoroughly verified. FP-based Zn^{2+} probes demonstrated strong functional properties in Zn^{2+} imaging. By fusing localizable signal peptides to these probes, it became possible to measure $[Zn^{2+}]$ within various subcellular compartments. Many FP probes exhibit ratiometric properties using two emission wavelengths, enabling the quantification of intracellular [Zn^{2+}]. Nonetheless, reported quantified values of [Zn^{2+}], especially in organelles, have varied significantly with different probes, and a consensus has not been reached for [Zn^{2+}] in some organelles. This discrepancy may stem from common limitations of FP probes, such as sensitivity to pH and/or redox environments and narrow DR. As a third approach, small-molecule–protein hybrid probes are emerging.

In this chapter, we introduced the molecular design, intracellular targeting, and quantification methods of ZnDA probes developed by the authors. Using these hybrid probes, subcellular [Zn^{2+}] in various organelles in HeLa cells were quantified. Our quantitative results surprisingly showed that [Zn^{2+}] among organelles differed over a range of several orders of magnitude. Using our hybrid probe technique, it was gradually revealed how such differences in the labile Zn^{2+} distribution at the organelle level are formed and what physiological significance these large differences hold. As reviewed in Chapter 6 by Amagai and Inaba, targeting ZnDA-1H in the Golgi apparatus through various signal peptides demonstrated that the sub-organelle-level distribution of labile Zn^{2+} in the Golgi is finely regulated by multiple ZnT transporters with different localizations.[33] Furthermore, the significant concentration gap between $[Zn^{2+}]_{ER}$ and $[Zn^{2+}]_{Golgi}$ plays a crucial role in the shuttling of a chaperone protein ERp44 between the ER and Golgi. Although numerous intracellular proteins possess peptide motifs predicted to coordinate Zn^{2+}, the physiological significance of these motifs in many proteins remains unclear. Investigating the functions of such proteins while quantifying [Zn^{2+}] in their local environment could reveal the molecular mechanisms underlying cellular homeostasis systems.

Nevertheless, the quantified values of [Zn^{2+}] often differ from those obtained by the previous probes, necessitating careful interpretation. To assess the reliability of these quantitative results, continuous optimization of quantification methods and studies across various cell types by multiple research groups are essential. It is also crucial to fully understand the limitations and uncertainties of the current methods to establish more reliable protocols. In our colocalization-based ratiometry, the fluorescence intensity ratio of the ZnDA probe to TMR varies across different cells due to differences in HaloTag protein expression levels in each cell. Consequently, this method has only enabled us to estimate the average concentration of [Zn^{2+}] on a per-cell basis and has not yet achieved sub-organellar Zn^{2+} mapping. However, for a better understanding of zinc biology, it is important to quantitatively monitor the Zn^{2+} dynamics at the sub-organelle level, in conjunction with pharmacological methods and optical manipulation techniques such as optogenetics. Therefore, development of quantitative time-lapse Zn^{2+} imaging in living cells is also expected.

Another critical aspect of quantitative Zn^{2+} imaging is the influence of pH and other metal ions. Although intracellular pH fluctuations have not been sufficiently investigated, simultaneous imaging of pH and Zn^{2+} is necessary for more accurate quantification of [Zn^{2+}], because the K_d values are affected by pH. Additionally, some transition metal ions competitively occupy the Zn^{2+}-binding site of the fluorescent probes, with copper ions (Cu^{2+} and Cu^{+}) being particularly significant. The effects of iron and manganese ions, which are thought to exist at higher levels in their free states within cells compared to free copper ions, also require further exploration. However, the development of imaging probes for these transition metal ions has lagged behind that for Zn^{2+}. Therefore, developing such transition metal ion-specific imaging probes would also accelerate the clarification of the overall picture of intracellular Zn^{2+} dynamics.

Acknowledgements

This work was supported by JSPS and MEXT KAKENHI (No. JP21H05252 and JP23H02077), Takeda Science Foundation, Nakatani Foundation, and AMED-CREST (21gm1410006h0001).

References

1. B. Chen, P. Yu, W. N. Chan, F. Xie, Y. Zhang, L. Liang, K. T. Leung, K. W. Lo, J. Yu, G. M. K. Tse, W. Kang and K. F. To, *Signal Transduction Targeted Ther.*, 2024, **9**, 6.
2. W. Maret, *Metallomics*, 2015, **7**, 202.
3. K. P. Kepp, *ChemBioChem*, 2022, **23**, e202100554.
4. S. Yamasaki, K. Sakata-Sogawa, A. Hasegawa, T. Suzuki, K. Kabu, E. Sato, T. Kurosaki, S. Yamashita, M. Tokunaga, K. Nishida and T. Hirano, *J. Cell Biol.*, 2007, **177**, 637.
5. K. M. Taylor, S. Hiscox, R. I. Nicholson, C. Hogstrand and P. Kille, *Sci. Signaling*, 2012, **5**, ra11.
6. A. Krężel and W. Maret, *Arch. Biochem. Biophys.*, 2016, **611**, 3.
7. T. Kambe, K. M. Taylor and D. Fu, *J. Biol. Chem.*, 2021, **296**, 100320.
8. M. C. Carpenter, M. N. Lo and A. E. Palmer, *Arch. Biochem. Biophys.*, 2016, **611**, 20.
9. A. Krężel and W. Maret, *J. Biol. Inorg. Chem.*, 2006, **11**, 1049.
10. Y. Qin, J. G. Miranda, C. I. Stoddard, K. M. Dean, D. F. Galati and A. E. Palmer, *ACS Chem. Biol.*, 2013, **8**, 2366.
11. Y. Han, J. M. Goldberg, S. J. Lippard and A. E. Palmer, *Sci. Rep.*, 2018, **8**, 15034.
12. G. Masanta, C. S. Lim, H. J. Kim, J. H. Han, H. M. Kim and B. R. Cho, *J. Am. Chem. Soc.*, 2011, **133**, 5698.
13. L. Xue, G. Li, C. Yu and H. Jiang, *Chem. – Eur. J.*, 2012, **18**, 1050.
14. Y. Shi, S. Wang, J. Wu, X. Jin and J. You, *J. Controlled Release*, 2021, **329**, 337.
15. L. Fang, G. Trigiante, R. Crespo-Otero, C. S. Hawes, M. P. Philpott, C. R. Jones and M. Watkinson, *Chem. Sci.*, 2019, **10**, 10881.
16. L. Xue, G. Li, D. Zhu, Q. Liu and H. Jiang, *Inorg. Chem.*, 2012, **51**, 10842.
17. B. T. Bajar, E. S. Wang, S. Zhang, M. Z. Lin and J. Chu, *Sensors*, 2016, **16**, 1488.

18. Y. Qin, P. J. Dittmer, J. G. Park, K. B. Jansen and A. E. Palmer, *Proc. Natl. Acad. Sci. U. S. A.*, 2011, **108**, 7351.
19. J. G. Park, Y. Qin, D. F. Galati and A. E. Palmer, *ACS Chem. Biol.*, 2012, 7, 1636.
20. J. L. Vinkenborg, T. J. Nicolson, E. A. Bellomo, M. S. Koay, G. A. Rutter and M. Merkx, *Nat. Methods*, 2009, **6**, 737.
21. P. Chabosseau, E. Tuncay, G. Meur, E. A. Bellomo, A. Hessels, S. Hughes, P. R. V. Johnson, M. Bugliani, P. Marchetti, B. Turan, A. R. Lyon, M. Merkx and G. A. Rutter, *ACS Chem. Biol.*, 2014, **9**, 2111.
22. A. M. Hessels, P. Chabosseau, M. H. Bakker, W. Engelen, G. A. Rutter, K. M. Taylor and M. Merkx, *ACS Chem. Biol.*, 2015, **10**, 2126.
23. Y. Xian, M. Zhou, S. Han, R. Yang and Y. Wang, *Chin. Chem. Lett.*, 2020, **31**, 468.
24. K. P. Carter, M. C. Carpenter, B. Fiedler, R. Jimenez and A. E. Palmer, *Anal. Chem.*, 2017, **89**, 9601.
25. E. Tomat, E. M. Nolan, J. Jaworski and S. J. Lippard, *J. Am. Chem. Soc.*, 2008, **130**, 15776.
26. A. Keppler, S. Gendreizig, T. Gronemeyer, H. Pick, H. Vogel and K. Johnsson, *Nat. Biotechnol.*, 2003, **21**, 86.
27. G. V. Los, L. P. Encell, M. G. McDougall, D. D. Hartzell, N. Karassina, C. Zimprich, M. G. Wood, R. Learish, R. F. Ohana, M. Urh, D. Simpson, J. Mendez, K. Zimmerman, P. Otto, G. Vidugiris, J. Zhu, A. Darzins, D. H. Klaubert, R. F. Bulleit and K. V. Wood, *ACS Chem. Biol.*, 2008, **3**, 373.
28. M. L. Zastrow, Z. Huang and S. J. Lippard, *ACS Chem. Biol.*, 2020, **15**, 396.
29. D. Li, L. Liu and W.-H. Li, *ACS Chem. Biol.*, 2015, **10**, 1054.
30. T. Kowada, T. Watanabe, Y. Amagai, R. Liu, M. Yamada, H. Takahashi, T. Matsui, K. Inaba and S. Mizukami, *Cell Chem. Biol.*, 2020, **27**, 1521.
31. R. Liu, T. Kowada, Y. Du, Y. Amagai, T. Matsui, K. Inaba and S. Mizukami, *ACS Sens.*, 2022, **7**, 748.
32. T. Kowada, T. Watanabe, R. Liu and S. Mizukami, *STAR Protoc.*, 2021, **2**, 100395.
33. Y. Amagai, M. Yamada, T. Kowada, T. Watanabe, Y. Du, R. Liu, S. Naramoto, S. Watanabe, J. Kyozuka, T. Anelli, T. Tempio, R. Sitia, S. Mizukami and K. Inaba, *Nat. Commun.*, 2023, **14**, 2683.

CHAPTER 2

Zinc Conditional Proteomics Using Zinc-responsive Protein Labelling Reagent and Recent Examples of Expansion to Other Biometals

TOMONORI TAMURA*[a,b] AND ITARU HAMACHI*[a,b]

[a] Department of Synthetic Chemistry and Biological Chemistry, Graduate School of Engineering, Kyoto University, Katsura, Nishikyo-ku, Kyoto 615-8510, Japan; [b] ERATO (Exploratory Research for Advanced Technology, JST), Sanbancho, Chiyoda-ku, Tokyo 102-0075, Japan
*Emails: tamura@sbchem.kyoto-u.ac.jp; ihamachi@sbchem.kyoto-u.ac.jp

2.1 Introduction

Zinc ions (Zn^{2+}) are essential metal cations for maintaining living systems and generally exist in mammalian cells in two forms. One is protein-bound Zn^{2+}, which serves as a structural element or catalytic cofactor of metalloproteins. The other form involves free or labile (loosely bound) Zn^{2+} functioning as a signalling molecule to control various cellular events.[1] Recent studies have revealed that the dynamics of such mobile Zn^{2+} vary dramatically depending on physiological responses and environmental changes inside and outside of cells.[2] For instance, Zn^{2+} is released along with the excitatory neurotransmitter glutamate from synaptic vesicles and regulates the activity of various neurotransmitter

Coordination Chemistry Perspectives No. 2
Zinc in Biology: Molecular Structures, Cellular Processes and Living Systems
Edited by Hiroyuki Yasui and Taiho Kambe

Published by the Royal Society of Chemistry, www.rsc.org

receptors located in the postsynaptic area, including *N*-methyl-D-aspartate (NMDA) receptors, α-amino-3-hydroxy-5-methyl-4-isoxazolepropionic acid (AMPA) receptors, voltage-dependent calcium channels (VDCC) and γ-aminobutyric acid ($GABA_A$) receptors. This clearly indicates that Zn^{2+} signalling is important for higher brain functions such as memory and learning. Additionally, dysregulation of Zn^{2+} levels has been observed in certain diseases, including neurodegenerative diseases, diabetes and cancers, suggesting a link between these diseases and abnormalities in Zn^{2+} homeostasis.

Conventionally, the fluctuation of Zn^{2+} levels in cells or tissues has been analysed by the fluorescence imaging of Zn^{2+}.[3–5] A variety of Zn^{2+} fluorescent sensors based on synthetic small molecules or genetically encodable fluorescent proteins have been developed (see Chapter 1), allowing direct observation of the changes in the concentration in living systems.[6] However, to obtain more detailed information about the molecular mechanisms involved in Zn^{2+} dynamics, imaging techniques alone might be insufficient. For example, it is known that, during the process of extracellular excretion of excess labile Zn^{2+}, vesicles containing accumulated Zn^{2+} are formed within the cell.[7] However, it is challenging to identify proteins related to this process and key components constituting these Zn^{2+}-rich vesicles using imaging-based analysis alone.

Our group has devised a novel chemical proteomics method aimed at specifically identifying and analysing proteomes in Zn^{2+}-rich environments within cells and that are potentially involved in Zn^{2+} homeostasis (Figure 2.1).[8] Using this method, called zinc conditional proteomics, we demonstrated that the increase in intracellular mobile Zn^{2+} concentration induced by nitric oxide (NO) stimulation dramatically alters the zinc-related proteome in a

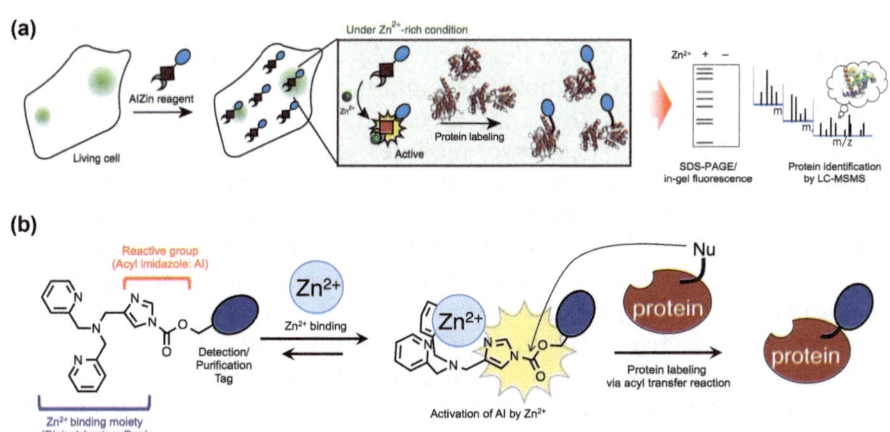

Figure 2.1 Conditional proteomics by Zn^{2+}-enhanced chemical labelling. (a) A schematic illustration of Zn^{2+}-conditional proteomics with AIZin (AI-based zinc-responsive reagent). (b) Molecular design of AIZin and its activation mechanism.

time-dependent manner. We also succeeded in identifying protein groups involved in forming Zn^{2+}-rich vesicles induced by NO stimulation by a liquid chromatography–tandem mass spectrometry (LC–MS/MS)-based chemical proteomics workflow. Furthermore, the obtained proteomic data revealed that these vesicles originate from the ER–Golgi intermediate compartment (ERGIC). Notably, since our initial report, several research groups including our own have successfully expanded this conditional proteomics concept to other metal ions such as iron and copper ions, offering powerful molecular tools in biometal research. This chapter describes the basic principles of our zinc conditional proteomics method and showcases several recent unique applications of it in biological research. We also highlight the expansion of this strategy to other metals and discuss its future potential.

2.2 Zinc Conditional Proteomics

2.2.1 Development of Zn^{2+}-responsive Protein Labelling Reagent

An overview of the Zn^{2+} conditional proteomics method is presented in Figure 2.1a.[8] This technique employs a protein modification reagent that exhibits enhanced reactivity upon binding to intracellular free or labile Zn^{2+}. This reagent can be activated specifically in the mobile Zn^{2+}-rich environment within the cell and reacts with proteins present there to incorporate a detection/purification tag into proteins. The labelled proteins can be visualised by confocal laser scanning microscopy (CLSM) after cell fixation and in-gel fluorescence analysis after cell lysis. Furthermore, if a pull-down compatible tag is added to proteins, the labelled proteins can be purified from the cell extract, which allows protein identification by LC–MSMS.

Based on this strategy, we designed a Zn^{2+}-responsive protein labelling reagent (Figure 2.1b). This molecule contains a dipicolylamine (Dpa) ligand that selectively binds to Zn^{2+}, and an acyl imidazole (AI) reactive group for protein labelling.[9–11] These two modules are connected such that the imidazole site can participate in Zn^{2+} complex formation. When Lewis acidic Zn^{2+} coordinates with this molecule, the electron density of the imidazole ring decreases, significantly enhancing the reactivity of the AI group. This activation mechanism accelerates the labelling reaction for proteins in the vicinity of a high concentration of Zn^{2+}, enabling Zn^{2+}-environment-selective proteomics. We named this Zn^{2+}-responsive protein modification reagent AIZin (AI-based zinc-responsive reagent).

X-ray crystallography of an AIZin–Zn^{2+} complex clearly showed that AIZin forms a 1 : 1 complex with Zn^{2+} in trigonal bipyramidal geometry (Figure 2.2a). As anticipated, it was confirmed that the nitrogen atom in the imidazole ring directly coordinates with Zn^{2+}. Additionally, ^{1}H-NMR analysis demonstrated that the electron density of the imidazole ring decreases upon Zn^{2+} coordination.

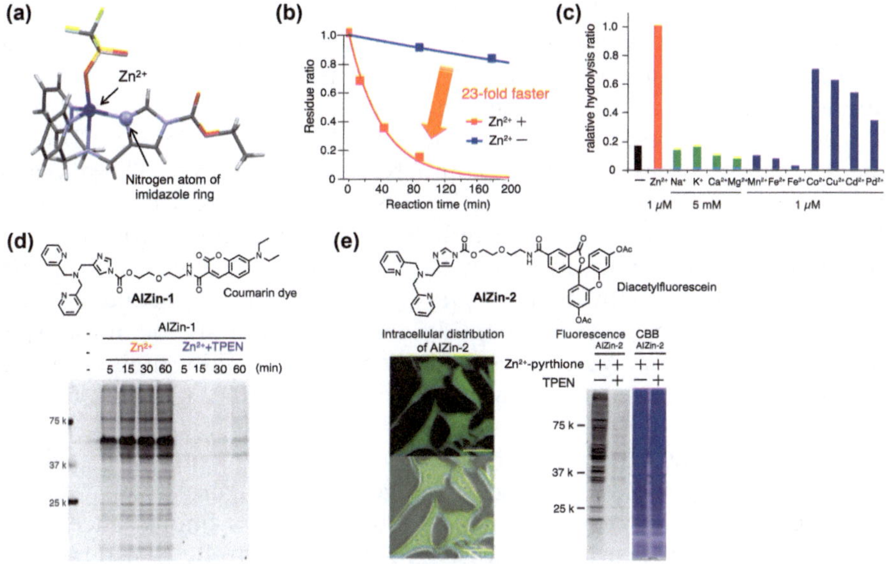

Figure 2.2 Characterisation of reaction properties of AIZin. (a) X-ray structure of the AIZin–Zn^{2+} complex. (b) Activation of AIZin by Zn^{2+}. Hydrolysis rate is 23-fold accelerated in the presence of Zn^{2+} compared to the absence of Zn^{2+}. (c) Metal ion selectivity of AIZin. (d) Zn^{2+}-responsive protein labelling in HeLa cell lysate using AIZin-1. (e) Intracellular distribution of AIZin-2 and Zn^{2+}-responsive protein labelling in live HeLa cells. Scale bar: 30 μm. CBB: Coomassie Brilliant Blue.

The activation of AIZin upon Zn^{2+} binding was evaluated in aqueous solution by measuring the hydrolysis rate. The results indicated that, in the presence of Zn^{2+}, the hydrolysis rate of AIZin was accelerated by 23 times compared with that in its absence (Figure 2.2b). This confirms that AIZin's reactivity is indeed enhanced by Zn^{2+}, as designed. The dissociation constant (K_d) of AIZin for Zn^{2+} was determined by similar hydrolysis assays to be 0.7 ± 0.3 nM. Given that the free Zn^{2+} concentration in cells is typically in the pM range but can rise to nM to μM levels in response to specific cellular stimuli or events, AIZin is expected to be unresponsive to basal intracellular free Zn^{2+} levels but activated only in specialised high-Zn^{2+} regions. Importantly, AIZin does not respond to other biologically abundant metal ions such as Na^+, K^+, Ca^{2+} and Mg^{2+}, or transition metals like Mn^{2+}, Fe^{2+} and Fe^{3+}, demonstrating very high selectivity for Zn^{2+} (Figure 2.2c). Although some transition metals like Co^{2+}, Cu^{2+}, Cd^{2+} and Pd^{2+} activated AIZin to some extent, in biological systems, these metal ions are present only at trace levels compared with Zn^{2+}, and thus their impact on the practical use of AIZin is expected to be minimal.

The reactivity of AIZin with proteins was evaluated *in vitro*. AIZin-1, which has a coumarin fluorescent dye as a detection probe, was added to HeLa cell lysate at a concentration of 1 μM, followed by the addition of $ZnSO_4$ (50 μM).

As shown in Figure 2.2d, under conditions with AIZin-1 and Zn^{2+}, several fluorescent bands corresponding to coumarin-labelled proteins were detected. In contrast, when *N*,*N*,*N*′,*N*′-tetrakis(2-pyridylmethyl)ethylenediamine (TPEN), a strong Zn^{2+} chelator, was present at a high concentration (1 mM), these fluorescent bands significantly decreased in intensity because the formation of AIZin-1–Zn^{2+} complexes was inhibited. These results clearly demonstrate that AIZin can label proteins in response to free Zn^{2+}. Notably, AIZin activation was not induced by zinc-binding enzymes such as carbonic anhydrase, indicating that AIZin is activated by free Zn^{2+} (or Zn^{2+} weakly interacting with biomolecules), but does not respond to Zn^{2+} strongly bound to proteins. Peptide mapping analysis by mass spectrometry revealed that AIZin reacts with diverse nucleophilic amino acids on the protein surface, including lysine, serine, threonine, tyrosine, cysteine and histidine.

Zn^{2+}-responsive protein labelling in live cell environments was tested with AIZin-2 bearing diacetylfluorescein as a probe (Figure 2.2e). Diacetylfluorescein is non-fluorescent outside cells but emits fluorescence inside cells owing to the cleavage of acetyl groups by intracellular esterases. As expected, AIZin-2 penetrated the cell membrane and was distributed uniformly within the cells approximately 15 min after being added to the culture medium. When zinc pyrithione was added as an exogenous Zn^{2+} source, protein labelling in response to the increase in intracellular Zn^{2+} was clearly observed (Figure 2.2e). Additionally, this labelling reaction was inhibited by TPEN. These results demonstrated that AIZin can modify proteins in a Zn^{2+}-responsive manner, even in live cells.

2.2.2 AIZin-2-mediated Profiling of Dynamic Changes in Proteins in the Zn^{2+}-rich Region upon NO Stimulation

With live cell-compatible AIZin-2 in hand, we attempted to label and identify proteins in Zn^{2+}-rich environments in live cells. As a proof-of-principle study, we induced oxidative stress in C6 glioma cells using nitric oxide (NO). Much of the intracellular Zn^{2+} is known to be strongly bound to metallothioneins, but when the thiol groups of cysteine in these proteins are *S*-nitrosylated by NO, their ability to bind to Zn^{2+} is lost, resulting in increased levels of free Zn^{2+} in the cell.[12] It has also been reported that, in C6 glioma cells, the excess free Zn^{2+} is temporarily sequestered in intracellular vesicles, thereby mitigating Zn^{2+} toxicity.[7] However, the protein components of these Zn^{2+}-rich vesicles, their origins and their relationship with other organelles are largely unknown, leaving many questions about the metabolism of free Zn^{2+} unanswered.

To characterise the Zn^{2+}-rich vesicles in C6 glioma cells, we used the developed AIZin-2 to identify proteins within them. First, we observed the distribution of Zn^{2+} in C6 cells under NO stimulation using a commercial fluorescent Zn^{2+} sensor, FluoZin-3. The results showed that Zn^{2+} levels temporarily increased throughout the cell (both in the cytoplasm and nucleus) for approximately 10 min after NO stimulation, and then Zn^{2+} became concentrated in specific small vesicles (Zn^{2+}-rich vesicles) over approximately 3 h (Figure 2.3a). AIZin-2 was then added to cells 10 min and 3 h after NO stimulation, and the labelled proteins were enriched by pull-down with anti-fluorescein antibody and subjected to tandem mass tag (TMT)-based quantitative mass spectrometry. Quantitative analysis of the approximately 300 identified proteins revealed that, 10 min after NO stimulation, proteins distributed in the cytoplasm and nucleus were primarily labelled (Figure 2.3b and c, group A). In contrast, 3 h after NO stimulation, the labelling of these proteins decreased, and instead, proteins associated with the endoplasmic reticulum (ER), Golgi apparatus and ER–Golgi intermediate vesicles were more enriched as the highly labelled proteins (Figure 2.3b and c, group B). Importantly, the information on intracellular Zn^{2+} distribution obtained from this proteomics-based analysis was well consistent with the Zn^{2+} localisation changes observed in the FluoZin-3 fluorescence imaging (Figure 2.3a). These results demonstrated that

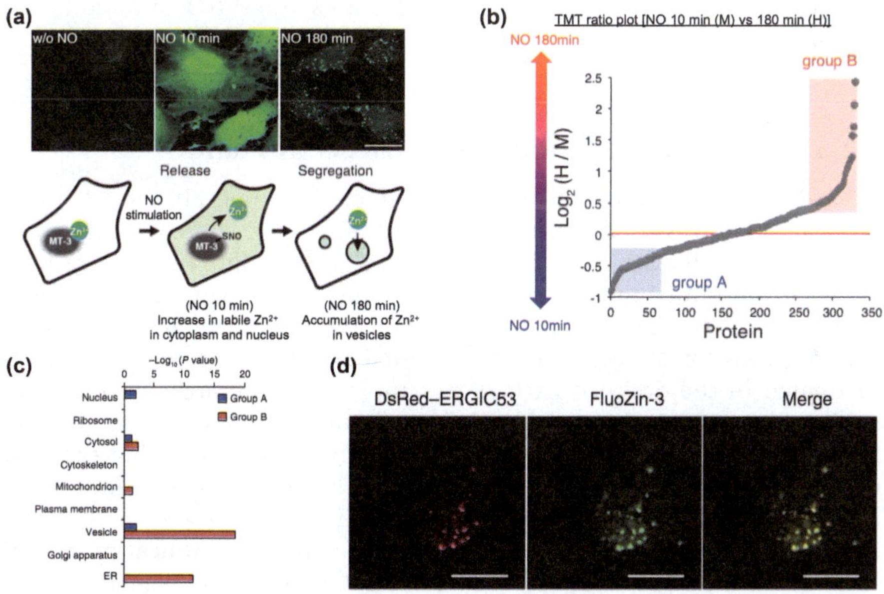

Figure 2.3 Conditional protein labelling in live C6 cells stimulated by NO. (a) Imaging of changes in intracellular Zn^{2+} distribution using FluoZin-3 upon NO stimulation. Scale bar: 20 μm. (b) Quantitative LC–MSMS analysis of the labelled proteins with AIZin-2 at 10 min and 3 h after NO stimulation. (c) Cellular component analysis of the subset enriched in NO-10 min over 180 min (group A, coloured in blue) and subset enriched in NO-180 min over 10 min (group B, coloured in red). Proteins in the nucleus and vesicle were enriched in group A, whereas proteins in the vesicle and endoplasmic reticulum (ER) were enriched in group B. (d) CLSM imaging of zinc localisation (FluoZin-3, green) in ERGIC vesicles (DsRed–ERGIC53; red). Scale bars, 10 μm.

combining AIZin-2 with an MS-based chemical proteomics approach can provide a snapshot of the array of proteins present around fluctuating Zn^{2+} levels in a spatiotemporal manner.

As described above, upon detailed analysis of the protein groups identified in the sample taken 3 h after NO stimulation, we found that a relatively large number of proteins were localised in the ER and Golgi apparatus. Additionally, many of these proteins were involved in folding and post-translational modifications, suggesting that the Zn^{2+}-rich vesicles observed by fluorescence imaging likely originate from the ER and Golgi apparatus. Specifically, the identified proteins included Sec proteins, G proteins such as Rab and ARF that regulate vesicle transport, and several COP proteins. These proteins are associated with COPI vesicles, COPII vesicles and the ERGIC, all of which are involved in vesicular transport between the ER and Golgi apparatus.[13] This suggested that free Zn^{2+} accumulates in these vesicles in response to NO stimulation. To test this hypothesis, we performed co-staining of the Zn^{2+}-rich vesicles visualised by FluoZin-3 and the ERGIC marker protein ERGIC53 fused with DsRed (Figure 2.3d). This revealed that the vesicles in which free Zn^{2+} accumulates after NO stimulation are intermediate vesicles that shuttle between the Golgi apparatus and the rough ER.

2.2.3 Detection of Mobile Zn^{2+} Levels in Plants

AIZin-2 is now commercially available under the name Zin-Pro Capture or Zinc-Stamp and is used for various applications by groups other than ours. For example, Shinozaki *et al.* used AIZin-2 (called Zin-Pro Capture in the paper) to determine mobile Zn^{2+} levels in wild-type and autophagy-deficient *Arabidopsis*.[14] In that work, AIZin-2 was vacuum-infiltrated into shoots of plants grown for 2 days under Zn^{2+}-deficient conditions, which were then incubated in the dark for 24 h. The samples were homogenised and then subjected to SDS-PAGE and in-gel fluorescence analysis. The mobile Zn^{2+} levels in these two samples could be determined by calculating the difference in fluorescence signal intensity of the labelled proteins. Indeed, this study showed that the level of fluorescently-labelled proteins was lower in autophagy-deficient *Arabidopsis* than in the wild type, indicating that the level of mobile Zn^{2+} inside plants was increased by autophagy under Zn^{2+}-deficient conditions. This study demonstrates that the protein labelling reaction with AIZin-2 proceeds in plants and highlights AIZin's unique ability to quantify zinc fluctuations based on the intensity of protein labelling, even in plant samples that are difficult to analyse by imaging-based techniques using conventional fluorescent zinc sensors.

2.3 Iron Conditional Proteomics

Ferroptosis is an iron-dependent form of programmed cell death characterised by the accumulation of oxidised lipids, which is known to be associated with various diseases. In particular, cancers resistant to existing anticancer drugs are highly susceptible to ferroptosis, and thus the artificial induction of ferroptosis is expected to be a new cancer treatment strategy. Recently, Renslo *et al.* developed an Fe^{2+}-responsive protein labelling reagent, named Ferroptosis Inducing Peroxide for Chemoproteomics-1 (FIPC-1) (Figure 2.4a), and used it to identify previously unknown ferroptosis regulators.[15] FIPC-1 is an artemisinin analogue with an ozonide and adamantane skeleton, which generates radical intermediates with high protein modification activity through an Fe^{2+}-mediated Fenton-like reaction (Figure 2.4a). Additionally, FIPC-1 itself was found to induce ferroptosis (although not mentioned in the paper, it is inferred that radicals generated by FIPC-1 lead to the reduction of oxygen molecules, producing superoxide with high lipid oxidation activity). Indeed, a compound with a structure nearly identical to FIPC-1 was shown to induce ferroptosis in HT-1080 cells ($EC_{50} = 9.7$ μM) with the accumulation of oxidised lipids inside cells.

Quantitative chemoproteomics analysis of ferroptosis-induced cells treated with FIPC-1 reproducibly identified proteins involved in redox homeostasis, such as thioredoxin, oxidoreductase and protein disulfide isomerase (PDI) (Figure 2.4b and c). The expression of these proteins is considered to increase during ferroptosis. Notably, proteins that had not previously been known as ferroptosis regulators, such as P4HB (a member of the PDI family)

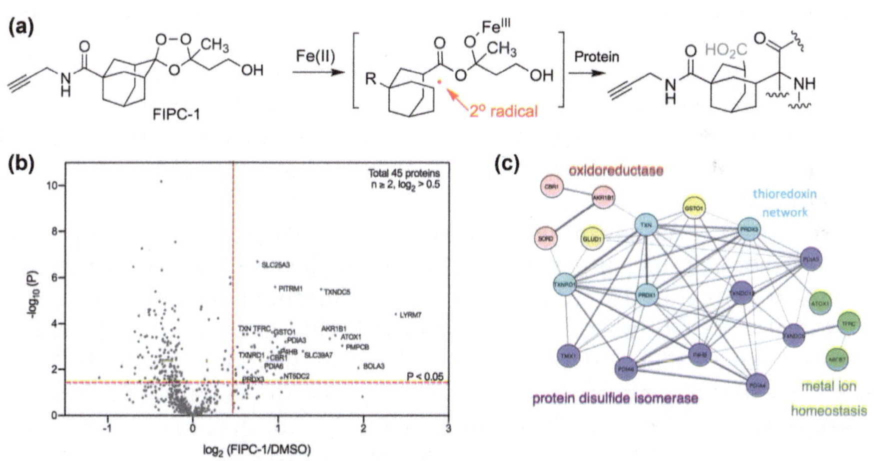

Figure 2.4 Fe^{2+}-responsive protein labelling and its application for conditional proteomics. (a) Molecular structure of FIPC-1 and putative mechanism of FIPC-1 mediated protein labelling. (b) Volcano plot of enriched FIPC-1 target proteins. Most enriched proteins by FIPC-1 mediated labelling were involved in the regulation of cellular redox homeostasis, including members of the thioredoxin network, the protein disulfide isomerase (PDI) family, and NADPH-dependent oxidoreductases. (c) Interactome network highlighting 19 proteins enriched by FIPC-1 and known to be involved in redox processes. Reproduced from ref. 15 with permission from American Chemical Society, Copyright 2020.

and the functionally unknown protein NT5DC2, were included among the identified proteins. The authors subsequently knocked down these genes and treated the cells with known ferroptosis inducers RSL3 (a GPX4 inhibitor) and erastin (a cystine transporter inhibitor). The results showed that the P4HB- and NT5DC2-knockdown cells were more sensitive to ferroptosis inducers than control cells were, indicating that these genes are involved in ferroptosis resistance. Moreover, the authors focused on PACMA31, an inhibitor of P4HB, and demonstrated that it can directly induce ferroptosis in HT1030 cells.

2.4 Copper Conditional Proteomics

2.4.1 AI-based Cu^+/Cu^{2+}-responsive Probes

Metal ion-responsive protein labelling reagents offer several advantages over conventional methods, not only for conditional proteomics but also for imaging applications. Conventional fluorescent chemosensors for the selective detection of intracellular metal ions diffuse freely within cells, so the spatial information of the observed fluorescent signals may not necessarily reflect the subcellular distribution of the metal ions. Additionally, most chemical probes are washed out from the cells after cell fixation, making it impossible to combine them with immunostaining commonly used in cell biology. In contrast, metal ion-responsive protein labelling reagents transfer the detection tag to fixable proteins upon binding to metal ions, thus preserving the information on the localisation of metal ions even after cell fixation. Moreover, when reacting with less diffusible proteins such as cytoskeletal proteins, the spatial information of the metal ions of interest is fixed, potentially providing higher spatial resolution than with conventional freely diffusible chemical probes.

Focusing on this advantage, Chang *et al.* developed a new copper ion-responsive protein labelling reagent, CD649, which connects a copper-selective bis-thioether ligand to an AI group.[16] *In vitro* experiments demonstrated that CD649 selectively binds to copper ions (both Cu^+ and Cu^{2+}) and becomes reactive (Figure 2.5a and b). When CD649 was introduced into HEK293T cells and external copper (Cu-gtsm complex) was added, strong fluorescence was observed inside the cells compared with the findings in untreated cells. This indicated that CD649 is activated in response to labile copper within the cells and covalently modifies proteins with fluorescent dye, while unreacted reagent is readily washed out from the cells. In addition, the increase in fluorescence intensity upon the addition of Cu-gtsm complexes was greater in cells in which ATP7A, which is responsible for the extracellular efflux of copper, was knocked out compared with that in the wild type. This result confirmed the validity of copper ion-responsive protein labelling by CD649. The authors subsequently applied CD649 to the three main cell types in the brain, namely, neurons, astrocytes and microglia, which revealed that copper pools in neurons migrate from the cell body to dendrites upon KCl

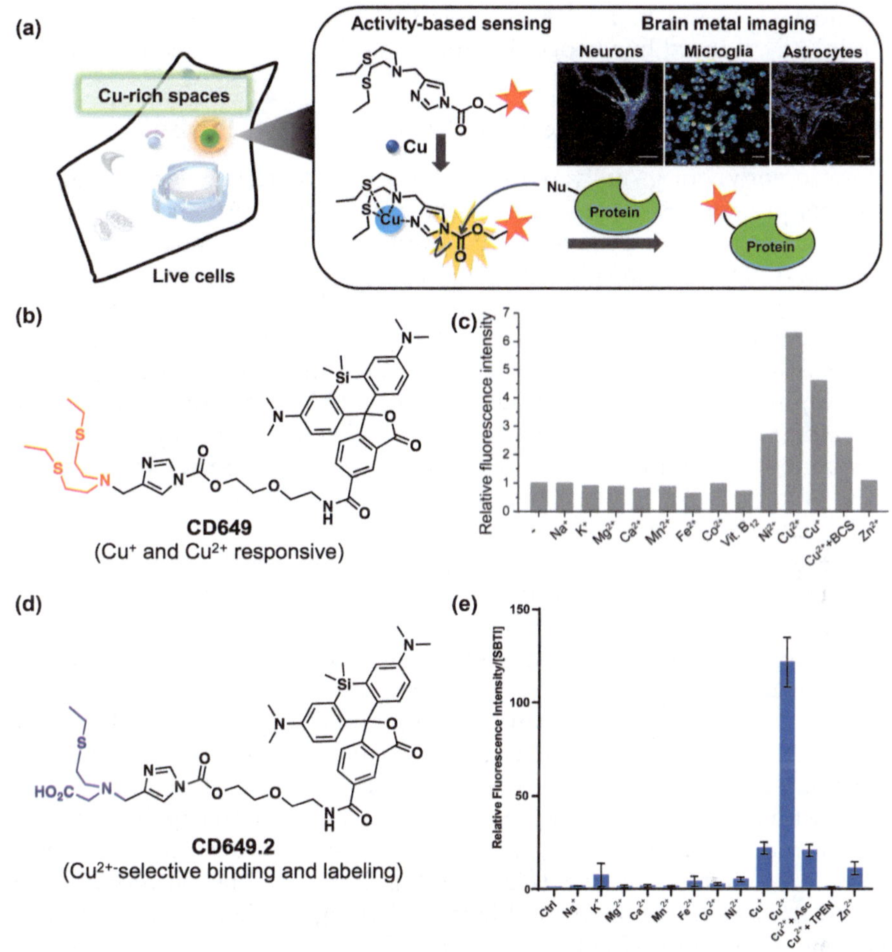

Figure 2.5 Copper ion-responsive protein labelling and imaging of intracellular copper pools. (a) Schematic illustration showing the working principle of copper responsive protein labelling reagent inside cells. (b) Molecular structure of CD649. (c) Metal ion selectivity of CD649. (d) Molecular structure of CD649.2. (e) Metal ion selectivity of CD649.2. Reproduced from ref. 16 with permission from American Chemical Society, Copyright 2020. Reproduced from ref. 17, https://doi.org/10.1073/pnas.2202736119, under the terms of the CC BY 4.0 license, https://creativecommons.org/licenses/by/4.0/.

stimulation, and that copper decreases in astrocytes but increases in microglia in response to inflammatory stimuli.

One important limitation of CD649 is its low selectivity for copper ion oxidation states (Cu^{+}/Cu^{2+}) (Figure 2.5c). To overcome this, the same group recently developed CD649.2, which shows higher selectivity for Cu^{2+} (Figure 2.5d and e).[17] This new probe not only provided fundamental information on the existence and regulation of labile Cu^{2+} pools, but also revealed reciprocal increases in labile Cu^{2+} accompanied by decreases in labile Cu^{+} induced by oncogenic mutations that promote oxidative stress.

2.4.2 Quinone Methide-based Cu^{+}-selective Probe

More recently, our group reported a novel design strategy for creating conditional proteomics probes that can be exclusively activated in response to labile Cu^{+}.[18] In our design of Cu^{+}-responsive protein labelling reagents, termed CuRs, a fluorescent dye and a copper chelator are concatenated through a latent quinone methide precursor with a benzyl ether linkage. The C–O bond of the benzyl ether is oxidatively cleaved upon Cu^{+} complexation of the reagent, which undergoes 1,4- or 1,6-elimination to generate a highly electrophilic quinone methide (Figure 2.6).[19] The released quinone methide reacts with nucleophilic amino acids (*e.g.*, Cys, Lys and His) of surrounding proteins to incorporate a fluorescent reporter into them. In this strategy, intracellular fluctuations in Cu^{+} can be quantified by analysing changes in the fluorescent signal intensity of the labelled protein using SDS-PAGE or fixed-cell imaging. Furthermore, the intracellular proteome with elevated labile Cu^{+} concentrations can be identified by LC–MSMS analysis.

We prepared a set of fluorescein-appended CuRs with latent *ortho*- or *para*-quinone methide precursor coupled with two different copper chelators, namely, tris(2-pyridylmethyl)amine (TPA) and 2-pyridylmethylbis(2-ethylthioethyl)amine (Figure 2.7a, O2, O3, P2, P3). Control CuRs without any chelating group were also prepared (Figure 2.7a, O1, P1). Gratifyingly, all CuRs with copper chelators were found to have activation capability for Cu^{+}, as evidenced by the labelling with BSA *in vitro*, while control compounds showed negligible reactivity (Figure 2.7b). Among the tested probes, we selected P2, which exhibited the highest labelling efficiency with BSA, as the optimal CuR, and characterised its selectivity toward Cu^{+} over other

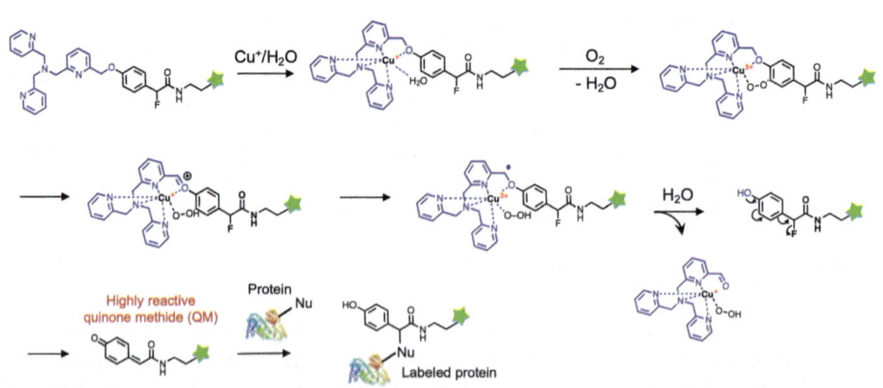

Figure 2.6 Reaction mechanism of the protein labelling reagent selectively activated by Cu^{+}.

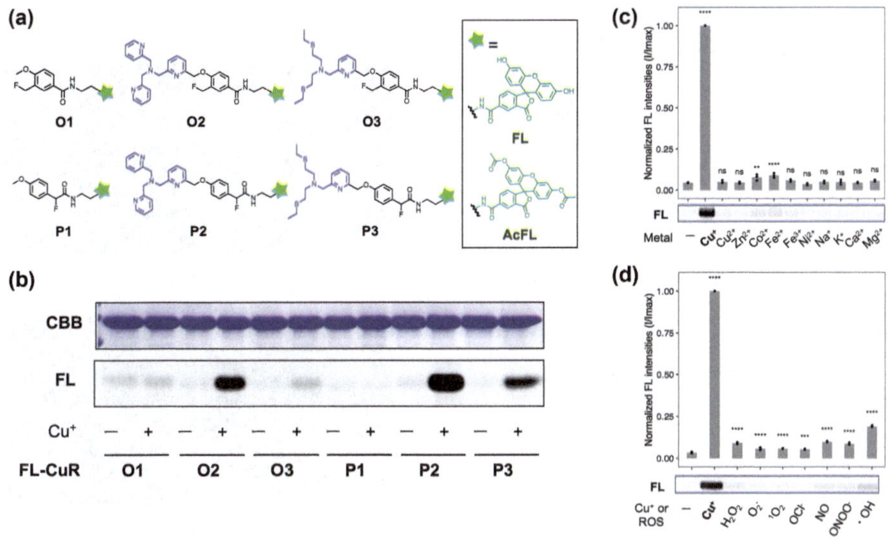

Figure 2.7 Development and characterisation of CuRs. (a) Molecular structures of Cu^+-responsive labelling reagents. FL, fluorescein; AcFL, diacetylated fluorescein. (b) SDS–PAGE and in-gel fluorescence analysis of BSA labelling with FL-tethered CuRs (FL-CuRs). The reagent (50 μM) was added to 50 mM HEPES buffer (pH 7.2) containing BSA (10 μM). (c) and (d) Selectivity of FL-P2 for several metal ions (panel c) and ROS (panel d).

biometals (Figure 2.7c). The results revealed that P2 was exclusively activated by Cu^+ among all tested metal ions, including Cu^{2+}, Zn^{2+} and redox active Fe^{2+}/Fe^{3+}. Importantly, P2 showed less sensitivity toward reactive oxygen species (ROS), such as H_2O_2, O_2^-, $^1O_2$, OCl^-, NO, $ONOO^-$ and $^{\bullet}OH$, which is an important feature for the specific detection of Cu^+ in cells because intracellular copper accumulation often generates a variety of ROS (Figure 2.7d).

Having confirmed the selectivity for Cu^+, we subsequently tested whether P2 can detect changes in labile copper levels inside cells. HeLa cells transiently transfected with copper transporter 1 (CTR1),[20] a membrane protein that selectively imports extracellular copper into cells, were preincubated in the presence or absence of 5 μM $CuCl_2$ for 8.5 h, followed by the administration of acetyl fluorescein-tethered P2 (AcFL-P2). SDS-PAGE and in-gel fluorescence analysis showed that many proteins were labelled in the CTR1-transfected cells treated with exogenous copper, whereas the labelling reaction scarcely occurred in the samples without CTR1 transfection or copper supplementation (Figure 2.8a). We also confirmed that treatment of the CTR1-overexpressing cells with *N*-methylmaleimide (*N*-MEM), a thiol alkylation reagent that can release protein-bound copper, caused a remarkable increase in labile Cu^+, resulting in enhancement of the labelling signals of in-gel fluorescence. In addition, CLSM imaging of fixed cells clearly showed

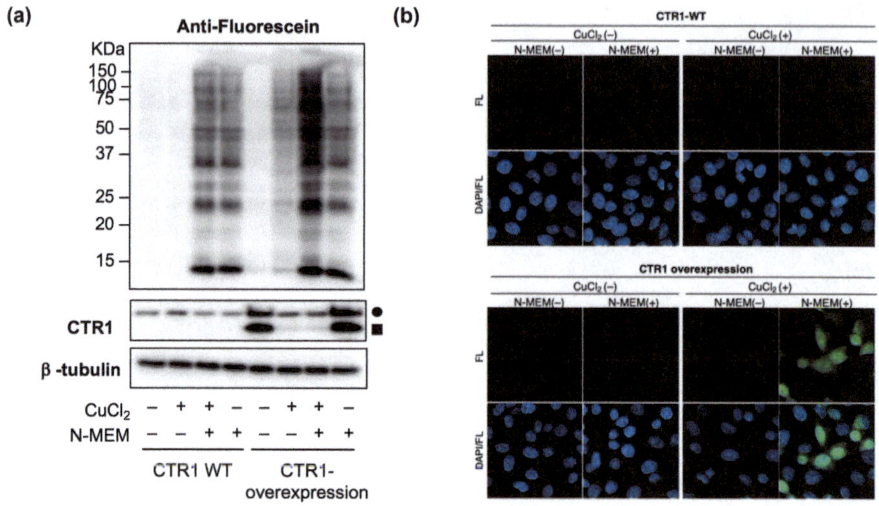

Figure 2.8 Enhancement of the labelling signal with AcFL-P2 by increasing free or labile Cu in CTR1-expressing cells. (a) Western blotting and (b) imaging analysis of the Cu^+-responsive protein labelling in cells. Nuclei were stained by 4′,6-diamidino-2- phenylindole (DAPI). ●, Mature CTR1; ■, immature CTR1.

strong fluorescence signals derived from the fluorescein-labelled proteins in CTR1-overexpressing cells (Figure 2.8b). All of these results clearly demonstrate that AcFL-P2 is activated and modifies intracellular proteins in response to the abnormal accumulation of labile Cu^+ transported into the cell by CTR1.

AcFL-P2 can also detect copper dysregulation induced by the genetic perturbation of ATP7A, the causative factor of Menkes disease.[21] ATP7A-knockout and wild-type human melanoma Mewo cells were preincubated with Cu-gtsm complex (2 μM) to increase intracellular copper levels, and then incubated with AcFL-P2 for 30 min. The elevated concentration of labile Cu^+ in ATP7A-knockout cells was successfully detected using both SDS-PAGE and CLSM, while wild-type cells exhibited a weaker signal because ATP7A rapidly excreted excess intracellular exogenous copper ions (Figure 2.9a and b).

Like the AIZin probe, CuR allows proteomics analysis in intracellular environments with elevated labile Cu^+ levels. Using a workflow similar to that for AIZin, we characterised the proteins modified with AcFL-P2 in Cu-gtsm-treated ATP7A-wild-type and -knockout cells (Figure 2.9c). The labelled proteins were enriched by immunoprecipitation with an anti-fluorescein antibody, digested using trypsin and analysed by LC–MSMS. In this study, label-free quantification was used in the mass analysis to detect the dynamic changes in protein labelling caused by exogenous copper relative to the basal (copper-untreated) condition. A larger number of proteins were identified in knockout cells than in wild-type ones, which was consistent with the gel- and

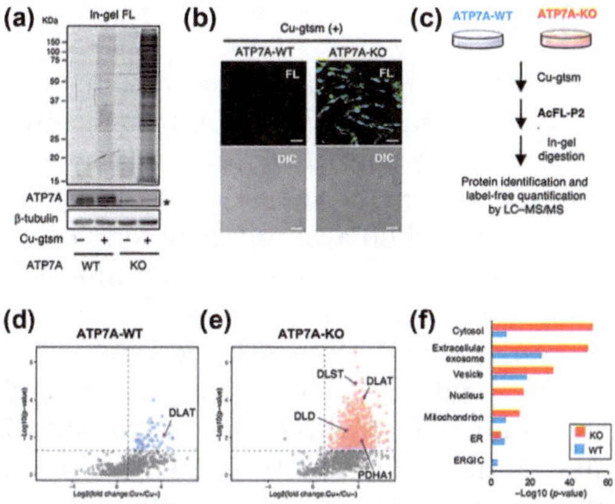

Figure 2.9 Cu^+-responsive protein labelling with AcFL-P2 in ATP7A-knockout (KO) cells. (a) In-gel fluorescence imaging of labelled proteins in wild-type (WT) and ATP7A-KO Mewo cells pretreated with Cu-gtsm (2 μM) for 30 min, then washed with medium and further incubated with AcFL-P2 (5 μM) for 30 min. The band marked with an asterisk results from nonspecific reactivity of the anti-ATP7A antibody. (b) CLSM images of Mewo cells after labelling. Scale bar, 40 μm. (c) Workflow for label-free quantitative analysis of labelled proteins using LC–MS/MS. (d) and (e) Volcano plots of proteins identified using Cu^+-responsive protein labelling in WT (panel d) and ATP7A-KO (panel e) Mewo cells. Significantly labelled proteins that met the criteria for hit proteins (fold change $(Cu(+)/Cu(-)) \geq 2$, number of identified peptides ≥ 2, and $p < 0.05$) in WT and KO cells are shown in blue and red plots, respectively. (f) Gene ontology analysis of significantly labelled proteins in WT (blue) and KO (red) cells. Cu-gtsm, glyoxal-bis(*N*(4)-methylthiosemicarbazonato)-copper(II).

imaging-based analyses (Figure 2.9d and e). We also found that proteins associated with exosomes and vesicles were more enriched in wild-type cells than other cellular components, suggesting that, in wild-type (normal) cells, copper is sequestered and concentrated in these intracellular compartments for subsequent excretion (Figure 2.9f). In contrast, cytosolic proteins were more enriched in ATP7A-knockout cells, indicating that abnormal copper efflux results in increased labile Cu^+ in the cytosol. We also noticed that mitochondrial proteins were more enriched in the knockout cells than in wild-type ones. Interestingly, our proteomics dataset contained DLAT, DLST, DLD and PDHA1, which were previously identified as proteins essential for cuproptosis.[22] In the proposed mechanism of cuproptosis, excess copper causes the aggregation of lipoylated proteins and the destabilisation of Fe–S cluster proteins, both of which are expressed in mitochondria, resulting in mitochondrial dysregulation and eventually cell death. However, there has been no direct evidence for the accumulation of Cu^+ in mitochondria. In this regard, our conditional proteomics study with AcFL-P2 provides the first experimental evidence to suggest the excess labile Cu^+ in mitochondria.

2.5 Conclusion

This chapter introduced recent research examples using metal ion-responsive protein modification reagents for a conditional proteomics approach. This new probe design concept provides a unique way not only of visualising the dynamics of intracellular metal ions, but also of characterising proteins present in metal ion-enriched environments at single-cell or organelle-level spatial resolution. Although we focused here on applications for trace metal elements, the conditional proteomics strategy should be extended to various intracellular environments, including ROS, reactive nitrogen species (RNS) and other redox active species. In fact, we have recently developed conditional proteomics probes for hydrogen peroxide and nitric oxide.[23,24] Because the abnormal production of ROS and RNS often correlates with the metabolic disruption of biometals, further development of these conditional proteomics probes should facilitate studies on both biometals and cellular redox homeostasis.[25,26] Looking ahead, as the range of target metals and biomolecules continues to grow, it is anticipated that previously unexplored biological phenomena that could not be captured by conventional imaging probes will be elucidated.

Acknowledgements

We thank Edanz (https://jp.edanz.com/ac) for editing a draft of this manuscript. This work was supported by the Japan Science and Technology Agency (JST) ERATO Grant JPMJER1802 and Japan Society for the Promotion of Science (JSPS) Grant-in-Aid for Specially Promoted Research (23H05405) to I. H.; Scientific Research (B) (21H02058) to T. T.

References

1. C. J. Frederickson, J. Y. Koh and A. I. Bush, *Nat. Rev. Neurosci.*, 2005, **6**, 449.
2. S. L. Sensi, P. Paoletti, A. I. Bush and I. Sekler, *Nat. Rev. Neurosci.*, 2009, **10**, 780.
3. K. P. Carter, A. M. Young and A. E. Palmer, *Chem. Rev.*, 2014, **114**, 4564.
4. T. Kowada, T. Watanabe, Y. Amagai, R. Liu, M. Yamada, H. Takahashi, T. Matsui, K. Inaba and S. Mizukami, *Cell Chem. Biol.*, 2020, **27**, 1521–1531.e8.
5. Y. Amagai, M. Yamada, T. Kowada, T. Watanabe, Y. Du, R. Liu, S. Naramoto, S. Watanabe, J. Kyozuka, T. Anelli, T. Tempio, R. Sitia, S. Mizukami and K. Inaba, *Nat. Commun.*, 2023, **14**, 2683.
6. D. W. Domaille, E. L. Que and C. J. Chang, *Nat. Chem. Biol.*, 2008, **4**, 168.
7. H. Haase and D. Beyersmann, *Biochem. Biophys. Res. Commun.*, 2002, **296**, 923.
8. T. Miki, M. Awa, Y. Nishikawa, S. Kiyonaka, M. Wakabayashi, Y. Ishihama and I. Hamachi, *Nat. Methods*, 2016, **13**, 931.

9. S. Fujishima, R. Yasui, T. Miki, A. Ojida and I. Hamachi, *J. Am. Chem. Soc.*, 2012, **134**, 3961.
10. T. Miki, S. Fujishima, K. Komatsu, K. Kuwata, S. Kiyonaka and I. Hamachi, *Chem. Biol.*, 2014, **21**, 1013.
11. T. Tamura and I. Hamachi, *J. Am. Chem. Soc.*, 2019, **141**, 2782.
12. R. A. Colvin, W. R. Holmes, C. P. Fontaine and W. Maret, *Metallomics*, 2010, **2**, 306.
13. F. Brandizzi and C. Barlowe, *Nat. Rev. Mol. Cell Biol.*, 2013, **14**, 382.
14. D. Shinozaki, E. A. Merkulova, L. Naya, T. Horie, Y. Kanno, M. Seo, Y. Ohsumi, C. Masclaux-Daubresse and K. Yoshimoto, *Plant Physiol.*, 2020, **182**, 1284.
15. Y. C. Chen, J. A. Oses-Prieto, L. E. Pope, A. L. Burlingame, S. J. Dixon and A. R. Renslo, *J. Am. Chem. Soc.*, 2020, **142**, 19085.
16. S. Lee, C. Y. S. Chung, P. Liu, L. Craciun, Y. Nishikawa, K. J. Bruemmer, I. Hamachi, K. Saijo, E. W. Miller and C. J. Chang, *J. Am. Chem. Soc.*, 2020, **142**, 14993.
17. A. T. Pezacki, C. D. Matier, X. Gu, E. Kummelstedt, S. E. Bond, L. Torrente, K. L. Jordan-Sciutto, G. M. DeNicola, T. A. Su, D. C. Brady and C. J. Chang, *Proc. Natl. Acad. Sci. U. S. A.*, 2022, **119**, e2202736119.
18. R. Cheng, Y. Nishikawa, T. Wagatsuma, T. Kambe, Y. K. Tanaka, Y. Ogra, T. Tamura and I. Hamachi, *ACS Chem. Biol.*, 2024, **19**, 1222.
19. K. K. Yu, K. Li, J. T. Hou and X. Q. Yu, *Tetrahedron Lett.*, 2013, **54**, 5771.
20. J. Lee, M. M. O. Peña, Y. Nose and D. J. Thiele, *J. Biol. Chem.*, 2002, **277**, 4380.
21. S. G. Kaler, *Nat. Rev. Neurol.*, 2011, **7**, 15.
22. P. Tsvetkov, S. Coy, B. Petrova, M. Dreishpoon, A. Verma, M. Abdusamad, J. Rossen, L. Joesch-Cohen, R. Humeidi, R. D. Spangler, J. K. Eaton, E. Frenkel, M. Kocak, S. M. Corsello, S. Lutsenko, N. Kanarek, S. Santagata and T. R. Golub, *Science*, 2022, **375**, 1254.
23. H. Zhu, T. Tamura, A. Fujisawa, Y. Nishikawa, R. Cheng, M. Takato and I. Hamachi, *J. Am. Chem. Soc.*, 2020, **142**, 15711.
24. Y. Nishikawa, T. Miki, M. Awa, K. Kuwata, T. Tamura and I. Hamachi, *ACS Chem. Biol.*, 2019, **14**, 397.
25. K. Jomova and M. Valko, *Toxicology*, 2011, **283**, 65.
26. S. J. Stohs and D. Bagchi, *Free Radical Biol. Med.*, 1995, **18**, 321.

CHAPTER 3

Structure and Biological Activity of Zinc Complexes Coordinated with Chalcogen Atoms

H. YASUI,*[a] T. NISHIGUCHI,[a] Y. NAITO[a] AND Y. YOSHIKAWA[b]

[a] Department of Analytical and Bioinorganic Chemistry, Division of Analytical and Physical Sciences, Kyoto Pharmaceutical University, Kyoto, Japan; [b] Department of Health, Sports, and Nutrition, Kobe Women's University, Kobe, Japan
*Email: yasui@mb.kyoto-phu.ac.jp

3.1 Introduction

According to a report by the International Diabetes Federation (IDF), the global diabetes population was 537 million in 2021 and is projected to reach approximately 700 million by 2045.[1] Diabetes is associated with increased risks of several serious complications such as retinopathy, renal disorders, and nerve disorders.[2] The long-term complications of diabetes also include the development of vascular disorders such as angiopathy.[3] Moreover, hypertension is a pathological state independently associated with an increased risk of cardiovascular diseases.[4,5] Many epidemiological studies have also demonstrated a correlation between insulin resistance and blood pressure in the state of diabetes, suggesting that insulin resistance is a risk factor in the development of hypertension in humans.[6]

Coordination Chemistry Perspectives No. 2
Zinc in Biology: Molecular Structures, Cellular Processes and Living Systems
Edited by Hiroyuki Yasui and Taiho Kambe

Published by the Royal Society of Chemistry, www.rsc.org

Diabetes is broadly divided into type 1 diabetes (IDDM) and type 2 diabetes (NIDDM).[7] IDDM is caused by an absolute lack of insulin secretion and is caused by an autoimmune mechanism triggered by T cells and other immune cells or, more recently, by viral infection, which is thought to cause damage to the pancreatic islets of Langerhans.[8] On the other hand, NIDDM is thought to be caused by a combination of environmental factors, such as overeating and lack of exercise, and genetic factors, leading to insulin resistance in the liver and muscles and loss of beta cells in the pancreatic islets, and accounts for around 90–95% of all diabetes cases.[7,9]

In addition to insulin injections to compensate for the lack of insulin action, synthetic hypoglycaemic agents are also used to treat diabetes. Examples include: (1) sulfonylurea (SU) drugs (such as glimepiride and gliclazide), which bind to the SU receptors on pancreatic beta cells, stimulate insulin secretion and lower blood glucose; (2) insulin secretagogues (glinides), which also stimulate pancreatic insulin secretion; (3) thiazolidinedione derivatives (pioglitazone), which mainly activate the PPARγ transcription factor and improve insulin resistance; (4) biguanide derivatives (metformin), which inhibit gluconeogenesis in the liver by a mechanism different from that of insulin and have antidiabetic effects; (5) α-glucosidase inhibitors (acarbose, voglibose, miglitol, *etc.*), which inhibit the action of disaccharide-decomposing enzymes (α-glucosidase) by competing with disaccharides in the small intestinal mucosa, thereby suppressing postprandial hyperglycaemia; (6) GLP-1 preparations (exenatide, liraglutide, *etc.*), which are glucagon-like peptide preparations that exert a hypoglycaemic effect by stimulating glucose-dependent insulin secretion through a mechanism distinct from that of sulfonylureas; (7) DPP-4 inhibitors (alogliptin, sitagliptin, omarigliptin, *etc.*), which inhibit dipeptidyl peptidase-4 (DPP-4), the enzyme responsible for degrading GLP-1; and (8) SGLT-2 inhibitors (dapagliflozin, empagliflozin, *etc.*), which promote glucose excretion by inhibiting its reabsorption in the kidney, and are used in clinical practice. However, these drugs all have mechanism-based side effects such as hypoglycaemia and hepatopathy. For example, the most serious side-effect of biguanides such as metformin is lactic acidosis.[10] Therefore, the development of treatments for diabetes is ongoing also because diabetes is difficult to completely cure.[11]

3.2 Metal-containing Clinical Medicines

On the other hand, to lead a healthy life, humans need to eat a balanced diet every day. This diet consists mainly of the three main nutrients (protein, fat and carbohydrates) and micronutrients.[12] Micronutrients include nutrients called essential trace elements, which are present in the body in amounts ranging from a few milligrams to a few grams. Essential trace minerals are defined as (1) when a deficiency of an element causes a deficiency disease; (2) supplementation with the deficient element improves the deficiency disease; (3) the element has been shown to be a component of a substance that is important to the body.[12] The nine elements currently reported as essential trace elements are iron (Fe), zinc (Zn), copper (Cu), selenium (Se), iodine (I), cobalt (Co), chromium (Cr), manganese (Mn) and molybdenum (Mo),[12] and it is worth noting that seven of the nine are metallic elements (Se is a semi-metallic element).

Medicines containing metal elements (metal-containing medicines) date back to arsphenamine (arsenic: As), a medicine for syphilis developed by Ehrlich and Hata in 1910.[13] Since then, metal-containing medicines have been used in clinical practice, including auranofin (gold: Au), a drug for rheumatoid arthritis; cyanocobalamin (Co), a vitamin B_{12} preparation used to treat peripheral neuropathy; cisplatin (platinum: Pt), an anticancer drug; and gastric ulcer drugs such as sucralfate (aluminium: Al) and polaprezinc (Zn).[13,14] Furthermore, in March 2017, the use of zinc acetate (brand name: Nobelzin), a treatment for Wilson's disease, was expanded to include the treatment of hypozincemia, and it is now used to treat zinc deficiency.[15] Recently, in August 2024, zinc histidine hydrate (brand name: Zintus) was introduced as a new medicine for the treatment of hypozincemia. Clinical trials have shown that zinc histidine hydrate actually increases the amount of zinc absorbed from the gastrointestinal tract more than inorganic zinc ion or zinc acetate (Nobelzin), and it also has fewer side effects than Nobelzin. Zinc histidine hydrate is therefore expected to be an important zinc supplement preparation for hypozincemia and zinc deficiency-related diseases in the future[16] (Figure 3.1).

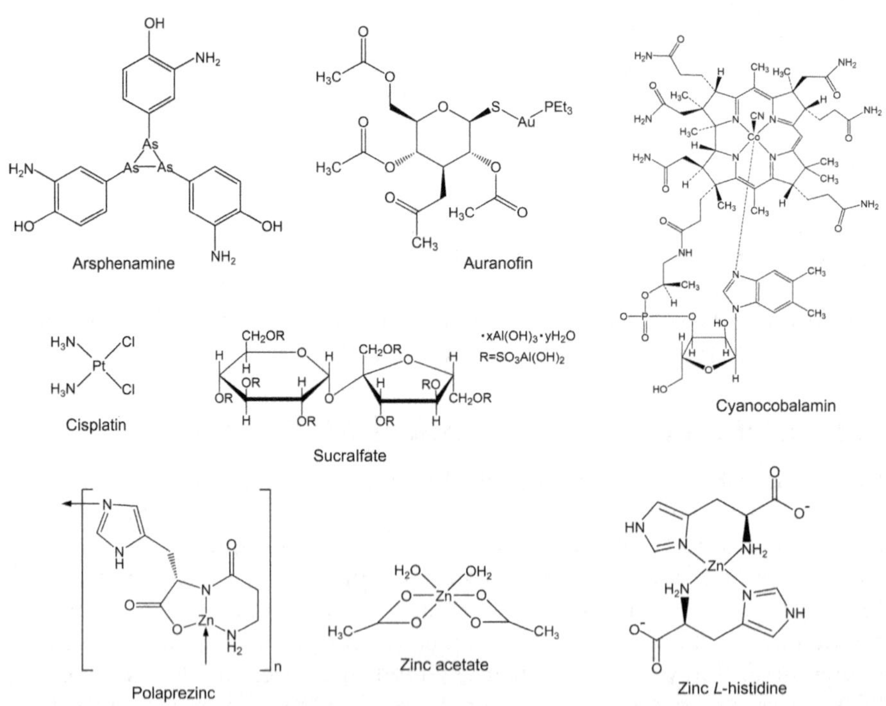

Figure 3.1 Clinically used metallo-pharmaceutical agents.

In this context, the use of metal elements to treat disease has become commonplace. Research and development into the use of metal elements as therapeutic agents continues, and it is expected that the number of metal-containing drugs will continue to increase.[17–19]

3.3 Relationship Between Diabetes and Metal Elements

There are several reports on the relationship between diabetes and metal elements. Chromium (Cr) has a long history as a metal element related to sugar metabolism in the body, and sugar metabolism abnormalities are included in the list of Cr deficiency symptoms. It has also been reported that copper (Cu) deficiency can also cause sugar metabolism abnormalities,[20] and it has also been reported that Cu complexes have anti-diabetic effects.[21] In a previous report from our laboratory, Yasumatsu *et al.* reported that a single intraperitoneal administration of bis(picolinato)Cu(II) ([Cu(pic)$_2$]) had a significant hypoglycaemic effect in an animal model of type 1 diabetes.[22] Furthermore, with regard to selenium, it has been reported that selenate (SeO_4^{2-}) increases the expression of both PPARγ and selenite (SeO_3^{2-}), the reduced form of SeO_4^{2-}, increases insulin sensitivity by inhibiting PTP-1B, a phosphatase that negatively regulates insulin signalling, and suppresses the increase in fasting blood glucose levels in db/db mice, a model of type 2 diabetes.[23] The history of the development of vanadium(V) as a diabetes treatment is very long, and there is a report from 1899 by the French physicians Lyonnet and others who administered sodium vanadate ($NaVO_3$), a pentavalent V ion, to type 2 diabetic patients and treated their diabetes.[24] Our research group has also been investigating the possibility of using V complexes to treat diabetes for many years. In the process of this research, we discovered that a once-daily oral administration of vanadyl (VO^{2+}) cysteine methyl ester complex, a complex of tetravalent V ions, to STZ rats as a model of type 1 diabetes, had a normalising effect on blood glucose levels.[25] In 1992, McNeill *et al.* reported that bis(maltolato)oxovanadium(IV) has insulin-like activity,[26] and there is also a report that it reduces blood glucose levels in STZ rats.[27] As such, V has been shown to have potential as a treatment for type 1 diabetes, for which there is currently only insulin therapy, however, due to its high toxicity,[28] it has not yet reached clinical application. In general, the use of metallic elements in pharmaceuticals is difficult due to their toxicity and side effects.

3.4 Insulin-like Activity of Zn and the Development of Anti-diabetic Zn Complexes

In our laboratory, Zn, which is an essential trace element in humans and is present in large quantities in the body, after Fe, was considered as a new drug candidate for the treatment of diabetes after V. The content of Zn is said to be about 2 g in a human with a body weight of 70 kg.[12] Zn is found in various proteins and enzymes and plays a very important role in the body. For example, it is found in oxidoreductases such as alcohol dehydrogenase and superoxide dismutase, transferases such as RNA polymerase and aspartate carbamoylase, and hydrolases such as carboxypeptidase A and matrix metalloproteinase. In addition to catalytic and structural roles in proteins, there are many Zn-essential nucleic acid recognition proteins that are important in the regulation of transcription and translation of gene information.[29]

3.4.1 Zn and Diabetes

The first report on Zn and diabetes was by Coulston and Dandona *et al.* in 1980, who found that $ZnCl_2$ (Zn ion) promoted lipid synthesis in rat adipocytes in the same way as insulin.[30] Furthermore, in 1982 James *et al.* reported that treatment of rat adipocytes with Zn ions promoted the uptake of 3-*O*-methylglucose and inhibited lipolysis by stimulating the adrenergic receptor for ritodrine.[31] These same activities as insulin shown by Zn ions are called insulin-like activity, and since this discovery, various groups have reported the antidiabetic effects of Zn ions in animal models of diabetes.[32,33]

3.4.2 Bis(maltrato)Zn(II) ([Zn(mal)$_2$])

Following several examples of the development of a highly active VO^{2+} complex by complexation with VO^{2+} in our laboratory, we have synthesised Zn complexes in which a Lewis acid Zn ion is bound to small organic compounds (ligands) with a coordinating atom such as oxygen (O), nitrogen (N) or sulphur (S), which are a Lewis base with a lone pair of electrons,[34] and tested the candidate compounds using an evaluation system for insulin-like activity in rat adipocytes.[35–37]

In 2000, we synthesised bis(maltrato)Zn(II) ([Zn(mal)$_2$]) (coordination mode: O_4) and evaluated its insulin-like activity, and reported for the first time in the world that it showed higher insulin-like activity than $ZnSO_4$ (Zn ion) as the control group.[38] Furthermore, when [Zn(mal)$_2$] was administered to KKA^y mice, a model of type 2 diabetes, at a dose of 2.0–4.5 mg Zn per kg body weight by intraperitoneal injection every day for 14 days, it was found to have potential as a diabetes treatment as it improved blood glucose-lowering effect, impaired glucose tolerance and insulin resistance.[39] Subsequently, we synthesised Zn complexes with different coordination modes, such as O_4, N_2O_2 and N_4, using natural products, amino acids and vitamins as ligands, and evaluated their *in vitro* and *in vivo* antidiabetic effects, but we have not been able to find any Zn complexes that show hypoglycaemic effects at low doses and by oral administrations.[40–42]

3.4.3 Organochalcogen–Zn Complexes

In this situation, in 2006–2007, Zn complexes with S_2O_2-type and S_4-type coordination modes showed very high *in vitro* insulin-like activity, and in a study of their *in vivo* antidiabetic effects in type 2 diabetes model animals, they not only showed hypoglycaemic effects when administered orally at low doses, but also promoted the secretion of adiponectin, one of the factors that alleviate insulin resistance, which is reduced in diabetic states, and improved the abnormal secretion of leptin (leptin resistance), a hormone related to feeding.[43,44] In 2013, a Se_2O_2-type Zn complex was synthesised and reported to reduce blood glucose levels in KKA^y mice, a model of type 2 diabetes, at the lowest oral dose to date (1.25–2.5 mg Zn per kg body weight).[45] Thus, Zn complexes with coordination atoms such as S and Se, both of which belong to the group of elements known as chalcogens, have very high anti-diabetic activity *in vitro* and *in vivo* and there is hope for their potential as future diabetes treatments.

On the other hand, although the fact that Zn complexes with coordination atoms of chalcogen elements show strong anti-diabetic effects, there is still no conclusion as to the organs in which Zn complexes act to treat diabetes *in vivo*, despite various discussions. Furthermore, it has been a matter of concern for many years as to which chemical species, Zn ions or Zn complexes, are actually acting to treat diabetes, and there is a need to develop a methodology that can easily measure the organ distribution of Zn complexes in the body. To solve the above problems, we planned to create highly active Zn complexes coordinated with chalcogen atoms that could be used to analyse their organ distribution in the body by quantifying both the Zn and chalcogen element of the ligand at the same time, using inductively coupled plasma mass spectrometry (ICP–MS). Then, we attempted to identify the organs in which Zn complexes exhibited anti-diabetic effects by synthesising Zn complexes containing Se as a coordinating atom in the structure of organic ligands and simultaneously evaluating the anti-diabetic effects and organ distribution of Zn complexes in the body.[46–48]

3.5 Action Mechanism of Zn Complexes Against Diabetes Mellitus

The majority of studies on the potential protective effects of Zn ions on the development of diabetes have been conducted in diabetic animal models. Some of the insulin-like effects of Zn ions can be explained by the influence on insulin signalling.[49,50] For example, it was observed that Zn ions induced tyrosine phosphorylation of the β-subunit of the insulin receptor, but to a lesser extent compared to insulin; in addition, the insulin receptor substrate (IRS) did not appear to play a role in enhancing glucose uptake in response to Zn stimulation.[49] This model proposed the activation of PI3K without the involvement of IRS.[49] Furthermore, Zn ions increased the phosphorylation of serine residues and subsequent activation of Akt in preadipocytes and adipocytes, thereby enhancing GLUT translocation.[49] Ezaki also showed that Zn ions induce GLUT translocation to the plasma membrane, resulting in increased glucose uptake by tissue cells.[50] This effect can be blocked by wortmannin, an inhibitor of PI3K, indicating the importance of PI3K in the activation of Akt by Zn ions. An alternative mechanism is that Zn ions affect carbohydrate metabolism *via* the insulin receptor, PTP1B, PTEN and other related proteins and enzymes.[51,52] However, these results always involved Zn ions. Furthermore, there have been few reports on the mechanisms of action of Zn complexes.

In our laboratory, we synthesised Zn complexes in which Zn ions were bound to low molecular weight organic compounds (ligands) and evaluated their *in vitro* insulin-like activity (activity to inhibit the release of free fatty acids or to promote glucose uptake) using the isolated rat adipocyte evaluation system[35] and found that they showed very strong activity compared with Zn ions. Therefore, this system was used to study the mechanisms underlying the insulin-like and anti-diabetic activities of Zn complexes in a whole cell.[36] Since Zn(mal)$_2$ was known to be taken up by the cell, the mechanisms underlying the effect of Zn(mal)$_2$ on the insulin receptor, PI3K, glucose transporter (GLUT) and phosphodiesterase (PDE) were investigated, using different inhibitors such as HNMPA-(AM)$_3$ (inhibitor of the tyrosine kinase activity of the insulin receptor *via* its 95 kDa β-subunit), wortmannin, cytochalasin B (GLUT inhibitor) and cilostamide (PDE inhibitor). Zn(mal)$_2$ was found to be a potent activator of GLUT4 and PDE. These results have helped to clarify the molecular basis of the effect of Zn complexes. In addition, the molecular mechanism underlying the effect of bis(hinokitiolato)Zn(II) ([Zn(hnk)$_2$]) was investigated and it was found that [Zn(hnk)$_2$] exhibited antidiabetic activities by activating the insulin signalling cascade through inhibition of PTP1B and PTEN and thus Akt/protein kinase B (Akt/PKB) or GSK3β phosphorylation.[37] Based on the above results, the proposed sites of action of Zn complexes are shown in Figure 3.2.

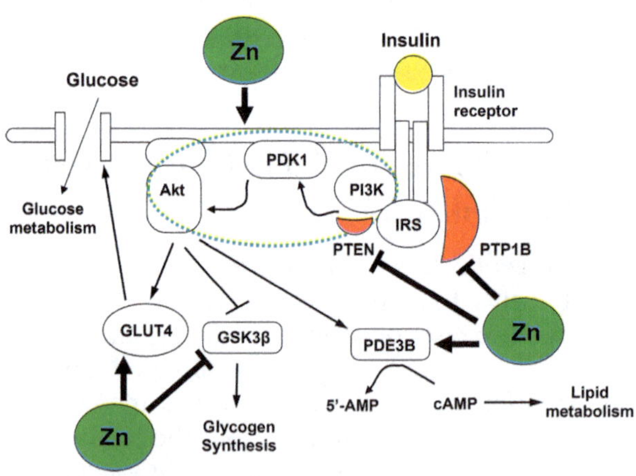

Figure 3.2 Potential action sites of Zn ions and Zn complexes in the insulin signal cascade. IRS; insulin receptor substrate, PI3K; phosphatidylinositol-3-kinase, PTP1B; protein tyrosine phosphatase 1B, PTEN; phosphatase and tensin homologue deleted from chromosome 10, PDK1; 3-phosphoinositide-dependent protein kinase 1, Akt; AKT8 virus oncogene cellular homologue, GLUT4; glucose transporter 4, GSK3β; glycogen synthase kinase 3β, PDE3B; cGMP-inhibited 3′,5′-cyclic phosphodiesterase B.

3.6 Physicochemical Properties of Zn Complexes with High *In Vitro* Insulin-like Activity

1. Amino acid and amine-related Zn complexes show higher insulin-like activity than Zn ions when the equilibrium constant (total stability constant; $\log\beta$) of the complex formation reaction is less than 10.5, whereas most Zn complexes with a total stability constant ($\log\beta$) of 11 or more do not show insulin-like activity.[53,54]
2. Although porphyrin-related Zn complexes are known to have very large total stability constants,[55] the concentration (IC_{50}) of [*meso*-tetrakis(4-sulfonato-phenyl)porphyrinato]Zn(II) ([Zn(tpps)$_2$]) that inhibits the release of free fatty acids by 50% is a relatively low 70 μM, and its insulin-like activity is exceptionally high.[55]
3. There is good correlation between the hydrophobicity (or lipophilicity) as the partition coefficient ($\log P_{\mathrm{org/water}}$) in *n*-octanol or chloroform/water of some Zn complexes and their insulin-like activity.[44]
4. Many Zn complexes with coordination modes of S_2O_2, S_4 and Se_2O_2 have high insulin-like activity.[43–45,56–58]

Among them, in relation to (4), Zn complexes with S or Se as coordinating atoms are bis(1,6-dimethyl-3-hydroxy-5-methoxy-2-pentyl-1,4-dihydropyridine-4-thionato)Zn(II) ([Zn(tanm)$_2$]),[43] bis(pyrrolidinone-dithiocarbamato)-Zn(II) ([Zn(pdc)$_2$]),[44] bis(1-oxy-2-pyridylthiolato)Zn(II) ([Zn(opt)$_2$]),[56,57] bis(2-mercaptotropolonato)Zn(II) ([Zn(mttr)$_2$]),[58] and bis(2-selenopyridin-*N*-oxide)Zn(II) ([Zn(spno)$_2$]).[45] It is particularly noteworthy that the IC_{50} of these Zn complexes is in the range of 1–10 μM and they exhibit very high *in vitro* insulin-like activity (Figure 3.3). Of these, the activity of the Zn complex with Se as the coordinating atom was investigated in one case of [Zn(spno)$_2$] and further investigation was considered necessary.

3.7 Synthesis of Hydroxypyrone Derivative Organochalcogen (O,S,Se)–Zn Complexes and Evaluation of *In Vitro* Insulin-like Activity

We therefore focused on maltol, the ligand of [Zn(mal)$_2$] because [Zn(mal)$_2$] was first shown to have higher insulin-like activity than Zn ions and to have an anti-diabetic effect in KKA^y mice, a model of type 2 diabetes.[38,39] Maltol is a hydroxy pyrone derivative known to have almost no physiological activity, and is also used as a food additive together with ethyl maltol, another hydroxy pyrone derivative, and is a compound with a high level of safety.[38,39] Therefore, the hydroxy pyrone derivatives (hydroxy pyrone, maltol, ethyl maltol) were thio- and seleno-carbonylated, respectively, and Zn complexes coordinated with chalcogen atoms such as the types O_4, S_2O_2 and Se_2O_2 were synthesised and their *in vitro* insulin-like activities were studied to search for highly active Zn complexes (Figure 3.4).[46–48]

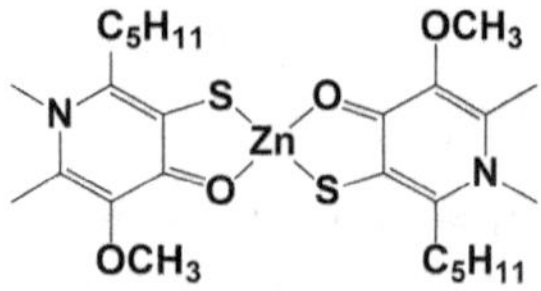

[Zn(tanm)$_2$]
IC$_{50}$=10.2 μM

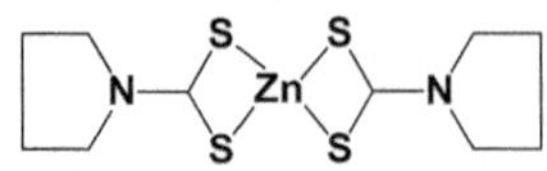

[Zn(pdc)$_2$]
IC$_{50}$=5.1 μM

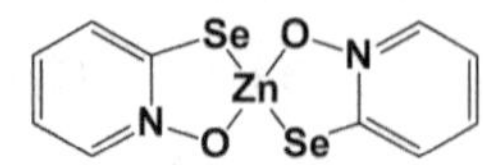

[Zn(spno)$_2$]
IC$_{50}$=7.4 μM

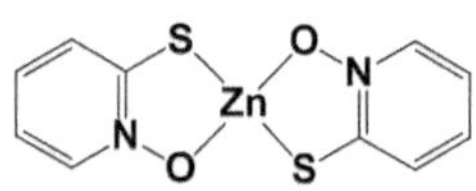

[Zn(opt)$_2$]
IC$_{50}$=8.9 μM

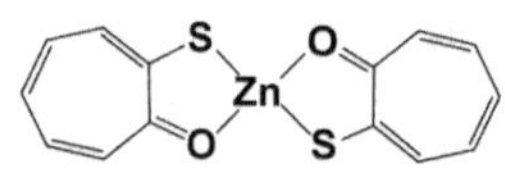

[Zn(mttr)$_2$]
IC$_{50}$=12.0 μM

Figure 3.3 Highly insulin-like active Zn complexes coordinated with chalcogen atoms and their IC$_{50}$ values for free fatty acid (FFA). Tanm; 1,6-dimethyl-3-hydroxy-5-methoxy-2-pentyl-1,4-dihydropyridine-4-thionato, pdc; pyrrolidinone-dithiocarbamate, spno; 2-selenopyridin-*N*-oxide, opt; 1-oxy-2-pyridylthiolate, mttr; 2-mercaptotropolonate. IC$_{50}$ (μM); 50% inhibitory concentrations of FFA release from isolated rat adipocytes.

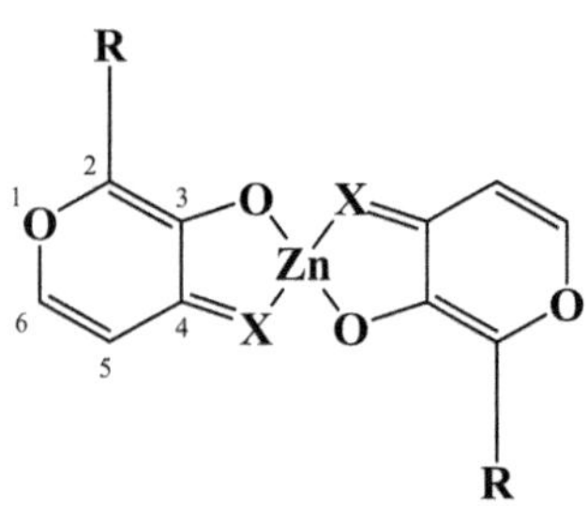

Compound	R	X
[Zn(hpo)$_2$] : bis(3-hydroxy-4*H*-pyran-4-ono)Zn	H	O
[Zn(hmpo)$_2$] : bis(3-hydroxy-2-methyl-4*H*-pyran-4-ono)Zn	-CH$_3$	O
[Zn(ehpo)$_2$] : bis(2-ethyl-3-hydroxy-4*H*-pyran-4-ono)Zn	-CH$_2$CH$_3$	O

Compound	R	X
[Zn(hpt)$_2$] : bis(3-hydroxy-4*H*-pyran-4-thiono)Zn	H	S
[Zn(hmpt)$_2$] : bis(3-hydroxy-2-methyl-4*H*-pyran-4-thiono)Zn	-CH$_3$	S
[Zn(ehpt)$_2$] : bis(2-ethyl-3-hydroxy-4*H*-pyran-4-thiono)Zn	-CH$_2$CH$_3$	S

Compound	R	X
[Zn(hps)$_2$] : bis(3-hydroxy-4*H*-pyran-4-seleno)Zn	H	Se
[Zn(hmps)$_2$] : bis(3-hydroxy-2-methyl-4*H*-pyran-4-seleno)Zn	-CH$_3$	Se
[Zn(ehps)$_2$] : bis(2-ethyl-3-hydroxy-4*H*-pyran-4-seleno)Zn	-CH$_2$CH$_3$	Se

Figure 3.4 Structures of organochalcogen–Zn complexes with hydroxypyrone derivatives. Hpo; 3-hydroxy-4*H*-pyran-4-one, hmpo; 3-hydroxy-2-methyl-4*H*-pyran-4-one, ehpo; 2-ethyl-3-hydroxy-4*H*-pyran-4-one, hpt; 3-hydroxy-4*H*-pyran-4-thione, hmpt; 3-hydroxy-2-methyl-4*H*-pyran-4-thione, ehpt; 2-ethyl-3-hydroxy-4*H*-pyran-4-thione, hps; 3-hydroxy-4*H*-pyran-4-selene, hmps; 3-hydroxy-2-methyl-4*H*-pyran-4-selene, ehps; 2-ethyl-3-hydroxy-4*H*-pyran-4-selene.

3.7.1 Chemical Structure of Hydroxypyrone Derivative Organochalcogen–Zn Complexes Based on ^{13}C-NMR

Regarding the synthesis of Zn complexes, the results of elemental analysis and EI-MS(+) for each Zn complex suggested that the Zn ion, which is the central metal, and the ligand were present in a 1 : 2 ratio. In the case of the organic O–Zn complexes, the IR results showed that the peaks for the stretching vibrations of the carbonyl groups observed in the ligands had shifted to lower wave numbers, suggesting that the carbonyl groups co-ordinate with the Zn ions {hmpo: 1655 ($\nu_{C=O}$), 3260 (ν_{OH}) and ehpo: 1647 ($\nu_{C=O}$), 3084 (ν_{OH})}. In addition, for the organic S–Zn and Se–Zn complexes, the results of the ^{1}H-NMR analysis of the ligands (hpt, hps, hmpo, hmpt, hmps, ehpo, ehpt and ehps) showed peaks for the hydroxy group at the 3-position (hpt: 7.73, hps: 7.80, hmpo: 8.85, hmpt: 7.78, hmps: 7.83, ehpo: 8.66, ehpt: 7.80 and ehps: 7.86 ppm) in the ligands, but disappeared in the Zn complexes. These results suggested that the hydroxy group (–OH) at the 3-position was converted to O– by a base and bound to the Zn ion. In addition, the ^{13}C-NMR results showed that the Zn complexes with O as the coordinating atom had a larger shift in C3 at the base of the O-group and C4 of the carbonyl group than other carbons compared to the ligands (Table 3.1[A]). Therefore, it

Table 3.1 ^{13}C-NMR chemical shift data for organochalcogen ligands and the corresponding Zn complexes; [A] organo-oxygen (O), [B] organo-sulfur (S), and [C] organo-selenium (Se) ligands and the corresponding Zn complexes.

	Chemical shift (ppm)				
Compound	C2	C3	C4	C5	C6
[A]					
hmpo	149.2	142.8	172.4	113.4	154.5
[Zn(hmpo)$_2$]	149.4	151.5	177.4	109.5	153.4
Δδ	+0.3	+8.7	+4.9	−3.9	−1.2
ehpo	153.3	142.3	172.8	113.5	154.8
[Zn(ehpo)$_2$]	154.0	151.0	177.8	109.6	153.6
Δδ	+0.6	+8.7	+5.0	−3.9	−1.2
[B]					
hpt	136.8	154.2	189.7	126.8	149.8
[Zn(hpt)$_2$]	140.5	162.0	182.9	123.4	148.9
Δδ	+3.7	+7.8	−6.8	−3.4	−0.9
hmpt	145.3	150.6	185.9	124.2	147.0
[Zn(hmpt)$_2$]	154.2	156.0	176.8	122.5	144.6
Δδ	+9.0	+9.4	−9.1	−1.7	−2.4
ehpt	149.6	149.9	186.0	124.2	147.1
[Zn(ehpt)$_2$]	158.3	159.5	176.9	122.5	144.6
Δδ	+8.6	+9.5	−9.1	−1.7	−2.5
[C]					
hps	132.8	156.7	189.1	130.6	146.0
[Zn(hps)$_2$]	140.6	164.4	182.0	126.9	145.1
Δδ	+7.8	+7.6	−7.1	−3.7	−0.9
hmps	145.0	154.2	185.8	129.6	145.5
[Zn(hmps)$_2$]	154.4	162.4	175.2	126.2	143.3
Δδ	+9.4	+8.2	−10.7	−3.4	−2.2
ehps	149.4	153.5	185.9	129.6	145.6
[Zn(ehps)$_2$]	158.4	161.8	175.1	126.1	143.2
Δδ	+9.0	+8.3	−10.8	−3.5	−2.3

was suggested that the O-group and the carbonyl group were bound to the Zn ion. In addition, in the Zn complexes with S and Se as coordinating atoms, the chemical shifts of C2 and C3 shifted to lower magnetic fields and the chemical shift of C4 shifted to higher magnetic fields compared to the ligands (Table 3.1[B] and [C]).

The low-field shift of C2 and C3 was thought to be caused by the resonance stabilisation of the phenolic OH group decreasing due to the O-group bound to the Zn ion, or by the electron density of C2 and C3 decreasing due to the electron being attracted by the Zn ion. The high field shift of C4 was thought to have occurred because the electron density of the S and Se atoms increased due to donor–acceptor interactions.[59] Donor–acceptor interaction refers to the effect of the electron density of the S and Se atoms increasing due to the spillover effect from the Zn atom caused by the formation of a polarised bond between the Zn atom (acceptor) and the S and Se atoms (donor). From the ^{1}H-NMR and ^{13}C-NMR results, it was suggested that the O-group of the ligand and the thiocarbonyl or selenocarbonyl group were bound to the Zn ion. Based on the above results for the IR and NMR analyses, the structure of the Zn complex was determined (Figure 3.4).

3.7.2 Single-crystal X-ray Structure of Hydroxypyrone Derivative Organochalcogen Zn–Complexes

Next, yellow plate-shaped single crystals of [Zn(hmps)$_2$] were obtained from a chloroform/methanol mixture, and prismatic and block-shaped single crystals of [Zn(ehpt)$_2$] and [Zn(ehps)$_2$] were obtained from ethyl acetate solutions, respectively, and single-crystal X-ray structural analysis was performed. It was found that in these three Zn complexes, the ligands were bound to the Zn ions as a distorted tetrahedral structure in the *trans* position by two S or Se and O atoms (Figure 3.5).

This structure was supported by the results of elemental analysis, EI-MS (+) and NMR. In comparison with the previously reported X-ray crystal structure of [Zn(thiomaltol)$_2$] (= [Zn(hmpt)$_2$]),[60] it was shown that in [Zn(thiomaltol)$_2$], the ligand thiomaltol was bound to the Zn ion as a distorted tetrahedral structure in the *trans* position with two S and O atoms, and its structure was indicated to be similar to our new three Zn complexes that were measured this time, and the similarity of chemical structure of Zn complexes with S_2O_2 or Se_2O_2 coordination modes was found as a distorted tetrahedral structure (Figure 3.5).

In order to obtain chemical structural information on organo-Se–Zn complexes, we compared the X-ray crystal structures of previously reported Se_2O_2-type Zn complexes with Se as the coordinating atom such as bis(*N*,*N*-diethyl-*N*′-benzoyl-selenoureato)Zn(II)[61] and bis(*N*,*N*-diethyl-*N*′-(2-fluoro)-benzoyl-selenoureato)Zn(II),[62] with those of [Zn(hmps)$_2$] and [Zn(ehps)$_2$]. The structures of the four Zn complexes had distorted tetrahedral coordination geometries with Se and O atoms in the *trans* positions of the two ligand molecules bound to the Zn ion (Figure 3.5). The bond lengths (Zn–Se, Zn–O)

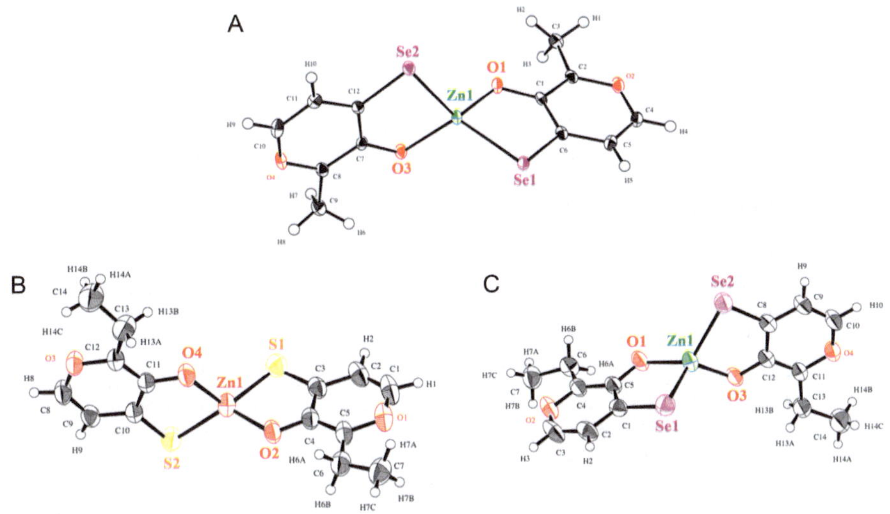

Figure 3.5 ORTEP (Oak Ridge Thermal-Ellipsoid Plot Program) drawing of [A] [Zn(hmps)$_2$], [B] [Zn(ehpt)$_2$] and [C] [Zn(ehps)$_2$]. hmps; 3-hydroxy-2-methyl-4*H*-pyran-4-selene, ehpt; 2-ethyl-3-hydroxy-4*H*-pyran-4-thione, ehps; 2-ethyl-3-hydroxy-4*H*-pyran-4-selene.

and angles (Se–Zn–Se, O–Zn–O, Se–Zn–O) were almost the same for the four complexes.[47,48,61,62]

3.7.3 Evaluation of the *In Vitro* Insulin-like Activity of Hydroxypyrone Derivative Organochalcogen–Zn Complexes

In order to investigate the *in vitro* insulin-like activity of the synthesised hydroxy pyrone derivative O–, S– and Se–Zn organic complexes, using both $ZnSO_4$ ionised in solution as a positive control and the synthesised Zn complexes, we evaluated the inhibitory activity (IC_{50}) of free fatty acid (FFA) release and the enhance activity (EC_{50}) of glucose uptake into adipocytes in rat adipocytes by means of the previously established scoring system.[35] As a result, all Zn complexes showed higher FFA release inhibitory activity and glucose uptake activity than $ZnSO_4$ (Table 3.2). Furthermore, the organic S– and Se–Zn complexes showed considerably higher activity than the organic O–Zn complexes (Table 3.2). This trend was consistent with previous reports from our laboratory,[43–45,56–58] and it was further found that among the Zn complexes, the Zn complexes with maltol as the basic skeleton showed stronger insulin-like activity than the Zn complexes with hydroxy pyrone and ethyl maltol as the basic skeleton (Table 3.2).

Zinc is classified as a relatively soft acid according to the HSAB principle. Therefore, it is expected that the stability of the Zn complex will improve as the binding force increases when the coordinating atoms are converted to

Table 3.2 IC_{50} values for free fatty acid release, EC_{50} values for glucose uptake, and $\log P_{org/water}$ of Zn complexes.[a]

Zn compounds	IC_{50} (μM)	EC_{50} (μM)	$\log P_{org/water}$
$ZnSO_4$	585 ± 305	N.D.	None
$[Zn(hpo)_2]$	166 ± 75	N.D.	None
$[Zn(hmpo)_2]$	166 ± 36	233 ± 7	$-1.94 \pm 0.06^{\ddagger}$
$[Zn(ehpo)_2]$	189 ± 31	110 ± 48	$-1.30 \pm 0.02^{\#\ddagger}$
$[Zn(hpt)_2]$	12 ± 1	5 ± 1	0.52 ± 0.02
$[Zn(hmpt)_2]$	2 ± 1	4 ± 1	0.62 ± 0.13
$[Zn(ehpt)_2]$	11 ± 2	3 ± 1	0.56 ± 0.10
$[Zn(hps)_2]$	108 ± 57	44 ± 7	1.11 ± 0.07
$[Zn(hmps)_2]$	$8 \pm 1^{*\#\dagger}$	$3 \pm 0^{\dagger}$	$1.39 \pm 0.13^{\#\dagger\ddagger}$
$[Zn(ehps)_2]$	$12 \pm 1^{*\#\dagger}$	$5 \pm 1^{\dagger}$	$1.43 \pm 0.08^{\#\dagger\ddagger}$

[a] Data are expressed as mean ± SDs. Statistical analysis was performed by Tukey test. Significance: $^{*}p < 0.01$ *vs.* $[Zn(hpo)_2]$, $^{\#}p < 0.01$ *vs.* $[Zn(hmpo)_2]$, $^{\dagger}p < 0.01$ *vs.* $[Zn(ehpo)_2]$, $^{\ddagger}p < 0.01$ *vs.* $[Zn(hps)_2]$.

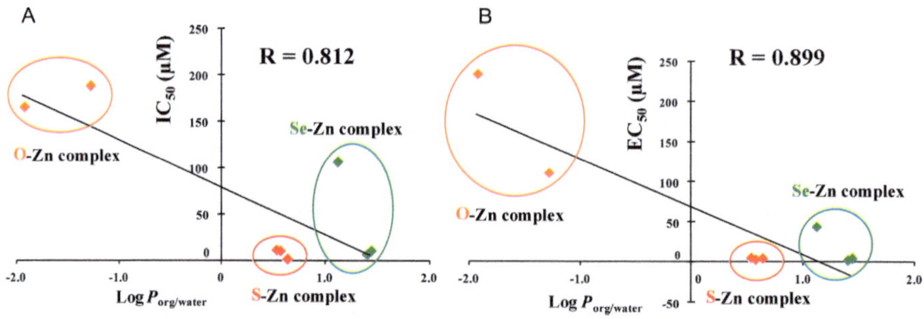

Figure 3.6 Relationships between [A] the FFA release inhibitory effect (IC_{50}), and [B] the glucose uptake effect (EC_{50}) and partition coefficient ($\log P_{org/water}$).

S and Se, which are bases that are softer than O atoms. Furthermore, Zn ions are divalent cations and generally have poor permeability across cell membranes. However, they become neutral Zn complexes by complexation, so it is thought that the permeability of cell membranes is improved by passive diffusion of Zn complexes and that this is why they show very high *in vitro* insulin-like activity, as in this case.

Furthermore, based on the results of our previous research, it has been suggested that there is a correlation between the $\log P_{org/water}$ of the Zn complex and the *in vitro* insulin-like activity of the dithiocarbamate derivative Zn complexes.[44] Therefore, we investigated whether there was a correlation between the $\log P_{org/water}$ of the newly synthesized Zn complexes[63] and the IC_{50} of FFA release and EC_{50} of glucose uptake, and there was a correlation between the $\log P_{org/water}$ and IC_{50} and EC_{50}. In particular, the R values for the Pearson's correlation coefficient with IC_{50} and EC_{50} was 0.812 and 0.889, respectively, indicating the strong negative correlations (Figure 3.6[A] and [B]).

However, the IC_{50} values of the Se_2O_2-type Zn complexes were almost the same as those of the previously synthesized S_2O_2-type $[Zn(opt)_2]$ ($IC_{50} = 8.9$ μM),[45,56]

[Zn(mttr)$_2$] ($IC_{50} = 12.0$ μM)[58] and Se_2O_2-type [Zn(spno)$_2$] ($IC_{50} = 7.4$ μM).[45] Therefore, it was found that the insulin-like activity of the Zn complex increases dramatically when the coordinating atoms are changed from O to S or Se, but that there is no further increase in activity associated with the change from S to Se (Table 3.2 and Figure 3.6[A], [B]).

3.8 Evaluation of the Biodistribution of Zn Complexes and Estimation of the Chemical Forms of [Zn(hmps)$_2$] in Organs

In order to investigate the organ distribution of Zn complex in the body, the concentrations of Zn and Se in organs of type 2 diabetic model KKA^y mice[64] were simultaneously quantified by ICP–MS, in which [Zn(hmps)$_2$] suspended in PEG-400 was administered orally to 9-week-old KKA^y mice at doses of 2.0–10 mg Zn per kg body weight for 28 consecutive days. The solvent, PEG-400, was administered to the control group of diabetic KKA^y mice.

3.8.1 *In Vivo* Anti-diabetic Effects of [Zn(hmps)$_2$] in KKA^y Mice

On the second day after administration, a rapid hypoglycaemic effect was observed in the [Zn(hmps)$_2$] group. Over the next 28 days, the [Zn(hmps)$_2$] group reduced blood glucose levels by approximately 100–150 mg dL^{-1} compared to the control group. HbA1c, an indicator of long-term blood glucose control, was significantly reduced by 3.4% compared to the control group. Fasting plasma insulin and leptin levels were significantly higher in the control group than in the normal state, indicating that the control KKA^y mice were in a state of insulin and leptin resistance. In the [Zn(hmps)$_2$] group, there was a significant decrease in plasma insulin and leptin concentrations to 60% and 38% of the control group, respectively. These results suggested that insulin and leptin resistance were improved in the [Zn(hmps)$_2$] group.[47] The results for AST and ALT, markers of liver damage, and BUN, a marker of kidney damage, showed that there was no significant increase in these plasma parameters in the [Zn(hmps)$_2$] group compared to the control group, suggesting that liver and kidney damage did not occur.[47] From these results, it was concluded that [Zn(hmps)$_2$] has a very potent anti-diabetic effect *in vivo*, in agreement with the *in vitro* insulin-like activity results.[47]

3.8.2 Biodistribution and Chemical Forms of [Zn(hmps)$_2$] in the Body

After oral administration of [Zn(hmps)$_2$] or vehicle (control) for 28 consecutive days, the plasma Zn concentration in the [Zn(hmps)$_2$] group was significantly higher than in the control group, suggesting that it was absorbed *via* the gastrointestinal tract and reached the systemic circulation.

Table 3.3 Increased Δ[Zn] (nmol g^{-1}), Δ[Se] (nmol g^{-1}), and Δ[Zn]/Δ[Se] molar ratios in organs of KKA^y mice treated with orally administered [Zn(hmps)$_2$], compared to non-treated control KKA^y mice.

	Plasma	Liver	Muscle	Spleen	Kidney	Adipose	Pancreas	Bone
Δ[Zn] (nmol g^{-1})	7.6	367	46	46	0	0	0	642
Δ[Se] (nmol g^{-1})	15.2	393	46	161	213	2	15	39
Δ[Zn]/Δ[Se]	0.50	0.93	1.0	0.29	0	0	0	16

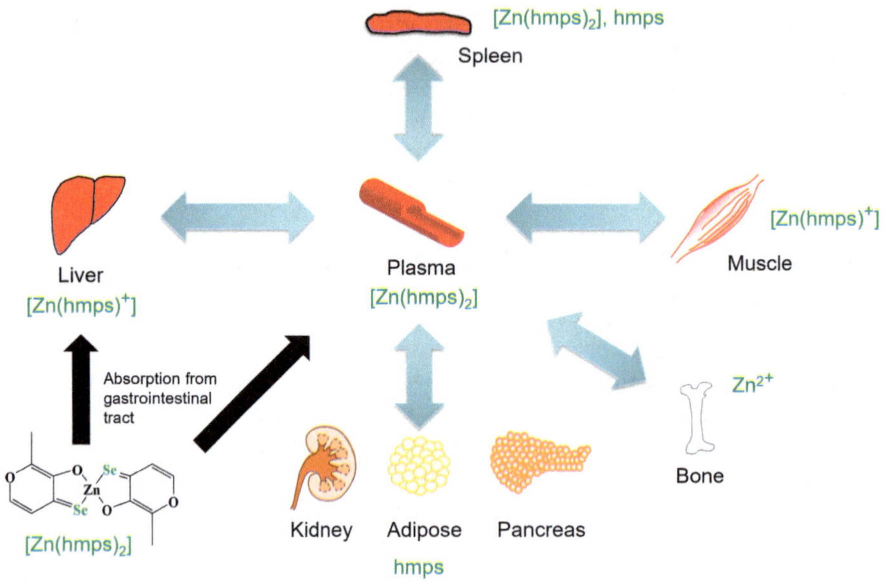

Figure 3.7 Biodistribution and estimated chemical forms of [Zn(hmps)$_2$] in organs after oral administration for 28 days to KKA^y mice.

Furthermore, since a significant increase in Zn concentration was observed in the liver and muscle, it was suggested that [Zn(hmps)$_2$] has an anti-diabetic effect in the liver and muscle. Therefore, the differences in molar concentration of Zn and Se (Δ[Zn] and Δ[Se]) between the control group and the [Zn(hmps)$_2$]-administered group were calculated, and the Δ[Zn]/Δ[Se] ratios were also calculated (Table 3.3).

Furthermore, before administration to animals, [Zn(hmps)$_2$] has a structure in which the Zn ion and ligand are in a 1 : 2 molar ratio, so it is estimated that if Δ[Zn]/Δ[Se] is equal to 0.5 in an organ, [Zn(hmps)$_2$] is present in the organ as its chemical form, and when Δ[Zn]/Δ[Se] is equal to 1, [Zn(hmps)$^+$] is present in the organ as its chemical form. In Table 3.3, the ratios of Δ[Zn]/Δ[Se] were 0.50 in plasma, 0.93 in liver, and 1.0 in muscle, respectively. Therefore, it was suggested that it exists as [Zn(hmps)$_2$] in plasma, and mainly [Zn(hmps)$^+$] form in liver and muscle (Figure 3.7).

In addition, the value of Δ[Zn]/Δ[Se] was equal to 0.29 in the spleen, suggesting that it exists as a mixture of the chemical forms of [Zn(hmps)$_2$] and hmps dissociated from [Zn(hmps)$_2$] in the spleen (Table 3.3 and Figure 3.7).

On the other hand, the values of Δ[Zn]/Δ[Se] equal to almost zero in the kidney, adipose, and pancreas suggested that most of the administered Zn is not distributed in the kidney, adipose, and pancreas although the ligand hmps dissociated from Zn complexes is distributed in those organs. And the value of Δ[Zn]/Δ[Se] equal to 16 in bone suggested that most of the administered Zn in bone was in the form of the Zn ions dissociated from the complex (Table 3.3 and Figure 3.7).

A significant increase in the concentration of Zn was observed in the liver and muscle, particularly in the liver. Furthermore, the inhibitory effect of [Zn(hmps)$_2$] on lipid accumulation in liver and muscle tissue sections was observed,[47] suggesting that [Zn(hmps)$_2$], in the chemical form of [Zn(hmps)$^+$], has an anti-diabetic effect in liver and muscle.

3.9 Conclusion and Perspectives

This review summarizes the research aimed at identifying the organs in which Zn complexes exert anti-diabetic effects, based on the synthesis of organochalcogen–Zn complexes using hydroxypyrone derivative ligands as lead compounds, the search for highly active Zn complexes, and the relationship between the anti-diabetic effects and organ distribution of Zn complexes in the body.

First, it was shown that Zn complexes with soft coordination atoms such as S and Se, which are governed by the HSAB principle, exhibited higher *in vitro* insulin-like activity than Zn complexes with hard O as the coordination atom, and a correlation was observed between the partition coefficient ($\log P_{org/water}$) and the *in vitro* insulin-like activity of the organochalcogen–Zn complexes. In addition, the single-crystal X-ray structure of [Zn(hmps)$_2$] and [Zn(ehps)$_2$], which have Se as the coordinating atom, have a distorted tetrahedral structure, and [Zn(hmpt)$_2$], which has S as the coordinating atom, was reported to be a similar structure, so it has also become clear that Zn complexes with distorted tetrahedral structures have very high insulin-like activity.

Second, it was shown that [Zn(hmps)$_2$] has hypoglycaemic, HbA1c improving and insulin/leptin resistance improving effects in type 2 diabetic model KKA^y mice. In addition, the [Zn(hmps)$_2$]-administered group was found to have an inhibitory effect on lipid accumulation in the liver, indicating that [Zn(hmps)$_2$] has the potential to inhibit the development of fatty liver, which is one of the pathological conditions of type 2 diabetes.

Third, we focused to estimate the target organs of the Zn complex with anti-diabetic effects by simultaneously evaluating the organ distribution of the Zn and Se-containing ligand in *in vivo* experiments. It was suggested that the chemical forms of [Zn(hmps)$_2$] in the organs can be estimated by quantifying Zn and Se concentrations in the organs using ICP–MS and calculating the Δ[Zn], Δ[Se] and Δ[Zn]/Δ[Se] ratios in each organ.

In the future, it will be necessary to carefully study the organ distribution of Zn complexes over time, and on a daily basis, during continuous administration. In addition, since it is expected that it will be difficult to elucidate the organs in which Zn complexes act *in vivo* by means of using only the conventional method of quantifying Zn concentrations in organs, it is believed that a new approach from a different perspective would be necessary to elucidate the detailed mechanism of action. As an example of a new approach, the spatio-temporal pharmacokinetics of Zn complexes using a radiotracer ^{65}Zn-labelled method has been proposed to monitor ^{65}Zn levels in the blood and organs of living animals that received a single intravenous injection or oral administration of Zn complexes.[65,66] In particular, this method will be important not only to confirm the proof of concept and the proof of mechanism for the *in vivo* pharmacological activity of Zn complexes, but also to predict new pharmaceutical designs of Zn complexes for target organs.

Abbreviations

Akt	AKT8 virus oncogene cellular homologue
DPP-4	Dipeptidyl peptidase-4
EC_{50}	50% glucose uptake concentration
EI	Electron ionisation
FFA	Free fatty acid
GLP-1	Glucagon like peptide-1
GLUT	Glucose transporter
GSK3β	Glycogen synthase kinase 3β
HbA1c	Haemoglobin A1c
IDF	International diabetes federation
IC_{50}	50% inhibitory concentration
IR	Infrared spectrometry
IRS	Insulin receptor substrate
MS	Mass spectrometry
NMR	Nuclear magnetic resonance
PDE3B	cGMP-inhibited 3′,5′-cyclic phosphodiesterase B
PDK1	3-Phosphoinositide-dependent protein kinase 1
PEG	Polyethylene glycol
PI3K	Phosphatidylinositol-3-kinase
PPARγ	Peroxisomal proliferator-activated receptor
PTP1B	Protein tyrosine phosphatase 1B
PTEN	Phosphatase and tensin homolog deleted from chromosome 10
SGLT-2	Sodium glucose co-transporter 2

Acknowledgements

This work was supported in part by a Grant-in-Aid from the Ministry of Education, Culture, Sports, Science, and Technology, Japan (MEXT)-Supported Program for the Strategic Research Foundation at Private Universities (S1201008) and a Grant-in-Aid for Scientific Research (C) (JSPS KAKENHI Grant Number JP26460636) from the Japan Society for the Promotion of Science (to H. Y.).

References

1. International Diabetes Federation, *IDF Diabetes Atlas*, IntDiabetesFed, 10th edn, 2021.
2. J. R. Gavin, K. G. M. M. Alberti, M. B. Davidson, R. A. DeFronzo, A. Drash, S. G. Gabbe, S. Genuth, M. I. Harris, R. Kahn, H. Keen, W. C. Knowler, H. Lebovitz, N. K. Maclaren, J. P. Palmer, P. Raskin, R. A. Rizza and M. P. Stem, *Diabetes Care*, 1997, **20**, 1183–1197.
3. D. Koya and G. L. King, *Diabetes*, 1998, **47**, 859–866.
4. C. Manrique, G. Lastra, A. Whaley-Connell and J. R. Sowers, *J. Clin. Hypertens.*, 2005, 7, 471–476.
5. N. M. Kaplan and L. H. Opie, *Lancet*, 2006, **367**, 168–176.
6. A. M. Sharma and V. T. Chetty, *Acta Diabetol.*, 2005, **42**, S3–S8.
7. American Diabetes Association, *Diabetes Care*, 2014, **37**, S81–S90.
8. K. T. Coppieters, F. Dotta, N. Amirian, P. D. Campbell, T. W. H. Kay, M. A. Atkinson, B. O. Roep and M. G. von Herrath, *J. Exp. Med.*, 2011, **209**, 51–60.
9. R. A. Defranzo, R. Eldor and M. A. bdul-Ghani, *Diabetes Care*, 2013, **36**, S127–S138.
10. M. Bodmer, C. Meier, S. Krähenbühl, S. S. Jick and C. R. Meier, *Diabetes Care*, 2008, **31**, 2086–2091.
11. F. Mittermayer, E. Caveney, C. D. Oliveira, L. Gourgiotis, M. Puri, L. J. Tai and J. R. Turner, *Curr. Diabetes Rev.*, 2015, **11**, 17–31.
12. E. J. Underwood, *Trace Elements in Human and Animal Nutrition*, Academic Press, New York, 4th edn, 1977.
13. H. Sakurai, Y. Yoshikawa and H. Yasui, *Chem. Soc. Rev.*, 2008, **37**, 2383–2392.
14. H. Yasui and Y. Yoshikawa, *Metallomics*, Springer Nature, Tokyo, 1st edn, 2017.
15. Nobelpharma Co., Ltd., *Nobelzin Package Insert*, 2017.
16. Nobelpharma Co., Ltd., *Zintus Package Insert*, 2024.
17. M. Zec, T. Srdic-Rajic, T. Todorovic, K. Andjelkovic, I. Filipovic-Ljeskovic and S. Radulovic, *Anti-Cancer Agents Med. Chem.*, 2012, **12**, 1071–1080.
18. J. R. Williams, E. Trias, P. R. Beilby, N. I. Lopez, E. M. Labut, C. S. Bradford, B. R. Roberts, E. J. McAllum, P. J. Crouch, T. W. Rhoads, C. Pereira, M. Son, J. L. Elliott, M. C. Franco, Á. G. Estévez, L. Barbeito and J. S. Beckman, *Neurobiol. Dis.*, 2016, **89**, 1–9.

19. J. Jia, Y. Zhang, M. Zheng, C. Shan, H. Yan, W. Wu, X. Gao, B. Cheng, W. Liu and Y. Tang, *Inorg. Chem.*, 2018, **57**, 300–310.
20. E. D. Harris, *Crit. Rev. Clin. Lab. Sci.*, 2003, **40**, 547–586.
21. J. R. J. Sorenson, *J. Prog Med. Chem.*, 1989, **26**, 437–468.
22. N. Yasumatsu, Y. Yoshikawa, Y. Adachi and H. Sakurai, *Bioorg. Med. Chem.*, 2007, **15**, 4917–4922.
23. A. S. Mueller and J. Pallauf, *J. Nutr. Biochem.*, 2006, **17**, 548–560.
24. B. Lyonnet, M. Martz and E. Martin, *La Presse Médicale*, 1899, **32**, 191–192.
25. H. Sakurai, K. Tsuchiya, M. Nukatsuka, J. Kawada, S. Ishikawa, H. Yoshida and M. Komatsu, *J. Clin. Biochem. Nutr.*, 1990, **8**, 193–200.
26. J. H. McNeill, V. G. Yuen, H. R. Hoveyda and C. Orvig, *J. Med. Chem.*, 1992, **35**, 1489–1491.
27. U. A. Shinde, A. A. Mehta and R. K. Goyal, *Indian J. Exp. Biol.*, 2001, **39**, 864–870.
28. A. Soussi, R. Abdennabi, F. Ghorbel, J. C. Murat and A. F. E. Feki, *Biol. Trace Elem. Res.*, 2017, **180**, 239–245.
29. R. R. Crichton, *Biological Inorganic Chemistry: A New Introduction to Molecular Structure and Function*, Academic Press, New York, 3rd edn, 2018.
30. L. Coulston and P. Dandona, *Diabetes*, 1980, **29**, 665–667.
31. J. M. May and C. S. Contoreggi, *J. Biol. Chem.*, 1982, **257**, 4362–4368.
32. A. Shisheva, D. Gefel and Y. Shechter, *Diabetes*, 1992, **41**, 982–988.
33. M. D. Chen, S. J. Liou, P. Y. Lin, V. C. Yang, P. S. Alexander and W. H. Lin, *Biol. Trace Elem. Res.*, 1998, **61**, 303–311.
34. H. Sakurai, Y. Kojima, Y. Yoshikawa, K. Kawabe and H. Yasui, *Coord. Chem. Rev.*, 2002, **226**, 187–198.
35. M. Nakai, H. Watanabe, C. Fujiwara, H. Kakegawa, T. Satoh, J. Takada, R. Matsushita and H. Sakurai, *Biol. Pharm. Bull.*, 1995, **18**, 719–725.
36. Y. Yoshikawa, E. Ueda, Y. Kojima and H. Sakurai, *Life Sci.*, 2004, **75**, 741–751.
37. Y. Naito, Y. Yoshikawa and H. Yasui, *J. Biol. Inorg. Chem.*, 2016, **21**, 537–548.
38. Y. Yoshikawa, E. Ueda, K. Kawabe, H. Miyake, H. Sakurai and Y. Kojima, *Chem. Lett.*, 2000, **29**, 874–875.
39. Y. Yoshikawa, E. Ueda, H. Miyake, H. Sakurai and Y. Kojima, *Biochem. Biophys. Res. Commun.*, 2001, **281**, 1190–1193.
40. Y. Yoshikawa, E. Ueda, K. Kawabe, H. Miyake, T. Takino, H. Sakurai and Y. Kojima, *J. Biol. Inorg. Chem.*, 2002, 7, 68–73.
41. Y. Yoshikawa, E. Ueda, H. Sakurai and Y. Kojima, *Chem. Pharm. Bull.*, 2003, **51**, 230–231.
42. E. Ueda, Y. Yoshikawa, N. Kishimoto, M. Tadokoro, N. Yanagihara, H. Sakurai and Y. Kojima, *Chem. Lett.*, 2001, **30**, 1184–1185.
43. Y. Adachi, J. Yoshida, Y. Kodera, T. Kiss, T. Jakusch, E. A. Enyedy, Y. Yoshikawa and H. Sakurai, *Biochem. Biophys. Res. Commun.*, 2006, **351**, 165–170.

44. Y. Yoshikawa, Y. Adachi and H. Sakurai, *Life Sci.*, 2007, **80**, 759–766.
45. S. Fujimoto, H. Yasui and Y. Yoshikawa, *J. Inorg. Biochem.*, 2013, **121**, 10–15.
46. T. Nishiguchi, Y. Yoshikawa and H. Yasui, *Int. J. Mol. Sci.*, 2017, **18**, 2647.
47. T. Nishiguchi, Y. Yoshikawa and H. Yasui, *J. Inorg. Biochem.*, 2018, **185**, 103–112.
48. T. Nishiguchi, M. Yamashita, Y. Yoshikawa and H. Yasui, *Crystallogr. Rep.*, 2019, **64**, 1099–1102.
49. X. H. Tang and N. F. Shay, *J. Nutr.*, 2001, **131**, 1414–1420.
50. O. Ezaki, *J. Biol. Chem.*, 1989, **264**, 16118–16122.
51. W. Wu, X. Wang, W. Zhang, W. Reed, J. M. Samet, Y. E. Whang and A. J. Ghio, *J. Biol. Chem.*, 2003, **278**, 28258–28263.
52. H. Hasse and W. Maret, *Biometals*, 2005, **18**, 333–338.
53. Y. Yoshikawa, E. Ueda, Y. Suzuki, N. Yanagihara, H. Sakurai and Y. Kojima, *Chem. Pharm. Bull.*, 2001, **49**, 652–654.
54. Y. Yoshikawa, M. Kondo, H. Sakurai and Y. Kojima, *J. Inorg. Biochem.*, 2005, **99**, 1497–1503.
55. T. K. Saha, Y. Yoshikawa and H. Sakurai, *ChemMedChem*, 2007, **2**, 218–225.
56. Y. Yoshikawa, A. Murayama, Y. Adachi, H. Sakurai and H. Yasui, *Metallomics*, 2011, **3**, 686–692.
57. T. Moroki, Y. Yoshikawa, K. Yoshizawa, A. Tsubura and H. Yasui, *Metallomics*, 2014, **6**, 1632–1638.
58. H. Murakami, H. Yasui and Y. Yoshikawa, *Chem. Pharm. Bull.*, 2011, **60**, 1096–1104.
59. G. Viktor, *The donor–acceptor approach to molecular interactions*, Plenum Press, New York, 1978.
60. J. A. Lewis, B. L. Tran, D. T. Puerta, E. M. Rumberger, D. N. Hendrickson and S. M. Cohen, *Dalton Trans.*, 2005, **15**, 2588–2596.
61. W. Beasch and M. Schuster, *Z. Anorg. Allg. Chem.*, 1993, **619**, 786–790.
62. M. Kampf, R. Richter, L. Hennig, A. Eidner, J. Baldamus and R. Kirmse, *Z. Anorg. Allg. Chem.*, 2004, **630**, 2677–2686.
63. R. C. Hider, L. Ejim, P. D. Taylor, R. Gale, E. Huehns and J. B. Porter, *Biochem. Pharmacol.*, 1990, **39**, 1005–1012.
64. H. Iwatsuka, A. Shino and Z. Suzuki, *Endocrinol. Jpn.*, 1970, **17**, 23–35.
65. H. Yasui, H. Nishimura, H. Haba, R. Hirunuma, S. Enomoto and H. Sakurai, *RIKEN Accel. Prog. Rep.*, 2005, **38**, 119.
66. H. Yasui, H. Nishimura, H. Haba, S. Enomoto and H. Sakurai, *RIKEN Accel. Prog. Rep.*, 2006, **39**, 118.

CHAPTER 4

Structure Activity Relationship of Insulin-mimetic Zinc Complexes with Hydroxyazine-type Heterocyclic Bidentate Ligands

R. SAITO[a,b]

[a] Department of Chemistry, Toho University, 2-2-1 Miyama, Funabashi, Chiba, 274-8510, Japan; [b] Research Center for Materials with Integrated Properties, Toho University, 2-2-1 Miyama, Funabashi, Chiba 274-8510, Japan
Email: saito@chem.sci.toho-u.ac.jp

4.1 Introduction

Diabetes mellitus (DM) is a chronic state of hyperglycemia caused by insufficient or reduced insulin action, leading to serious complications without symptoms, and in the worst cases, death. According to a survey conducted by the International Diabetes Federation, the prevalence of diabetes in the global population aged 20–79 exceeded 10% (511 million people) in 2021, and is expected to increase to more than 12% (770 million people) by 2024,[1] thereby rendering the treatment of DM a significant global challenge. Depending on its pathogenesis, DM can be broadly divided into two types: type 1 and type 2. Type 1 DM is caused by the destruction of beta cells in the pancreas and can only be treated with insulin injections, whereas type 2 DM is

Coordination Chemistry Perspectives No. 2
Zinc in Biology: Molecular Structures, Cellular Processes and Living Systems
Edited by Hiroyuki Yasui and Taiho Kambe

Published by the Royal Society of Chemistry, www.rsc.org

caused by a deficiency in insulin action due to genetic predisposition, as well as environmental factors such as aging and lifestyle-related disorders such as being overweight and obesity. More than 90% of patients with diabetes suffer from type 2 DM.[2] The treatment of type 2 diabetes involves the administration of synthetic drugs along with exercise and diet therapy. Currently, the most commonly used hypoglycemic agents include alpha-glycosidase inhibitors, which suppress the postprandial increase in blood glucose levels by degrading carbohydrates to delay their absorption; sulfonylureas, which stimulate insulin secretion; and thiazolidinediones, which improve insulin resistance by acting on insulin receptors. However, even with these drugs, there are cases in which blood glucose control becomes difficult and requires insulin injections, such as in the treatment of type 1 DM, which causes significant mental and physical distress. Therefore, it is desirable to develop a drug that works *via* a mechanism different from that of conventional drugs and can be administered orally without the need for insulin injections. Therefore, zinc ions have recently attracted considerable attention as novel drug candidates for DM. Zinc ions are trace metal elements that are essential for life, and are the second most abundant metal in the human body, following iron.[3] The optimal concentration range of zinc ions in the human body is wide and the risk of overdose poisoning is low,[4–6] making the development of drugs utilizing zinc ions a highly attractive field of research.

Zinc ions play various roles in numerous biological processes, including protein and DNA synthesis, enzyme function, bone structure, nerve function, and cognitive function.[6,7] Additionally, zinc ions are involved in the formation and structural stability of insulin.[8] Zinc ions also play a role in the insulin signaling pathway, stimulating lipogenesis and glucose uptake into adipocytes,[9] suggesting that abnormal zinc metabolism is involved in the pathogenesis and progression of diabetes.

In 1980, it was demonstrated that zinc ions stimulated lipogenesis in rat adipocytes in a manner analogous to the action of insulin.[10] In practice, the oral administration of $ZnCl_2$ can effectively regulate plasma glucose concentration without the assistance of insulin injection.[11–13] However, the low bioavailability of the inorganic zinc salts necessitated the administration of increased doses, which in turn resulted in the emergence of adverse effects, probably combined with the high reactivity of $ZnCl_2$ as a Lewis acid.[14] It was thus imperative to develop methodologies to enhance the absorption efficacy of zinc ions. Based on the idea that an organozinc complex might improve the rate of absorption of zinc ions, organozinc complexes were orally administered to rats, which resulted in increased bioavailability compared to inorganic $ZnCl_2$.[14,15] These findings prompted the synthesis of various organozinc complexes, which were then investigated to determine their *in vitro* insulin-mimetic activities, hypoglycemic effects, and mechanisms of action. In early studies on the insulin-mimetic activity of organozinc complexes, naturally occurring zinc chelators and their derivatives, such as picolinic acids,[16] lactic and quinic acids,[17] maltol,[18] carnitine,[19] and allixin,[20] as well as commercially available dithiocarbamates[21] and amino acids[22] were mainly used as ligands. Some of these complexes demonstrated *in vivo* potential as chemotherapeutic agents for DM. For example, bis(6-methylpicolinato)zinc(II) and bis(maltolato)zinc(II) complexes exhibited insulin-mimetic activity and effectively lowered the blood glucose level of type-2-diabetic KK-A^y mice.[16,23] Yoshikawa *et al.* demonstrated that zinc complexes with maltol ligands activated the glucose transporter protein type-4 (also known as GLUT4), which mediated insulin-stimulated glucose transport in adipocytes, to exhibit insulin-mimetic activity.[24] They also reported that a zinc complex with hinokitiol ligands exhibited high insulin-mimetic activity by promoting Akt/PKB (protein kinase B) phosphorylation.[25] Therefore, the utilization of organozinc complexes as therapeutic agents for DM has been attracting considerable research attention in the fields of medicinal chemistry, since such drugs can replace conventional oral hypoglycemic drugs for type 2 DM and can be used as orally available hypoglycemic agents for type 1 DM without the assistance of insulin injections. To develop more potent organozinc complexes, detailed studies on the relationship between the structure of the ligands and the insulin-mimetic activity of their zinc complexes are required. For this purpose, it is necessary to expand the compound library; therefore, the development of new synthetic ligands is highly desirable. However, only a limited number of commercially available compounds can be used as ligands for zinc complexes. In addition, the utilization of naturally-occurring compounds as ligands has remained challenging due to the difficulty in isolating the large quantities necessary for the comprehensive studies of the compounds. From this perspective, producing new ligands by organic synthesis can greatly contribute to the advancement of the research and development of new complexes, and synthetic ligands have thus been used to expand the library of insulin-mimetic organozinc complexes. Among the synthetic ligands, those containing heterocyclic structures are particularly advantageous due to the following characteristics: (i) even low molecular weight monocyclic compounds have a considerably greater variety of derivatives than benzene derivatives, depending on the number of ring members and number and type of heteroatoms, (ii) the synthesis of compounds with oxygen or nitrogen functional groups that are capable of coordinating with metal ions is a more straightforward process than that of benzene derivatives, (iii) this ease of introducing substituents also allows for providing various water- or fat-soluble compounds as well as alteration of their metal coordination modes , and (iv) compounds with widely different acidities (pK_a values) can be synthesized more easily than benzene derivatives, providing complexes with different stability constants. Over the last two decades, hydroxyazine-type heterocyclic bidentate ligands with diverse structures have been synthesized, and the insulin-mimetic activities of their zinc complexes have been investigated. This section describes recent achievements associated with insulin-mimetic zinc complexes with hydroxyazine-type heterocyclic bidentate ligands, elucidates the correlation between their structural characteristics and insulin-mimetic activity, and discusses the structural factors crucial for obtaining high insulin-mimetic activity.

4.2 Recent Achievements in Insulin-mimetic Zinc Complexes in Terms of the Structures of Hydroxyazine-type Heterocyclic Ligands

4.2.1 Hydroxypyridinones and Their Thione Derivatives as Ligands

After the discovery of the insulin-mimetic activity of maltol-zinc complexes,[18] the utilization of 3-hydroxypyridin-4(1*H*)-ones (3,4-HOPOs) as ligands became a research focus, as 3,4-HOPOs are readily prepared by reacting maltol with appropriate amines. Thus, 3-hydroxy-1,2-dimethylpyridin-4(1*H*)-one (3,4-HOPO-Me) was synthesized. Naturally occurring allixin, which has the same 3-hydroxy-4*H*-pyranon skeleton as maltol, was similarly converted to the corresponding pyridinone, 1-(methylamino)allixin (maaxn). The insulin-mimetic activities of these complexes were evaluated by estimating the 50% inhibition concentrations of the complexes (IC_{50} values) on free fatty acid (FFA) release from isolated rat adipocytes.[26] However, this transformation did not significantly improve in the insulin-mimetic activity of the corresponding zinc complexes (see Figure 4.1).[27,28]

Subsequently, 3-hydroxypyridine-4(1*H*)-thiones (3,4-HOPTs), which are the thiocarbonyl derivatives of the corresponding 3,4-HOPOs, were prepared *via* a thiocarbonylation reaction of the corresponding carbonyl compounds with phosphorus pentoxide.[29] Among them, the 1-phenyl analogue exhibited significantly higher insulin-mimetic activity ($IC_{50} = 40$ μM) than vanadyl sulfate ($IC_{50} = 1.0$ mM), as a positive control (see Figure 4.2).[29]

In association with the use of pyridinetione derivatives as ligands, zinc pyrithione, a commercially available antibacterial and antiseborrheic agent with a structural isomer of 3,4-HOPTs, 1-hydroxypyridine-2(1*H*)-thione, exhibited insulin-mimetic activity approximately 30-fold higher than that of zinc sulfate, a positive control (see Figure 4.2).[30] Despite this fascinating activity, few zinc pyrithione derivatives have been studied because of the difficulty of their synthesis. Therefore, future research on zinc pyrithione

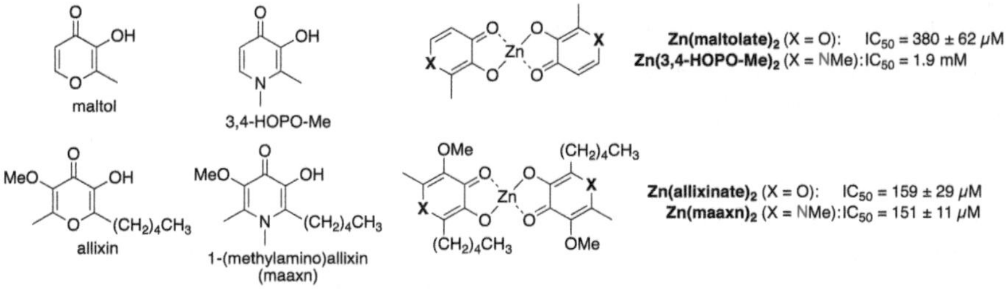

Figure 4.1 Structures of maltol, allixin, 3-hydroxy-1,2-dimethylpyridin-4(1*H*)-one (3,4-HOPO-Me), 1-(methylamino)allixin (maaxn), and their zinc complexes with their insulin-mimetic activity described as the IC_{50} values.

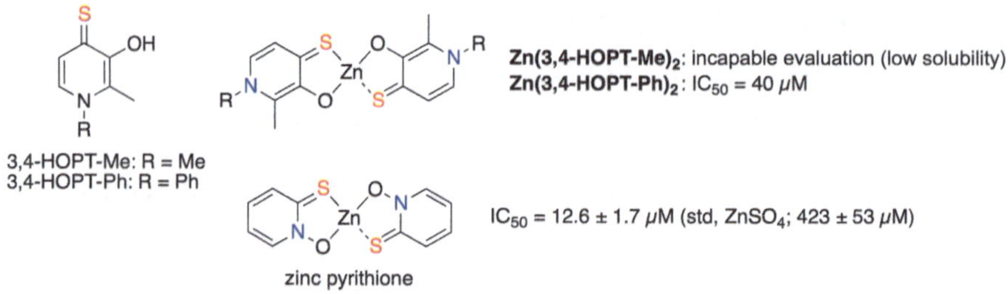

Figure 4.2 Structures of 3-hydroxypyridine-4(1*H*)-thiones (3,4-HOPTs), zinc complexes of 3,4-HOPTs, and zinc pyrithione with their insulin-mimetic activity (IC_{50} values).

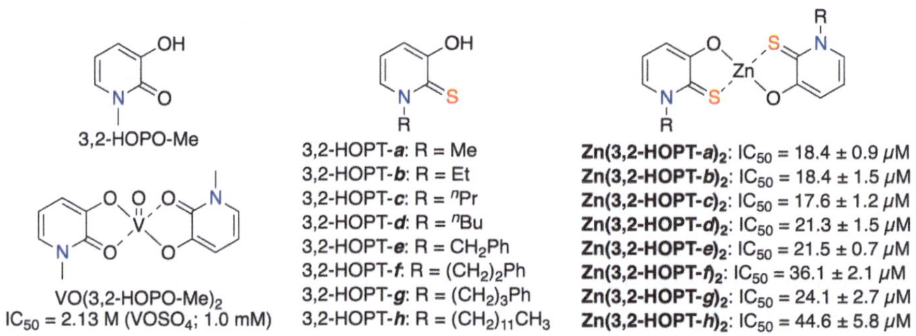

Figure 4.3 Structures of 3-hydroxy-1-methylpyridin-2(1*H*)-one (3,2-HOPO-Me), vanadyl complex of 3,2-HOPO-Me, 1-alkyl-3-hydroxypyridine-2(1*H*)-thiones (3,2-HOPT-***a–h***), and zinc complexes of 3,2-HOPTs accompanied by their insulin-mimetic activity (IC_{50} values).

derivatives, including the establishment of methods for the synthesis of 3-hydroxypyridine-2(1*H*)-thiones, is highly anticipated.

4.2.2 1-Alkyl-3-hydroxypyridine-2(1*H*)-thiones

Due to the low bioactivity of the vanadyl complex of 3-hydroxy-1-methylpyridin-2(1*H*)-one (3,2-HOPO-Me), zinc complexes of 3,2-HOPOs have received little attention. However, the above results prompted the synthesis of zinc complexes with thiocarbonyl derivatives of 3,2-HOPOs, namely 1-alkyl-3-hydroxypyridine-2(1*H*)-thiones (3,2-HOPTs), which exhibited 6- to 17-fold higher insulin-mimetic activities than that of the positive control, zinc sulfate (see Figure 4.3).[31]

Regarding the effect of the substituent (R) on the activity of the compound, complexes with shorter alkyl chains, such as methyl, ethyl, and propyl groups, at the N1 position showed higher activities than the complexes with phenyl and dodecyl groups. These results suggest a correlation between the hydrophobicity of the complex and its activity. This is discussed later in Section 4.3.

HOPyO-***a***: R_1 = H, R_2 = H
HOPyO-***b***: R_1 = Me, R_2 = Me
HOPyO-***c***: R_1 = H, R_2 = NHC_2H_5
HOPyO-***d***: R_1 = H, R_2 = $NH(CH_2)_3CH_3$
HOPyO-***e***: R_1 = H, R_2 = $NH(CH_2)_5CH_3$
HOPyO-***f***: R_1 = H, R_2 = (*S*)-$NH(CH_3)CO_2CH_3$
HOPyO-***g***: R_1 = H, R_2 = (*R*)-$NH(CH_3)CO_2CH_3$

Zn(HOPyO-*a*)$_2$: IC_{50} = 5.45 ± 2.08 mM
Zn(HOPyO-*b*)$_2$: IC_{50} = 4.54 ± 1.64 mM
Zn(HOPyO-*c*)$_2$: no activity was measuerd
Zn(HOPyO-*d*)$_2$: IC_{50} = 6.06 ± 1.01 mM
Zn(HOPyO-*e*)$_2$: IC_{50} = 2.58 ± 0.00 mM
Zn(HOPyO-*f*)$_2$: IC_{50} = 1.49 ± 0.03 mM
Zn(HOPyO-*g*)$_2$: IC_{50} = 1.05 ± 0.15 mM

Figure 4.4 Structures of 1-hydroxypyrimidin-2(1*H*)-ones (HOPyOs) and their zinc complexes with their insulin-mimetic activity (IC_{50} values).

4.2.3 1-Hydroxypyrimidin-2(1*H*)-ones

1-Hydroxypyrimidin-2(1*H*)-ones (HOPyOs) have previously been employed as ligands for a range of metal ions, such as iron, aluminum, gallium, and oxovanadium.[32–35] In 2000, the vanadyl (VO^{2+}) complex with 4,6-dimethyl-HOPyO was found to have high insulin-mimetic activity compared to the bis(picolinolato)oxovanadium complex, which was already known to be a potent antidiabetic agent.[36] Based on this finding, zinc complexes of HOPyOs with various substituents on the C4 and C6 positions were synthesized (see Figure 4.4).[34] The insulin-mimetic activities, that is, the IC_{50} values, were observed in the range of 1.05–6.06 mM, and were larger than that of zinc sulfate ($IC_{50} = 0.70$ mM) as the positive control.[34] Thus, the zinc complexes of HOPyOs were found to be less potent than zinc sulfate.

This result also indicated that these complexes were less potent than those having a 3,4-HOPO derivative as a ligand.[27] HOPyOs with an amino group at the C4 position, such as **Zn(HOPyO-*c*)$_2$–Zn(HOPyO-*g*)$_2$**, exist in tautomeric equilibrium between the three isomers.[37] It is conceivable that this structural uncertainty is a contributing factor to the observed low activity. Interestingly, the vanadyl complexes of all but two of these ligands exhibited higher insulin-mimetic activities than vanadyl sulfate, a positive control.[34] This suggests that the structure–activity relationships obtained for the vanadyl complexes cannot be directly applied to the development of zinc complexes. No examples of zinc complexes of 1-hydroxypyrimidin-2(1*H*)-thiones have been reported because of the difficulty of their synthesis.

4.2.4 3-Hydroxy-4-(4-substituted)phenylthiazole-2(3*H*)-thiones

All of the ligands described above are six-membered heterocycles and have previously been demonstrated to form complexes with metal ions, such as ferrous ion.[38] Whereas, there are few studies on the insulin-mimetic activity of zinc complexes with five-membered heterocyclic ligands. In this

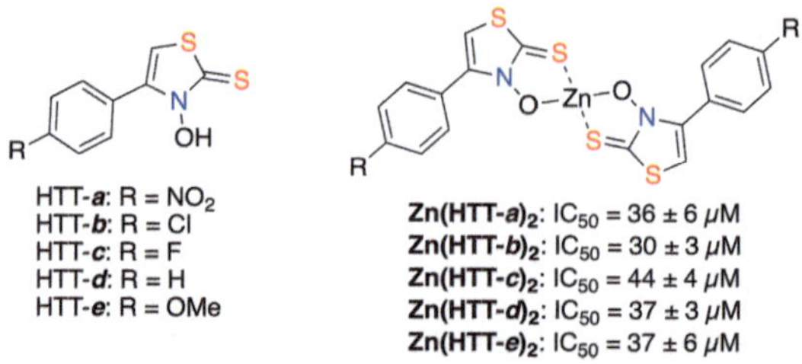

Figure 4.5 Structures of 3-hydroxy-4-(4-substituted)phenylthiazole-2(3*H*)-thiones (HTTs) and their zinc complexes with their insulin-mimetic activity (IC_{50} values).

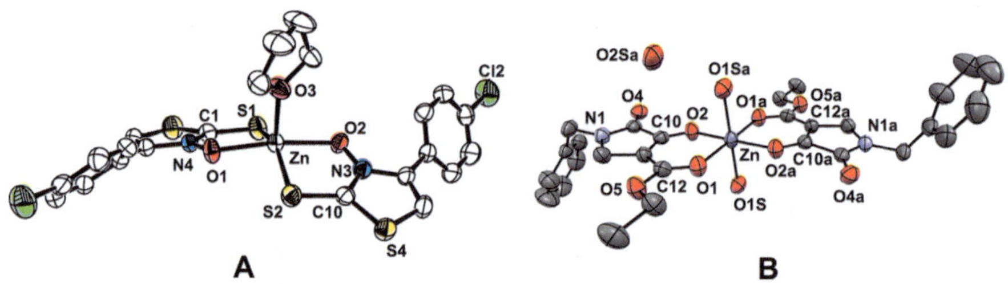

Figure 4.6 ORTEP diagrams of (A) **Zn(HTT-*b*)$_2$** and (B) **Zn(HOPE-*a*)$_2$**. The thermal ellipsoids of non-H atoms are depicted at the 50% probability level. Hydrogen atoms are omitted for clarity.

context, 3-hydroxy-4-(4-substituted)phenylthiazole-2(3*H*)-thiones (HTTs)[39] were identified as new ligand candidates because HTTs have a five-membered thiazolidinone skeleton, which is found in thiazolidinedione-based antidiabetic drugs such as pioglitazone, and a metal-coordinating thiohydroxamate group is incorporated into the ring system. Thus, several zinc complexes of HTTs have been synthesized (see Figure 4.5).[40] X-ray crystallographic analysis revealed the structural features of the complexes, forming a slightly disordered square-pyramidal geometry around zinc with two HTT ligands coordinated in *trans* configuration, and a solvent (tetrahydrofuran) molecule binding to the central metal from the apical position (see Figure 4.6A). This distorted bipyramidal geometry, in which zinc ions are coordinated by five donor atoms, has also been observed in the zinc complex of maltol[18] and in zinc pyrithione,[41] which has high insulin-mimetic activity. In several cases, zinc complexes recrystallized from organic solvents that do not have atoms capable of coordinating metals adopt a tetrahedral geometry.[42] In contrast, when a coordinating solvent is present in the medium, Zn(HTTs)$_2$ has one apical coordinating solvent molecule. These results suggest that water molecules coordinate apically and assume a square pyramidal state when HTTs/Zn complexes are administered to the body.

The insulin-mimetic activities of these complexes ($IC_{50} = 30–44$ μM) were ten-fold higher than that of the positive control, zinc sulfate (470 μM). The two most active complexes, **Zn(HTT-*a*)$_2$** ($R = NO_2$) and **Zn(HTT-*b*)$_2$** (R = Cl), were further evaluated for their antidiabetic effects *in vivo*. In an oral glucose tolerance test using KK-A^y mice, which are model animals for type 2 DM, these complexes substantially lowered blood glucose levels is diabetic mice. In addition, these complexes effectively improved the HbA1c values, blood pressure, and insulin levels in diabetic mice. Based on these results, zinc complexes with HTTs have been identified as a new class of chemotherapeutic agents for the treatment of DM. This was the first example of a zinc complex with five-membered heterocycles as ligands exhibiting high insulin-mimetic activities and provided an impetus for subsequent studies on zinc complexes with five-membered heterocyclic ligands.

4.2.5 Pyrrole-based Bidentate Ligands

In the search for new five-membered heterocyclic ligands for antidiabetic zinc complexes, Saito *et al.* focused on 1-arylmethyl-2,5-dihydro-4-hydroxy-5-oxo-1*H*-pyrrole-3-carboxylic acid ethyl esters (HOPEs; see Figure 4.7) as HOPEs have a structure similar to that of HTTs, as well as functional groups that are capable of coordinating to zinc ions.[43] Moreover, HOPEs have inhibitory activity against aldose reductase,[44] which is associated with the onset of diabetic complications,[45] suggesting that zinc complexes with these ligands may lead to the development of chemotherapeutics that simultaneously treat DM and diabetic complications. Thus, HOPEs with various substituents on the phenyl groups were synthesized. Interestingly, this ligand coordinated to zinc *via* hydroxy oxygen atoms and ester carbonyl oxygen, forming a η^6-type complex, in contrast to the previously reported organozinc complexes, where the ligands coordinated to zinc in the η^5 form. Furthermore, this complex exhibits an octahedral six-coordinate geometry (see Figure 4.6B), which differs from the tetrahedral or square pyramidal geometries observed in the previously reported zinc complexes.[18,41]

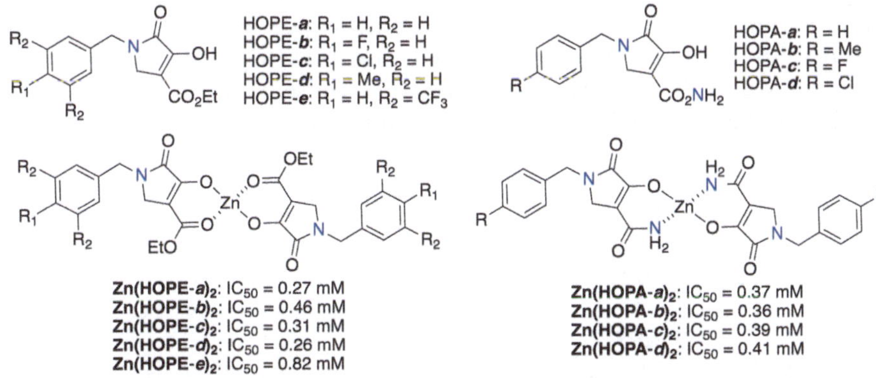

Figure 4.7 Structures of 1-arylmethyl-2,5-dihydro-4-hydroxy-5-oxo-1*H*-pyrrole-3-carboxylic acid ethyl esters (HOPEs) and their zinc complexes with their insulin-mimetic activity (IC_{50} values).

The insulin-mimetic activities (IC_{50} values) of the complexes were in the range of 0.26–0.82 mM, and all but one were more potent than standard zinc sulfate ($IC_{50} = 0.75$ mM). Based on these results, it can be said that these complexes are potential dual chemotherapeutic agents for both DM and diabetic complications. One exception was a derivative in which four trifluoromethyl groups were introduced at the R_2 positions. In this complex, four large highly hydrophobic substituents were introduced into an already sufficiently large and hydrophobic complex with two phenyl groups. This significantly increased the hydrophobicity and molecular length diameter.

To examine the effect of altering the coordination mode from O_4 to N_2O_2 in HOPEs, 4-hydroxy-5-oxo-2,5-dihydro-1*H*-pyrrole-3-carboxamides (HOPAs) were synthesized, and their zinc complexes were prepared (see Figure 4.7).[46]

The coordination mode of the ligand to zinc was confirmed by nuclear magnetic resonance (NMR) spectroscopy as being coordinated to zinc with the η^6 form *via* the oxygen of the hydroxyl group at the 4-position and the nitrogen of the carbamoyl group at the 3-position. The derivative containing a trifluoromethyl group was insoluble in the assay buffer; therefore, its bioactivity could not be determined. Considering the results for Zn–HOPE complexes, the introduction of trifluoromethyl groups was not advantageous, at least for pyrrole-based ligands. All of the other complexes exhibited higher insulin-mimetic activity ($IC_{50} = 0.36$–0.41 mM) than that of standard zinc sulfate ($IC_{50} = 0.44$ mM).

Saito *et al.* reported the synthesis and the insulin-mimetic activities of zinc complexes with 1-arylmethyl-5-oxo-2-thioxopyrrolidines (OTPs) as ligands (see Figure 4.8).[47] These compounds were not synthesized with a specific objective in mind and were unexpectedly obtained in an attempt to synthesize the SH analogs of HOPEs to study the effect of changing the coordination mode from $Zn(O_4)$ to $Zn(S_2O_2)$ in the Zn-HOPE complexes.[47] The zinc complexes were successfully obtained through the conventional method of reacting the ligands, 5-oxo-2-thioxopyrrolidines, with zinc sulfate. In these complexes, the ligands were coordinated to zinc in the η^6 form, and the coordination mode was identified as the S_2O_2-type by NMR spectroscopy. The insulin-mimetic activities of these complexes (IC_{50} values) were in the range of 0.178–0.237 mM, indicating that these complexes had higher insulin-mimetic activities than those of Zn-HOPEs.

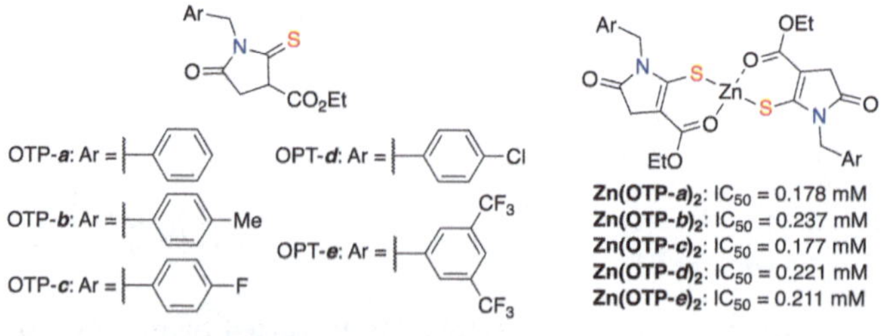

Figure 4.8 Structures of 1-arylmethyl-5-oxo-2-thioxopyrrolidines (OTPs) and their zinc complexes with their insulin-mimetic activity (IC_{50} values).

4.3 Relationship Between the Physicochemical Properties of the Zinc Complexes and Their Insulin-mimetic Activities

In the development of new drugs, it is important to understand the correlation between the physicochemical properties and bioactivity of drug candidates to consider their pharmacokinetics and pharmacodynamics. This section describes the significant physicochemical properties of zinc complexes that influence the insulin-mimetic activity based on the above results.

4.3.1 Coordinating Atoms to the Central Zinc Ions

Studies with vanadyl (VO^{2+}) complexes of heterocyclic bidentate ligands have indicated that the set of coordinating atoms, that is, the coordination mode; $VO(O_4)$, $VO(N_4)$, $VO(N_2O_2)$, $VO(S_2O_2)$, and $VO(S_4)$, is an important factor affecting the insulin-mimetic activity.[48] The coordination mode is also a key factor in the insulin-mimetic activity of zinc complexes of non-heterocyclic ligands.[49] Therefore, it is reasonable to investigate whether the coordination mode is also affected in the zinc complexes with hydroxyazine-type heterocyclic ligands.

The insulin-mimetic activity of 3,4-HOPO and its derivatives was dramatically enhanced when the carbonyl oxygen at the C4 position was replaced by sulfur (see Figure 4.9). A significant increase in the insulin-mimetic activity by converting the carbonyl (C=O) to the corresponding thiocarbonyl (C=S) groups was also observed in 3,2-HOPTs. A similar activity enhancement by chemical transformation of C=O to C=S has been observed in other non-hydroxyazine-type bidentate ligands such as

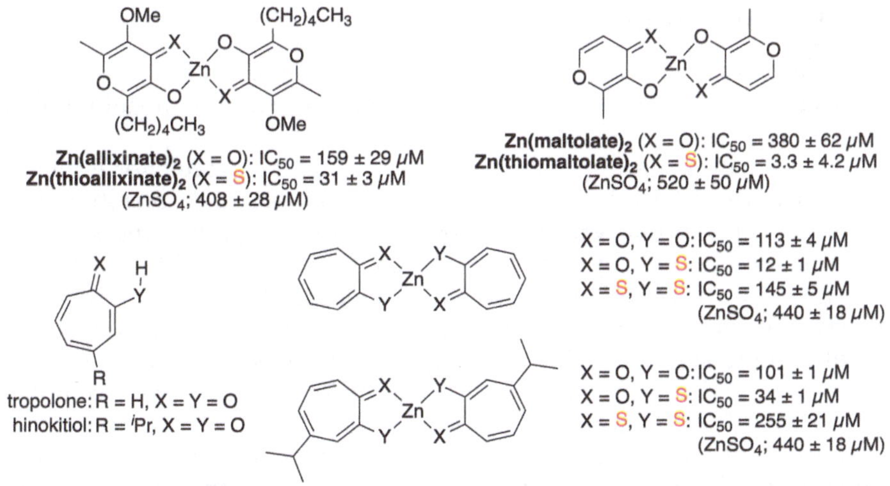

Figure 4.9 Structures of zinc complexes of 3-hydroxy-4*H*-pyrrones and tropolones as ligands accompanied by their insulin-mimetic activity (IC_{50} values).

3-hydroxy-4-pyrones (see Figure 4.9).[27,28] Furthermore, the insulin-mimetic activity of the zinc complex with the 4-isopropyltropolone (also known as hinokitiol) ligand was enhanced approximately 3-fold upon conversion of one of the two oxygen atoms in hinokitiol into a sulfur atom. In the case of tropolone, in which the isopropyl group of hinokitiol was replaced with a hydrogen atom, a 9-fold increase in insulin-mimetic activity was observed with the same transformation. In both cases, the conversion of the all oxygen atoms into sulfur atoms resulted in a reduction in activity (see Figure 4.9).[50]

Studies of 3,2-HTTs focused on novel heterocyclic bidentate ligands for developing antidiabetic zinc complexes. These complexes showed high levels of insulin-mimetic and antidiabetic activities, comparable to those of the aforementioned zinc complexes with six-membered heterocyclic bidentate ligands, which are already known as potent antidiabetic agents.

Comparing the IC_{50} values for the zinc complexes of OTPs with those of HOPEs and HOPAs revealed that the zinc complexes of OTPs were more potent, indicating that even for zinc complexes with pyrrole-based ligands coordinating with the central zinc ion in the η^6 form, higher insulin-mimetic activities were observed in the complexes with the S_2O_2-type coordination mode. There was no significant difference in the insulin-mimetic activity between the O_4- and N_2O_2-type coordination modes, at least for the zinc complexes with pyrrole-based ligands.

These results imply that zinc complexes with S_2O_2-type coordination mode exhibit higher insulin-mimetic activity than those with O_4- and N_2O_2-type coordination modes, without exception, irrespective of the structure of the heterocyclic ring and the form of coordination to the central zinc ion (η^5 or η^6 form). In other words, the S_2O_2-type coordination mode is a significant factor for achieving high insulin-mimetic activity.

4.3.2 Effect of Lipophilicity (log *P*)

In addition to the coordination mode, it is also important to understand the physicochemical properties of the ligand that influence its insulin-mimetic activity.

Lipophilicity is an important physicochemical parameter for evaluating the bioactivity of drug candidates because it is related to membrane permeability, absorption, and distribution of compounds in living systems.[51] Commonly, lipophilicity is expressed quantitatively as $\log P$, which is the common logarithm of the ratio of the concentration of a compound at equilibrium between the *n*-octanol and water phases. In the evaluation of insulin-mimetic activity, zinc complexes are assumed to be taken up by rat adipocytes, and it has been suggested that the lipophilicity of zinc complexes affects their insulin-mimetic activities.[16,23,24]

In the series of zinc complexes of 3,2-HOPTs, there may be a correlation between the lipophilicity of the complex and its activity because the lipophilicity of the complexes is thought to vary depending on the length of the alkyl chain introduced and the presence or absence of a phenyl group.

It would be best to measure the lipophilicity of all complexes; however, this is challenging because of the solubility limitations of the complexes in the measurement solvents. Thus, the log *P* values of the ligands were measured instead because there exists a positive correlation between log *P* of a complex and log *P* of a ligand.[52,53] All values obtained were greater than zero, indicating that these ligands were lipophilic.[31] Comparing the activity between these compounds, the complexes with less lipophilic ligands tended to exhibit better insulin-mimetic activity, although the correlation was not particularly strong. In this series of complexes, more lipophilic complexes were modified with longer or larger substituents, suggesting that the size of the complexes, rather than their lipophilicity, influenced their activity.

It is also important to understand that the relationship between calculated log *P* (Clog *P*) values and insulin-mimetic activity is as important as in the case of using experimental log *P* values because the activity of a zinc complex can be predicted at the design stage of its synthesis. Although Clog *P* values may not exactly match the experimental log *P* values, there is a correlation between these two parameters,[54,55] enabling the utilization of Clog *P* values for the expeditious development of new drugs. Based on this idea, Saito *et al.* investigated the relationship between the insulin-mimetic activities and the Clog *P* values of the zinc complexes of OTPs.[47] The Clog *P* values of potent **Zn(OTP-*a*)$_2$** and **Zn(OTP-*c*)$_2$** were estimated to be 4.30 and 4.62, respectively, and those of the other less potent complexes were greater than 5. A similar trend was observed for the zinc complexes of HOPEs, with Clog *P* values ranging from 4.43 to 8.10. Moreover, the Clog *P* values of the HOPE-based complexes were holistically larger than those of the Zn-OTP complexes. Considering that the Zn-HOPE complexes were less potent than the Zn-OTP complexes, it can be concluded that the less lipophilic complexes were more potent. However, Zn-HOPyO complexes exhibited insulin-mimetic activity ranging from 2.58 to 6.06 mM, which increased with the increasing lipophilicity of the complexes.[34] To address the apparent inconsistencies in the observed outcomes, the same Clog *P* calculations were conducted for a subset of the pyrimidine-based complexes. Their Clog *P* values were obtained in the range of −0.72 to 4.09,[47] indicating that Zn-HOPyO complexes were less lipophilic than the Zn-OTP complexes. These results led to the following conclusion: for complexes with Clog *P* values less than 5, the insulin-mimetic activity increases with increasing lipophilicity of the complex, whereas for the complexes with Clog *P* values more than 5, the activity diminishes with increasing lipophilicity. This corroborates the previous findings that the hydrophilicity/lipophilicity balance plays an important role in determining the insulin-mimetic activities of zinc complexes with heterocyclic bidentate ligands.[34,40]

4.3.3 Other Physicochemical Properties

High insulin-mimetic activity is observed for zinc complexes with stability constants (log β values) not exceeding 11, whereas the activity tends to be

Table 4.1 Structures, IC_{50} values, and physicochemical properties of insulin-mimetic zinc complexes with heterocyclic bidentate ligands.

Complex	Coordination mode	$IC_{50}{}^{a}$/μM	Positive control	$IC_{50,rel}{}^{b}$/ μM	log P	pK_a of ligand	CpK_a of ligand	Ref.
Zn(maltolate)$_2$	O_4	590	0.81 mM ($ZnSO_4$)	728	0.60 ± 0.01	8.45	8.41[i]	Ref. 18, ref. 62 (pK_a), ref. 20 (log P)
Zn(3,4-HOPT-Me)$_2$	S_2O_2	nd[c]	1.0 mM ($VOSO_4$)			9.47	9.39[i]	Ref. 29, ref. 42 (pK_a)
Zn(3,4-HOPT-Ph)$_2$	S_2O_2	40	1.0 mM ($VOSO_4$)	40			8.29[i]	Ref. 29
Zn(etylmaltolate)$_2$	O_4	440 ± 50	1.00 ± 0.07 mM ($ZnSO_4$)	440	1.24 ± 0.03		8.38[i]	Ref. 20
Zn(allixinate)$_2$	O_4	370 ± 30	1.00 ± 0.07 mM ($ZnSO_4$)	370	1.65 ± 0.03		9.27[i]	Ref. 20
Zn(thioallixinate)$_2$	S_2O_2	31 ± 3	408 ± 28 μM ($ZnSO_4$)	76	(1.36 ± 0.06)[d]		7.34[i]	Ref. 27
Zn(maaxn)$_2$	O_4	159 ± 26	408 ± 28 μM ($ZnSO_4$)	390	(1.73 ± 0.01)[d]		9.66[i]	Ref. 27
Zn(HOPyO-***a***)$_2$	O_4	5450 ± 2080	0.70 ± 0.05 mM ($ZnSO_4$)	7786	nd[c]	6.1	6.00[i]	Ref. 34, ref. 32 (pK_a)
Zn(HOPyO-***b***)$_2$	O_4	4540 ± 1640	0.70 ± 0.05 mM ($ZnSO_4$)	6486	−0.30		3.27[i]	Ref. 34
Zn(HOPyO-***c***)$_2$	O_4	None	0.70 ± 0.05 mM ($ZnSO_4$)		nd[c]		3.27[i]	Ref. 34
Zn(HOPyO-***d***)$_2$	O_4	6060 ± 1010	0.70 ± 0.05 mM ($ZnSO_4$)	8657	-0.74		3.09[i]	Ref. 34
Zn(HOPyO-***e***)$_2$	O_4	2580 ± 0	0.70 ± 0.05 mM ($ZnSO_4$)	3686	0.55		3.10[i]	Ref. 34
Zn(HOPyO-***f***)$_2$	O_4	1490 ± 30	0.70 ± 0.05 mM ($ZnSO_4$)	2129	$-\infty$[e]		3.04[i]	Ref. 34
Zn(HOPyO-***g***)$_2$	O_4	1050 ± 150	0.70 ± 0.05 mM ($ZnSO_4$)	1500	$-\infty$[e]		3.04[i]	Ref. 34
Zn(HTT-***a***)$_2$	S_2O_2	36 ± 6	0.47 ± 0.03 mM ($ZnSO_4$)	77		3.92	4.22[i]	Ref. 40
Zn(HTT-***b***)$_2$	S_2O_2	30 ± 3	0.47 ± 0.03 mM ($ZnSO_4$)	64		4.91	4.40[i]	Ref. 40
Zn(HTT-***c***)$_2$	S_2O_2	44 ± 4	0.47 ± 0.03 mM ($ZnSO_4$)	94		4.80	4.44[i]	Ref. 40
Zn(HTT-***d***)$_2$	S_2O_2	37 ± 3	0.47 ± 0.03 mM ($ZnSO_4$)	79		4.01	4.46[i]	Ref. 40
Zn(HTT-***e***)$_2$	S_2O_2	37 ± 6	0.47 ± 0.03 mM ($ZnSO_4$)	79		4.63	4.48[i]	Ref. 40
Zn(thiomaltolate)$_2$	S_2O_2	3.3 ± 0.4	520 ± 50 μM ($ZnSO_4$)	6	0.54	8.06	7.58[i]	Ref. 28, ref. 62 (pK_a)
Zn(3,4-HOPO-Me)$_2$	O_4	1900 ± 41	520 ± 50 μM ($ZnSO_4$)	3668	−0.85	9.82	9.90[i]	Ref. 28, ref. 63 (pK_a)
Zn(HOPT-***a***)$_2$	S_2O_2	18.4 ± 0.9	292 μM ($ZnSO_4$)	63	(0.3)[d]	8.97	8.31[i]	Ref. 31, ref. 42 (pK_a)

Zn(HOPT-***b***)$_2$	S_2O_2	18.4 ± 1.5	292 μM ($ZnSO_4$)	63	(1.2)[d]		8.27[i]	Ref. 31
Zn(HOPT-***c***)$_2$	S_2O_2	17.6 ± 1.2	292 μM ($ZnSO_4$)	60	(1.3)[d]		8.28[i]	Ref. 31
Zn(HOPT-***d***)$_2$	S_2O_2	21.3 ± 1.5	292 μM ($ZnSO_4$)	73	(1.4)[d]		8.27[i]	Ref. 31
Zn(HOPT-***e***)$_2$	S_2O_2	21.5 ± 0.7	292 μM ($ZnSO_4$)	74	(1.7)[d]		8.16[i]	Ref. 31
Zn(HOPT-***f***)$_2$	S_2O_2	36.1 ± 2.0	292 μM ($ZnSO_4$)	124	(∞)[d]		8.16[i]	Ref. 31
Zn(HOPT-***g***)$_2$	S_2O_2	24.1 ± 2.7	292 μM ($ZnSO_4$)	83	(∞[f])[d]		8.23[i]	Ref. 31
Zn(HOPT-***h***)$_2$	S_2O_2	44.6 ± 5.8	292 μM ($ZnSO_4$)	153	(∞[f])[d]		8.26[i]	Ref. 31
Zinc pyrithione	S_2O_2	12.6 ± 1.7	423 ± 53 μM ($ZnSO_4$)	30	0.48 ± 0.04		4.70[i]	Ref. 30
Zn(HOPE-***a***)$_2$	O_4	270	0.75 mM ($ZnSO_4$)	613	0.62	3.87[g]	4.50[j]	Ref. 43
Zn(HOPE-***b***)$_2$	O_4	460	0.75 mM ($ZnSO_4$)	413	0.66	3.73[g]	4.50[j]	Ref. 43
Zn(HOPE-***c***)$_2$	O_4	310	0.75 mM ($ZnSO_4$)	347	nd[c]	3.44[g]	4.50[j]	Ref. 43
Zn(HOPE-***d***)$_2$	O_4	260	0.75 mM ($ZnSO_4$)	1093	nd[c]	3.61[g]	4.50[j]	Ref. 43
Zn(HOPE-***e***)$_2$	O_4	820	0.75 mM ($ZnSO_4$)	841	nd[c]	nd[c]	4.50[j]	Ref. 43
Zn(HOPA-***a***)$_2$	O_4	370	0.44 mM ($ZnSO_4$)	841			4.50[j]	Ref. 46
Zn(HOPA-***b***)$_2$	N_2O_2	360	0.44 mM ($ZnSO_4$)	818			4.50[j]	Ref. 46
Zn(HOPA-***c***)$_2$	N_2O_2	390	0.44 mM ($ZnSO_4$)	886			4.50[j]	Ref. 46
Zn(HOPA-***d***)$_2$	N_2O_2	410	0.44 mM ($ZnSO_4$)	932			4.50[j]	Ref. 46
Zn(OTP-***a***)$_2$	S_2O_2	178	0.44 mM ($ZnSO_4$)	405	4.30[h]		—[k]	Ref. 47
Zn(OTP-***b***)$_2$	S_2O_2	237	0.44 mM ($ZnSO_4$)	539	5.17[h]		—[k]	Ref. 47
Zn(OTP-***c***)$_2$	S_2O_2	177	0.44 mM ($ZnSO_4$)	502	5.62[h]		—[k]	Ref. 47
Zn(OTP-***d***)$_2$	S_2O_2	221	0.44 mM ($ZnSO_4$)	480	5.54[h]		—[k]	Ref. 47
Zn(OTP-***e***)$_2$	S_2O_2	211	0.44 mM ($ZnSO_4$)	480	7.98[h]		—[k]	Ref. 47

[a] 50% inhibitory concentration of a complex on epinephrine-stimulated FFA release from isolated rat adipocytes.
[b] Values relative to the positive control of which the IC_{50} value is converted to 1.0 mM.
[c] Incapable of evaluation because of low solubility.
[d] Values for ligands.
[e] Highly hydrophilic (water soluble).
[f] Highly lipophilic (insoluble in water).
[g] Measured in 50% aqueous *N*,*N*-dimethylformamide.
[h] Calculated using Chem3D software.
[i] Calculated using Advanced Chemistry Development (ACD/Labs) Software (©1994–2024 ACD/Labs).
[j] Calculated using MolGpKa.[64]
[k] Incorrect prediction of dissociating acidic protons.

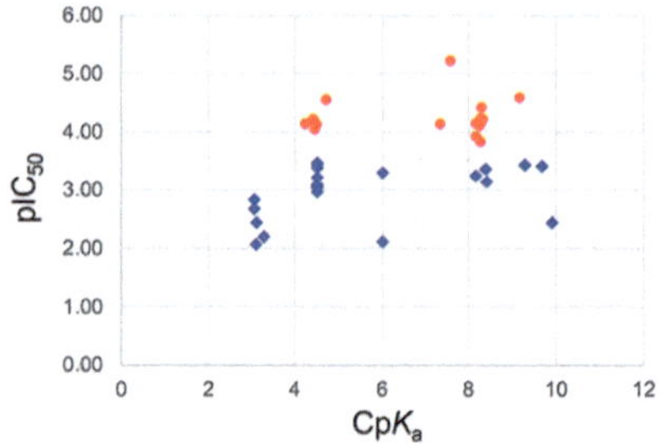

Figure 4.10 Correlation between the calculated pK_a (CpK_a) values and the insulin-mimetic activities (expressed as pIC_{50} values) of the zinc complexes with hydroxyazine-type heterocyclic bidentate ligands with S_2O_2 (closed circle) and O_4 (closed diamond) coordination modes.

lower for complexes with $\log\beta$ values exceeding 11,[22,56] and so far no significant deviations from this trend have been observed. Therefore, the $\log\beta$ of a complex is an important factor that affects its insulin-mimetic activity, and it is advantageous to estimate it beforehand. However, it is difficult to predict the stability constants of these zinc complexes. One alternative is to utilize the pK_a values of the ligands, because the stability constants of the metal complexes are approximately proportional to the pK_a values of the non-bulky ligands measured under identical conditions.[57,58] In the molecular design phase, the calculated values are used. Table 4.1 lists the calculated pK_a (CpK_a) values of the ligands. Figure 4.10 shows the insulin-mimetic activities ($pIC_{50,rel}$ values, defined by eqn (4.1)) of the zinc complexes plotted against their CpK_a values.

$$pIC_{50,rel} = -\log_{10}(IC_{50,rel}), \tag{4.1}$$

where $IC_{50,rel}$ represents the value relative to the positive control.

From these results, there appears to be no clear correlation between the CpK_a and pIC_{50} values, making it difficult to estimate the stability of the zinc complexes based on the CpK_a values of the ligands. However, it is evident that the S_2O_2 type is more potent than the O_4 type. This demonstrates that the coordination mode has a more pronounced effect on the insulin-mimetic activity of the zinc complexes with hydroxyazine-type heterocyclic bidentate ligands than the other physicochemical parameters.

Molecular chirality can be an important factor in controlling drug discovery because the biological environment is chiral. To determine whether chirality control is necessary for the development of insulin-mimetic zinc complexes, chiral substituents were introduced into some complexes. Consequently, no significant difference in the activity between the enantiomers was observed for complexes with chiral carbons in the ligand side chain.[34,59] This is beneficial for future drug development because it will help reduce the cost of purifying new compounds.

4.4 Conclusion

Although the insulin-mimetic activity of zinc complexes with six-membered heterocyclic bidentate ligands has been extensively investigated, the results presented here indicate that five-membered hydroxyazine-type heterocycles can also be used as ligands to design zinc complexes with good insulin-mimetic activities.

The hydrophilic/lipophilic balance and, in particular, the coordination mode were noted as important physicochemical properties to consider when developing new zinc complexes. O_4- and N_2O_2-type ligands are easier to obtain commercially and synthesize compared with S_2O_2-type ligands. Even when natural products or food components are used as ligands, O_4 and N_2O_2 type ligands are more readily available than S_2O_2-type ligands. For this purpose, one can envision the direct conversion of alcohols to thiols,[60] direct thioesterification of carboxylic acids,[61] and one-step conversion of carbonyl to thiocarbonyl groups,[29,31] as accessible strategies for obtaining S_2O_2-type ligands.

It is also noteworthy that good insulin-mimetic activity was observed for zinc complexes with HOPE ligands that have aldose reductase inhibitory activity. These complexes are potential drug candidates for the simultaneous treatment of DM and diabetic complications. Thus, the use of compounds with established bioactivities as ligands represents a pivotal strategy for the development of less toxic and more multipotent zinc complexes, and an increasing number of analogous cases will emerge in the future.

Acknowledgements

The author would like to thank Editage (www.editage.jp) for the English language editing.

References

1. H. Sun, P. Saeedi, S. Karuranga, M. Pinkepank, K. Ogurtsova, B. B. Duncan, C. Stein, A. Basit, J. C. N. Chan, J. C. Mbanya, M. E. Pavkov, A. Ramachandaran, S. H. Wild, S. James, W. H. H. Herman, P. Zhang, C. Bommer, S. H. Kuo, E. J. J. Boyko and D. J. Magliano, *Diabetes Res. Clin. Pract.*, 2022, **183**, 109119.
2. A. F. Amos, D. J. McCarty and P. Zimmet, *Diabetic Med.*, 1997, **14**, S7–S85.
3. W. Kaim and B. Schwederski, *Bioinorganic Chemistry: Inorganic Elements in the Chemistry of Life: An Introduction and Guide*, John Wiley & Sons Ltd, Chichester, UK, 1994.
4. G. J. Fosmire, *Am. J. Clin. Nutr.*, 1990, **51**, 225–227.
5. J. Wolf, H. H. Sandstead and L. Rink, in *Handbook on the Toxicology of Metals*, ed. G. F. Nordberg and M. Costa, Academic Press, 5th edn, 2022, pp. 963–984.
6. L. I. Stiles, K. Ferrao and K. J. Mehta, *Clin. Exp. Med.*, 2024, **24**, 38–56.
7. M. Maares and H. Haase, *Nutrients*, 2020, **12**, 762.

8. G. Dodson and D. Steiner, *Curr. Opin. Struc. Biol.*, 1998, **8**, 189–194.
9. A. Fukunaka and Y. Fujitani, *Int. J. Mol. Sci.*, 2018, **19**, 476.
10. L. Coulston and P. Dandona, *Diabetes*, 1980, **29**, 665–667.
11. J. M. May and C. S. Contoreggi, *J. Biol. Chem.*, 1982, **257**, 4362–4368.
12. A. Shisheva, D. Gefel and Y. Shechter, *Diabetes*, 1992, **41**, 982–988.
13. M. D. Chen, S. J. Liou, P. Y. Lin, V. C. Yang, P. S. Alexander and W. H. Lin, *Biol. Trace Elem. Res.*, 1998, **61**, 303–311.
14. M. Nishida, H. Sakurai, J. Kawada, M. Koyama and J. Takada, *Naturwissenschaften*, 1989, **76**, 220–222.
15. J. Fugono, K. Fujimoto, H. Yasui, K. Kawabe, Y. Yoshikawa, Y. Kojima and H. Sakurai, *Drug Metab. Pharmacokinet.*, 2002, **17**, 340–347.
16. Y. Yoshikawa, E. Ueda, K. Kawabe, H. Miyake, T. Takino, H. Sakurai and Y. Kojima, *J. Biol. Inorg. Chem.*, 2002, **7**, 68–73.
17. Y. Kojima, Y. Yoshikawa, E. Ueda, R. Ueda, S. Yamamoto, K. Kumekawa, N. Yanagihara and H. Sakurai, *Chem. Pharm. Bull.*, 2003, **51**, 1006–1008.
18. Y. Yoshikawa, E. Ueda, K. Kawabe, H. Miyake, H. Sakurai and Y. Kojima, *Chem. Lett.*, 2000, 874–875.
19. Y. Yoshikawa, E. Ueda, H. Sakurai and Y. Kojima, *Chem. Pharm. Bull.*, 2003, **51**, 230–231.
20. Y. Adachi, J. Yoshida, Y. Kodera, A. Kato, Y. Yoshikawa, Y. Kojima and H. Sakurai, *J. Biol. Inorg. Chem.*, 2004, **9**, 885–893.
21. Y. Yoshikawa, Y. Adachi and H. Sakurai, *Life Sci.*, 2007, **80**, 759–766.
22. Y. Yoshikawa, E. Ueda, Y. Suzuki, N. Yanagihara, H. Sakura and Y. Kojima, *Chem. Pharm. Bull.*, 2001, **49**, 652–654.
23. Y. Yoshikawa, E. Ueda, H. Miyake, H. Sakurai and Y. Kojima, *Biochem. Biophys. Res. Commun.*, 2001, **281**, 1190–1193.
24. Y. Yoshikawa, E. Ueda, Y. Kojima and H. Sakurai, *Life Sci.*, 2004, **75**, 741–751.
25. Y. Naito, Y. Yoshikawa and H. Yasui, *Bull. Chem. Soc. Jpn.*, 2011, **84**, 298–305.
26. M. Nakai, H. Watanabe, C. Fujiwara, H. Kakegawa, T. Satoh, J. Takada, R. Matsushita and H. Sakurai, *Biol. Pharm. Bull.*, 1995, **18**, 719–725.
27. Y. Adachi, J. Yoshida, Y. Kodera and H. Sakurai, *Chem. Lett.*, 2005, **34**, 656–657.
28. S. Chaves, R. Jelic, C. Mendonça, M. Carrasco, Y. Yoshikawa, H. Sakurai and M. A. Santos, *Metallomics*, 2010, **2**, 220–227.
29. A. Katoh, T. Tsukahara, R. Saito, K. K. Ghosh, Y. Yoshikawa, Y. Kojima, A. Tamura and H. Sakurai, *Chem. Lett.*, 2002, 114–115.
30. Y. Yoshikawa, A. Murayama, Y. Adachi, H. Sakurai and H. Yasui, *Metallomics*, 2011, **3**, 686–692.
31. A. Katoh, H. Yokoyama, Y. Matsumura, Y. Yoshikawa, H. Yasui and H. Sakurai, *Heterocycles*, 2010, **81**, 585–600.
32. J. Ohkanda, T. Tokumitsu, K. Mitsuhashi and A. Katoh, *Bull. Chem. Soc. Jpn.*, 1993, **66**, 841–847.
33. T. Masuda, J. Hayashi and S. Tamagaki, *J. Chem. Soc., Perkin Trans. 2*, 2000, 161–167.

34. M. Yamaguchi, K. Wakasugi, R. Saito, Y. Adachi, Y. Yoshikawa, H. Sakurai and A. Katoh, *J. Inorg. Biochem.*, 2006, **100**, 260–269.
35. S. Chaves, A. Capelo, L. Areias, S. M. Marques, L. Gano, M. A. Esteves and M. A. Santos, *Dalton Trans.*, 2013, **42**, 6033–6045.
36. A. Katoh, K. Taguchi, H. Okada, M. Harata, Y. Fujisawa, T. Takino and H. Sakurai, *Chem. Lett.*, 2000, 866–867.
37. A. Katoh, Y. Hida, J. Kamitani and J. Ohkanda, *J. Chem. Soc., Dalton Trans.*, 1998, 3859–3864.
38. J. Ohkanda and A. Katoh, *Rev. Heteroat. Chem.*, 1998, **18**, 87–118.
39. J. Härtung, M. Schwarz, R. S. Tangirala and D. P. Curran, *Org. Synth.*, 2002, **79**, 228–235.
40. M. Yamaguchi, R. Saito, Y. Adachi, Y. Yoshikawa, H. Sakurai and A. Katoh, *Heterocycles*, 2007, **73**, 603–615.
41. J. Kladnik, A. Dolinar, J. Kljun, D. Perea, J. Grau-Expósito, M. Genescà, M. Novinec, M. J. Buzon and I. Turel, *J. Chem. Inf. Model.*, 2022, **37**, 2158–2168.
42. J. A. Lewis, B. L. Tran, D. T. Puerta, E. M. Rumberger, D. N. Hendrickson and S. M. Cohen, *Dalton Trans.*, 2005, 2588–2596.
43. H. Kawarada, Y. Yoshikawa, H. Yasui, S. Kuwahara, Y. Habata and R. Saito, *Metallomics*, 2011, **3**, 675–679.
44. B. L. Mylari, T. A. Beyer and T. W. Siegel, *J. Med. Chem.*, 1991, **34**, 1011–1018.
45. K. H. Gabbay, *N. Engl. J. Med.*, 1973, **288**, 831.
46. R. Saito, M. Tamura, S. Kawano, Y. Yoshikawa, A. Kato, K. Sasaki and H. Yasui, *New J. Chem.*, 2017, **41**, 5572–5581.
47. S. Kawano, Y. Yoshikawa, A. Kato, S. Higashi, K. Mitani, H. Yasui, Y. Habata, S. Kuwahara, K. Sasaki and R. Saito, *Tetrahedron Lett.*, 2019, **60**, 1534–1537.
48. H. Sakurai, Y. Kojima, Y. Yoshikawa, K. Kawabe and H. Yasui, *Coord. Chem. Rev.*, 2002, **226**, 187–198.
49. H. Sakurai and Y. Adachi, *Biometals*, 2005, **18**, 319–323.
50. H. Murakami, H. Yasui and Y. Yoshikawa, *Chem. Pharm. Bull.*, 2012, **60**, 1096–1104.
51. S. Hongmao, in *A Practical Guide to Rational Drug Design*, ed. S. Hongmao, Woodhead Publishing, 2016, pp. 193–223.
52. Y. Adachi, J. Yoshida, Y. Kodera, A. Katoh, J. Takada and H. Sakurai, *J. Med. Chem.*, 2006, **49**, 3251–3256.
53. H. Yasui, Y. Adachi, A. Katoh and H. Sakurai, *J. Biol. Inorg. Chem.*, 2007, **12**, 843–853.
54. F. A. A. Núñez and S. H. Yalkowsky, *J. Pharm. Sci.*, 1997, **86**, 1187–1189.
55. D. J. Lu, P. Chambers, P. Wipf, X. Q. Xie, D. Englert and S. Weber, *J. Chromatogr. A*, 2012, **1258**, 161–167.
56. H. Sakurai, A. Katoh and Y. Yoshikawa, *Bull. Chem. Soc. Jpn.*, 2006, **79**, 1645–1664.
57. H. M. Irving and H. S. Rossotti, *J. Chem. Soc.*, 1941, 2904–2910.
58. H. Irving and H. Rossotti, *Acta Chem. Scand.*, 1956, **10**, 72–93.

59. A. Katoh, Y. Matsumura, Y. Yoshikawa, H. Yasui and H. Sakurai, *J. Inorg. Biochem.*, 2009, **103**, 567–574.
60. T. Nishio, *J. Chem. Soc., Perkin Trans. 1*, 1993, 1113–1117.
61. S. Iimura, K. Manabe and S. Kobayashi, *Chem. Commun.*, 2002, 94–95.
62. É. A. Enyedy, O. Dömötör, E. Varga, L. Kiss, R. Trondl, C. G. Hartinger and B. K. Keppler, *J. Inorg. Biochem.*, 2012, **117**, 189–197.
63. V. M. Nurchi, G. Crisponi, T. Pivetta, M. Donatoni and M. Remelli, *J. Inorg. Biochem.*, 2008, **102**, 684–692.
64. X. L. Pan, H. Wang, C. Y. Li, J. Z. H. Zhang and C. G. Ji, *J. Chem. Inf. Model.*, 2021, **61**, 3159–3165.

Section II: Regulation of Systemic and Cellular Zinc Homeostasis and Its Importance in Cellular Functions

CHAPTER 5

Regulation of the Zn Absorption Process in the Small Intestine: Functional Insights into Zn Transporters ZIP4 and ZNT1

A. HASHIMOTO,[a,†] Y. NISHITO,[b,†] K. NAKAKITA[c] AND T. KAMBE*[c]

[a] Department of Food and Nutrition, Faculty of Home Economics, Kyoto Women's University, Kyoto, Japan; [b] Department of Analytical and Bioinorganic Chemistry, Division of Analytical and Physical Sciences, Kyoto Pharmaceutical University, Kyoto, Japan; [c] Division of Integrated Life Science, Graduate School of Biostudies, Kyoto University, Kyoto, Japan
*Email: kambe.taiho.7z@kyoto-u.ac.jp

5.1 Introduction

Dietary Zn is absorbed through the small intestinc, primarily through the duodenum and jejunum. In general, the absorption rate of Zn from dietary sources is approximately 30%.[1] The absorbed Zn is delivered to peripheral tissues *via* blood circulation, where it is bound to albumin (80%) and α-macroglobulin (20%). The adult human body contains 2–3 g of Zn, most of which is found in the bones (60% of the body's Zn) and skeletal muscle (30%),

[†] These authors contributed equally to this work.

Coordination Chemistry Perspectives No. 2
Zinc in Biology: Molecular Structures, Cellular Processes and Living Systems
Edited by Hiroyuki Yasui and Taiho Kambe

Published by the Royal Society of Chemistry, www.rsc.org

followed by the liver and skin.[1–3] Zn, which is not absorbed by the gastrointestinal tract, along with the Zn present in gastrointestinal secretions and desquamated mucosal epithelial cells, is excreted in feces.

The gastrointestinal Zn absorption system is crucial for regulating systemic Zn homeostasis. It is mainly divided into three parts: (1) uptake of dietary Zn into the enterocytes at the apical membrane, (2) intracellular transfer of Zn from the apical membrane to the basolateral membrane in the enterocytes, and (3) excretion from the basolateral side of the enterocytes into the portal blood for circulation. Recent studies have indicated that the uptake and excretion processes are conducted by specific Zn transporters, which enable the unidirectional transport of Zn in the enterocytes.

Zn transporters are divided into two families: ZIP (Zrt- and Irt-like protein, also known as solute carrier 39A (SLC39A)) and ZNT (Zn transporter, also known as SLC30A). At present, with respect to mammals, the ZIP family contains 14 members and thc ZNT family contains 10 members.[4,5] Two Zn transporters, ZIP4 and ZNT1, play essential roles in Zn absorption by the small intestine. ZIP4 is located on the apical plasma membrane of enterocytes facing the lumen and uptakes Zn into the cells,[6,7] whereas ZNT1 is located on the basolateral membrane and releases Zn into the portal blood.[8,9] This chapter summarizes the pivotal roles of ZIP4 and ZNT1 in Zn absorption and discusses unsolved questions regarding the Zn absorption process in enterocytes.

5.2 Zn Deficiency

Zn functions as a cofactor important for the activity of over 300 Zn-containing enzymes and as a structural component essential for protein folding and signaling factors.[2,5,10,11] Hence, Zn is important for regulating many cellular processes, including cell proliferation and differentiation, gene transcription, and signal transduction,[2,3,12] and its deficiency causes various symptoms, such as taste disorders, dermatitis, and alopecia, as well as decreased appetite, impaired growth, impaired immune function, and reduced wound healing ability.[13,14] Although several mechanisms have been reported,[15–18] the reasons for the numerous symptoms associated with Zn deficiency have not yet been fully elucidated. Dietary Zn deficiency is one of the most common forms of micronutrient malnutrition worldwide; however, it remains one of the least well recognized deficiencies studied.[19,20] Moreover, two types of inherited Zn deficiency were identified.

5.2.1 Dietary Zn Deficiency

Generally, the recommended daily intake for adults is usually set at 8–19 mg per day.[20] Animal-based foods generally contain more Zn than plant-based foods and are known to be more bioavailable; the proportion of total dietary Zn available from animal foods is inversely correlated with the estimated prevalence of Zn insufficiency.[21] Based on an analysis using estimates of Zn intake and bioavailability obtained from food balance data of the Food and Agriculture Organization of the United Nations, approximately 20% of the world's population may be at risk of Zn deficiency.[20,21] To prevent dietary Zn deficiency, it is important to increase Zn intake, which requires Zn fortification and active intake of foods with high Zn content.

The rate of Zn absorption typically ranges from 16 to 50%,[1] with an average of approximately 30%. This rate varies depending on the dietary Zn intake.[22] For example, Zn absorption increases to approximately 90% in Zn-deficient individuals.[23] Unlike Fe, the absorption of which is facilitated by ascorbic acid, no absorption enhancer for Zn is known. However, several other potential food components have been suggested.[24,25] Similar to Fe absorption, Zn absorption can be easily inhibited by various dietary compounds, such as phytate and polyphosphate, as well as by the consumption of medicines that inhibit Zn absorption.

5.2.2 Inherited Zn Deficiency

Two types of inherited Zn deficiencies have been identified.[26,27] One is acrodermatitis enteropathica (AE), and the other is transient neonatal Zn deficiency (TNZD) (Table 5.1). Both Zn deficiencies are identified in infancy

Table 5.1 Zn transporters whose mutations can cause inherited Zn deficiency.

Gene/protein (cytogenetic location)	Diseases	MIM no.	Clinical features
SLC39A4/ZIP4 (8q24.3)	Acrodermatitis enteropathica (AE)	201100	Eczematous dermatitis around the body orifices and extremities; total alopecia; diarrhea; growth delay; mental slowing; poor wound healing, anemia, diminished taste, anorexia, delayed puberty, and hypogonadism in males. Clinical symptoms improve with Zn supplementation (1–3 mg per kg per day), which should be continued into adulthood.
SLC30A2/ZNT2 (1p36.11)	Transient neonatal Zn deficiency (TNZD)	608118	Erythematous and erosive dermatitis localized around the body orifices and on the extremities, along with diarrhea, hair loss, and growth delay. Clinical symptoms improve with Zn supplementation during nursing; however, infants with TNZD do not require Zn supplementation after weaning.

and are characterized by acral dermatitis, diarrhea, and alopecia.[28,29] TNZD is caused by a mutation in the *SLC30A2/ZNT2* gene, resulting in the secretion of Zn-deficient breast milk during lactation, which causes severe Zn deficiency in infants nourished only with breast milk.[30,31] In contrast, AE is caused by a mutation in the *SLC39A4/ZIP4* gene and is a rare autosomal recessive disorder with an incidence of approximately 1 in 500 000 births. To date, more than 50 mutations and unclassified variants of ZIP4 have been documented.[32] Defective intestinal Zn absorption occurs in patients with AE, resulting in severe Zn deficiency. In patients with both TNZD and AE, Zn deficiency symptoms improved with Zn supplementation.

5.3 Zn Transporters Involved in Zn Absorption

ZIP4 and ZNT1 play vital roles in Zn absorption. The expression of both the transporters is tightly regulated by the Zn status, which is essential for controlling the rate of Zn absorption.[1,33] This section provides an overview of the Zn-responsive properties of each transporter and highlights the significance of their interactions.

5.3.1 ZIP4

ZIP4 is an essential molecule for Zn absorption in humans,[28,29] cattle[34] and mice.[35] In mice, the loss of intestinal ZIP4 results in rapid wasting and death unless provided with excess Zn,[36] and ZIP4 haploinsufficiency causes hypersensitivity to Zn deficiency,[35] both of which clearly demonstrate the importance of ZIP4 in Zn absorption in mammals. ZIP4 is a primary regulator of Zn absorption, and its expression is finely tuned by the Zn status through both post-transcriptional and post-translational mechanisms[6,7,37,38] (Figure 5.1). Recent study estimates the apparent turnover number of ZIP4 to be in the range of 0.08 to 0.2 s^{-1}.[39] When Zn is present in sufficient amounts, ZIP4 constantly undergoes endocytosis and is rapidly degraded *via* the lysosomal and proteasomal pathways.[7,40] In contrast, when Zn is deficient, endocytosis and degradation of ZIP4 are inhibited, resulting in its accumulation on the cell surface. *ZIP4* transcription is not enhanced, but its mRNA is stabilized during Zn deficiency, contributing to the accumulation of ZIP4 on the cell surface.[6,7,37,40,41] ZIP4 undergoes proteolytic cleavage (processing) of its extracellular N-terminal region during prolonged Zn deficiency.[41,42] As the N-terminal portion of ZIP4 plays a crucial role in Zn transport,[39] the physiological significance of processing of the N-terminal region of ZIP4 remains unclear.

Understanding the structure of ZIP4 can provide clarity regarding its endocytosis and degradation processes. While the complete structure of human ZIP4 has not yet been crystallized, the extracellular N-terminal portion of ZIP4 from *Pteropus Alecto* ortholog has been reported.[43] The unique feature of ZIP4 is a large extracellular N-terminal region containing a highly conserved proline–alanine–leucine (PAL) motif and a histidine-rich

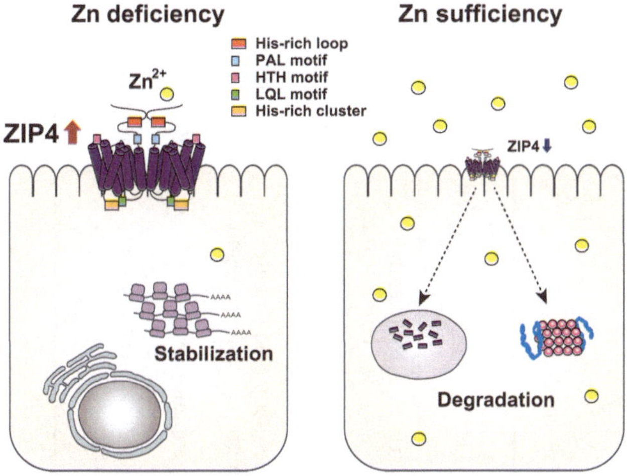

Figure 5.1 Expression regulation of ZIP4 on the apical membrane of enterocytes by Zn status. ZIP4 protein accumulates on the apical surface in Zn deficient condition (left), whereas it is lost under the Zn sufficient condition (right). The responses occur through Zn-induced endocytosis followed by degradation of ZIP4 *via* lysosomal and proteasomal pathways. The stabilization of *ZIP4* mRNA might contribute to its accumulation on the apical surface under the Zn deficient condition. The PAL motif and His-rich loop, both of which are present in the extracellular N-terminal portion, the HTH motif in the extracellular loop between TM2 and TM3, and the cytoplasmic His-rich cluster and LQL motif between the TM3–TM4 loops are also described. The details of the motifs are referred to in the main text and Table 5.2.

(His-rich) loop that are putatively involved in dimerization and a Zn-binding site, respectively.[43,44] Moreover, the structure of *Bordetella bronchiseptica* (BbZIP), a bacterial homolog of ZIP, has been elucidated,[45] which is now being used as a model to characterize the properties of ZIP4, including Zn-responsive expression, using computational modeling. Based on the structure of BbZIP, ZIP4 likely possesses eight transmembrane segments (TMs) that form homodimers.[45,46] The Zn-binding site within TMs is formed as a binuclear metal center per monomer, and several amino acids, such as His and Asn, which are semi-conserved among ZIP family proteins, are critical for the Zn transport activity of ZIP4.[45,47]

Several sequences responsible for Zn sensing have been proposed for ZIP4, indicating that multiple Zn-sensing pathways may be involved in the fine-tuned regulation of Zn-responsive ZIP4 expression (Table 5.2). One is a histidine-rich (His-rich) cluster in the cytosolic loop between TM3 and TM4, which binds to Zn with nanomolar affinity and is crucial for the degradation of ZIP4 under conditions of excess Zn.[40,48–50] Specifically, Zn bound to this motif induces structural remodeling and likely causes the ubiquitination of at least one lysine residue. The conserved leucine–glutamine–leucine (LQL) motif in the same loop contributes to the Zn-dependent endocytosis.[49]

Table 5.2 Sequences involved in Zn-dependent expression regulation of ZIP4.

Sequence (human)	Extracellular/ intracellular	Location	Sensitivity	Response	Conservation (between human and mouse)	Ref.
His-rich loop: EAHSDHSHRHRGAS SRDPVPLISSS (236-260)	Extracellular	N-terminus	Low μM	May contribute to Zn sensing	No (highly homologous)	43 and 44
HTH (398–400: mouse)	Extracellular	Between TM2 and TM3	Low μM	Zn sensing Zn-dependent endocytosis	Yes	51
HSSHSHGGHSH (438-448)	Intracellular	Between TM3 and TM4	nM	Required for Zn-dependent degradation (promote ubiquitination)	No (highly homologous)	40 and 48–50
LQL (452–454)	Intracellular	Between TM3 and TM4	—	Required for Zn-dependent endocytosis	Yes	49
D (511)	TM domain	TM4	—	Bind to Zn, maintain Zn transport, and surface expression	Yes	49

Moreover, the 511th Asp residue (D511) in TM4, which constitutes the binuclear metal center and is required for Zn transport, regulates the ZIP4 protein expression on the cell surface; the ZIP4 mutant of this site is endocytosed even in the absence of Zn.[49] In addition, the aforementioned His-rich loop in the extracellular N-terminal portion may contribute to Zn sensing, which allosterically regulates Zn transport function.[43,44] Analysis of mouse Zip4 showed a His-containing motif (HTH motif) in the extracellular loop between TM2 and TM3 is important for Zn-dependent endocytosis.[51] Further studies are needed to elucidate the regulation and Zn-sensing mechanism of ZIP4.

Recent reports have indicated that ZIP transporters play a role in the elevator transport mechanism,[50,52–54] and thus the Zn transport mode of ZIP4 should be described as a carrier rather than an ion channel.[39] A Zn transport mode of ZIPs is suggested, such as Zn^{2+}/HCO^{3-} symport mechanism,[55] but that of ZIP4 is proposed as a Zn^{2+}/H^{+} symport mechanism.[56] The detailed mechanism underling it has not yet been elucidated.

5.3.2 ZNT1

ZNT1 was the first mammalian Zn transporter to be identified by expression cloning in 1995 as a molecule that rescued a Zn-sensitive cell line.[57] Many studies have indicated the important roles of ZNT1 in various physiological and pathophysiological processes through its Zn efflux function in cells.[58–63] The expression of ZNT1 on the basolateral membrane in the enterocytes suggests its role in releasing Zn into the portal blood, which was confirmed in recent studies using intestine-specific *Znt1*-knockout mice.[64,65] The knockout mice showed loss of body weight and lethality unless supplemented with excess Zn. The Zn-releasing function to the portal blood by ZNT1 is supported by the lethal phenotype of Drosophila lacking the Znt1 ortholog (dZnt1), with Zn accumulation in the gastric ceca.[66]

ZNT1 expression is also regulated by the Zn status in a manner opposite to that of ZIP4[9,63,67] (Figure 5.2). Specifically, ZNT1 expression on the cell surface is upregulated under Zn-sufficient conditions *via* both transcriptional and post-translational mechanisms. Transcriptional expression regulation is mediated by the binding of metal responsive transcription factor 1 (MTF-1) to the metal responsive element (MRE) in the *ZNT1* promoter region, which is paralleled by the expression of the cytosolic Zn-binding protein metallothionein (MT) because its expression is also controlled by the MTF-1-MRE pathway.[63,67] Moreover, ZNT1 expression on the cell surface is facilitated by the post-translational mechanism. Specifically, cell surface ZNT1 expression decreased through endocytosis, followed by degradation by proteasomal or lysosomal pathways under Zn-deficient conditions.[9] The molecular mechanism underlying these responses remains unclear; however, *N*-glycosylation at the 299th Asn residue (N299) in extracellular loop 3 (EL3) between TM5 and TM6 is involved in ZNT1 expression on the cell surface under Zn-deficient conditions.[9]

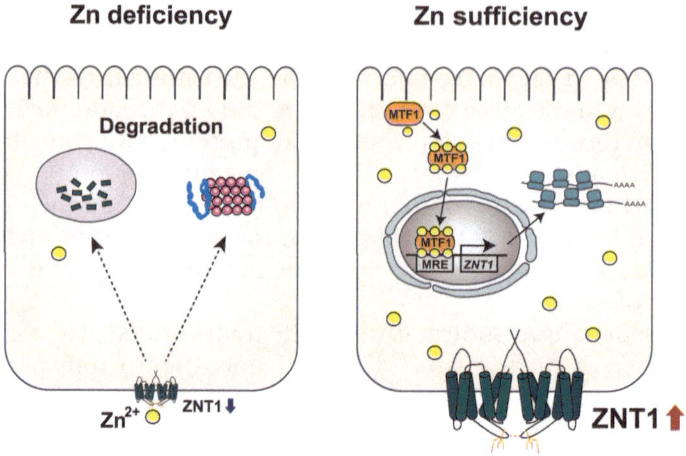

Figure 5.2 Expression regulation of ZNT1 on the basolateral membrane of the enterocytes by Zn status. ZNT1 protein on the basolateral surface decreased under the Zn deficient condition (left), whereas it accumulated under the Zn sufficient condition (right). Zn-induced transcriptional upregulation, through the binding of MTF-1 to the MRE in the *ZNT1* promoter contributes to the accumulation of ZNT1 on the basolateral surface under the Zn sufficient condition. The responses are post-translationally regulated, either through Zn deficiency-induced endocytosis followed by degradation of ZNT1 *via* lysosomal and proteasomal pathways, or through facilitation of Zn induced trafficking to the plasma membrane. The *N*-glycosylation on the extracellular loop between TM 5 and 6 is involved in this regulation. Dashed line in red: the intermolecular disulfide bond. For details refer to the main text.

Recent reports on the structure of ZNT1 indicate that ZNT1 forms homodimers, with each protomer containing six TMs that have a metal-binding site consisting of TM2 and TM5,[64,65,68,69] similar to those found in ZNT7,[70] ZNT8,[71] and the bacterial homolog YiiP.[72–74] Generally, Zn transport among ZNT family members operates as a Zn^{2+}/H^{+} antiporter in the rocker-switch alternating-access mechanism,[54,64,69–74] and ZNT1 is believed to function similarly.[75] However, recent reports indicate that ZNT1 is uniquely driven by the Zn^{2+}/Ca^{2+} antiport mechanism.[68,76] Given the localization of ZNT1 on the cell surface, the latter mechanism may be a reasonable mode for Zn transport by ZNT1. ZNT10, which is localized to the cell surface and is the close homolog of ZNT1, is also thought to operate *via* an Mn^{2+}/Ca^{2+} antiport mechanism.[77] The ZNT1 structure shows the formation of a disulfide bond between the protomer at the 291st Cys residue (C291) in the EL3,[64,65] close to the *N*-glycosylation site. Since the *N*-glycosylation site is involved in the regulation of cell-surface ZNT1 expression during Zn deficiency, the disulfide bond may also be involved in this regulation. Both the *N*-glycosylation site and disulfide bond are highly conserved among vertebrate orthologs of ZNT1,[9] suggesting their potential importance in ZNT1 function.

5.3.3 Unidirectional Transport of Zn in the Enterocytes

As mentioned earlier, the expression of ZIP4 and ZNT1 exhibits completely opposite patterns in terms of localization and response to Zn status. These distinct expression patterns play a role in the regulation of Zn absorption. We recently presented a model suggesting that the coordinated, opposite expression of ZIP4 and ZNT1 is important for facilitating the unidirectional transport of Zn in the enterocytes during the absorption process[33] (Figure 5.3). This model showed that apically localized ZIP4 increased Zn uptake, which induced the expression of basolaterally localized ZNT1 in the polarized cells. In this model, ZIP4 and ZNT1 expression was regulated through Zn-dependent or Zn-deficiency-dependent degradation in response to the Zn status (Figure 5.3).

In the process of the unidirectional transport of Zn in the enterocytes, the mechanism by which imported Zn is shuttled to ZNT1 on the basolateral membrane has not yet been elucidated. MTs may be functional in this role because they can function as a chaperone for Zn supply to Zn proteins or enzymes.[78,79] In the small intestine of humans, MT2A and MT1X are highly expressed among the 11 isoforms of MTs (according to data from the Human

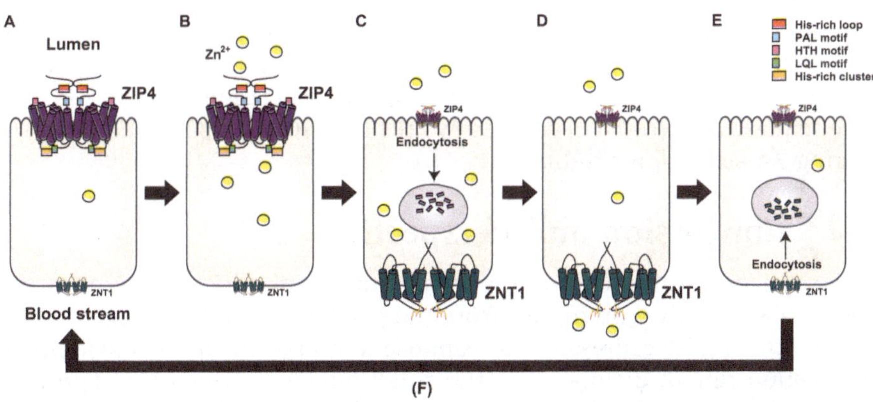

Figure 5.3 ZIP4-driven ZNT1 expression regulation is important for the unidirectional transport of Zn in the enterocytes in Zn absorption. ZIP4-driven ZNT1 expression regulation is accomplished by coordinated expression of ZIP4 and ZNT1 as follows. (A) ZIP4 accumulation on the apical surface, whereas ZNT1 expression is decreased in Zn deficient conditions (*ex*, before meals). (B) Zn is taken up into the enterocytes *via* ZIP4 in Zn sufficient conditions (*ex*, during a meal). (C) ZIP4 is degraded following endocytosis and ZNT1 accumulates on the basolateral membrane. (D) The basolaterally accumulated ZNT1 facilitates Zn efflux into the blood. (E) ZNT1 expression decreases following endocytosis under Zn deficient conditions (*ex*, after meals), and (F) ZIP4 expression on the apical surface increases (*ex*, between meals).

Protein Atlas (https://www.proteinatlas.org)), suggesting that MT2A and MT1X may play a significant role in this process.

5.3.4 Possible Involvement of Other Zn Transport Proteins in the Regulation of Zn Absorption

ZIP4 and ZNT1 play essential roles in mammalian Zn absorption. However, other Zn transporters may also be involved in the fine-tuning of the Zn absorption rate. The primary candidate is ZNT4, which is located in the endosomal compartment or *trans*-Golgi network,[80] because the lethal milk mouse, in which *Znt4* is mutated, is reported to show severe dermatitis after 8 months of age.[81] Moreover, Golgi-resident ZNT7 may be involved,[82] because *Znt7*-deficient mice showed a mildly reduced Zn absorption rate in a radioactive feeding study.[83] Thus, ZNT transporters located in intracellular organelles and vehicles might fine-tune Zn absorption efficiency.

ZIP5 and ZIP14 are both localized to the basolateral membrane and may play a role in the reabsorption of Zn from the blood.[7,84–86] ZIP5 expression is oppositely regulated by Zn status compared to ZIP4; ZIP5 expression is upregulated by Zn and downregulated by Zn deficiency similar to ZNT1.[9,87] Therefore, the expression responses may contribute to the regulation of Zn absorption or reabsorption processes. The intestine-specific *Zip5*-knockout mice showed an increase in the abundance of *Zip4* mRNA, suggesting such functions.[88] However, *Zip5* knockout mice showed almost normal phenotypes. ZIP14 may also play a role because *Zip14* knockout mice showed increased intestinal Zn concentrations, especially accumulating in vesicles.[85] The sequestration of Zn in vesicles by ZIP14 might be important as a Zn store during Zn starvation conditions.

5.4 Conclusion and Perspectives

This review summarizes the current understanding of Zn absorption mechanisms in the enterocytes from the perspective of Zn transporters. The regulation of ZIP4 expression in response to Zn is critical for controlling the absorption rate of dietary Zn in the small intestine, which is supported by the regulation of ZIP4-driven ZNT1 expression. However, the mechanism of shuttling the imported Zn from the apical side to the basolateral side, and thus from ZIP4 to ZNT1, remains unknown. Recently, Zn-regulated GTPase metalloprotein activator 1 (ZNG1) was identified as a Zn chaperone[89,90] that delivers Zn to metalloprotease methionine aminopeptidase 1 (METAP1) for its enzyme activity, suggesting that other Zn chaperones are operative in many physiological functions in mammals. Certain Zn chaperone proteins may play a role in the shuttling functions in enterocytes.

We recently reported a close interaction between Zn and Mn homeostasis,[91] suggesting that the competition between Zn and other trace elements during Zn absorption needs to be studied further. It is well-established that intestinal Zn absorption competes with Cu absorption.[92] The competition between Zn and Cu may be mediated by MTs, which bind to both Zn and Cu.[93,94] However, the molecular mechanism involved in this competition is still unknown. A recent study reported that ZNT1 acts as a Cu^{2+} importer under high Cu conditions.[64] This suggests that unexpected novel interactions between Zn and Cu may occur during Zn absorption and metabolism after absorption. However, this needs to be clarified in future studies.

Abbreviations

ZIP	Zrt- and Irt-like protein
ZNT	Zn transporter
AE	Acrodermatitis enteropathica
TNZD	Transient neonatal Zn deficiency
BbZIP	*Bordetella bronchiseptica*
EL	Extracellular loop
His-rich	Histidine-rich
TMs	Transmembrane segments
MT	Metallothionein
MTF-1	Metal responsive transcription factor 1
MRE	Metal responsive element

Acknowledgements

This work was supported by a Grant-in-Aid for Scientific Research on Innovative Areas "Integrated Bio-metal Science" (MEXT KAKENHI Grant Number JP19H05768) from the Ministry of Education, Culture, Sports, Science, and Technology, Japan, a Grant-in-Aid for Scientific Research (B) (JSPS KAKENHI Grant Number JP22H02257) from the Japan Society for the Promotion of Science, KOSÉ Cosmetology Research Foundation, and Tojuro Iijima Foundation for Food Science and Technology (to T. K.).

References

1. M. Maares and H. Haase, *Nutrients*, 2020, **12**, 762.
2. T. Kambe, T. Tsuji, A. Hashimoto and N. Itsumura, *Physiol. Rev.*, 2015, **95**, 749–784.
3. T. Hara, T. A. Takeda, T. Takagishi, K. Fukue, T. Kambe and T. Fukada, *J. Physiol. Sci.*, 2017, **67**, 283–301.
4. T. Kambe, K. M. Taylor and D. Fu, *J. Biol. Chem.*, 2021, **296**, 100320.
5. B. Chen, P. Yu, W. N. Chan, F. Xie, Y. Zhang, L. Liang, K. T. Leung, K. W. Lo, J. Yu, G. M. K. Tse, W. Kang and K. F. To, *Signal Transduction Targeted Ther.*, 2024, **9**, 6.
6. J. Dufner-Beattie, F. Wang, Y. M. Kuo, J. Gitschier, D. Eide and G. K. Andrews, *J. Biol. Chem.*, 2003, **278**, 33474–33481.
7. B. P. Weaver, J. Dufner-Beattie, T. Kambe and G. K. Andrews, *Biol. Chem.*, 2007, **388**, 1301–1312.

8. R. J. McMahon and R. J. Cousins, *Proc. Natl. Acad. Sci. U. S. A.*, 1998, **95**, 4841–4846.
9. Y. Nishito and T. Kambe, *J. Biol. Chem.*, 2019, **294**, 15686–15697.
10. A. Kocyla, J. B. Tran and A. Krezel, *Trends Biochem. Sci.*, 2021, **46**, 64–79.
11. T. Fukada, S. Yamasaki, K. Nishida, M. Murakami and T. Hirano, *J. Biol. Inorg. Chem.*, 2011, **16**, 1123–1134.
12. A. Fukunaka and Y. Fujitani, *Int. J. Mol. Sci.*, 2018, **19**, 476.
13. A. S. Prasad, *Adv. Nutr.*, 2013, **4**, 176–190.
14. H. Kodama, M. Tanaka, Y. Naito, K. Katayama and M. Moriyama, *Int. J. Mol. Sci.*, 2020, **21**, 2941.
15. I. Wessels, H. J. Fischer and L. Rink, *Annu. Rev. Nutr.*, 2021, **41**, 133–175.
16. I. Wessels, M. Maywald and L. Rink, *Nutrients*, 2017, **9**, 1286.
17. T. A. Takeda, S. Miyazaki, M. Kobayashi, K. Nishino, T. Goto, M. Matsunaga, M. Ooi, H. Shirakawa, F. Tani, T. Kawamura, M. Komai and T. Kambe, *Commun. Biol.*, 2018, **1**, 113.
18. H. Yuasa, K. Miyazaki, T. Kido, K. Nishino, M. Shiotsu, T. Fukuwatari, M. Suka, M. Nagao and T. Kambe, *FASEB J.*, 2024, **38**, e70025.
19. A. S. Prasad, *J. Trace Elem. Med. Biol.*, 2014, **28**, 357–363.
20. D. B. Kumssa, E. J. Joy, E. L. Ander, M. J. Watts, S. D. Young, S. Walker and M. R. Broadley, *Sci. Rep.*, 2015, **5**, 10974.
21. K. R. Wessells and K. H. Brown, *PLoS One*, 2012, **7**, e50568.
22. B. Sandstrom, A. Almgren, B. Kivisto and A. Cederblad, *J. Nutr.*, 1989, **119**, 48–53.
23. C. M. Taylor, J. R. Bacon, P. J. Aggett and I. Bremner, *Am. J. Clin. Nutr.*, 1991, **53**, 755–763.
24. S. Gautam, K. Platel and K. Srinivasan, *J. Agric. Food Chem.*, 2010, **58**, 8426–8429.
25. A. Hashimoto, K. Ohkura, M. Takahashi, K. Kizu, H. Narita, S. Enomoto, Y. Miyamae, S. Masuda, M. Nagao, K. Irie, H. Ohigashi, G. K. Andrews and T. Kambe, *Biochem. J.*, 2015, **472**, 183–193.
26. T. Kambe, A. Hashimoto and S. Fujimoto, *Cell. Mol. Life Sci.*, 2014, **71**, 3281–3295.
27. S. Kasana, J. Din and W. Maret, *J. Trace Elem. Med. Biol.*, 2014, **29C**, 47–62.
28. S. Kury, B. Dreno, S. Bezieau, S. Giraudet, M. Kharfi, R. Kamoun and J. P. Moisan, *Nat. Genet.*, 2002, **31**, 239–240.
29. K. Wang, B. Zhou, Y. M. Kuo, J. Zemansky and J. Gitschier, *Am. J. Hum. Genet.*, 2002, **71**, 66–73.
30. W. Chowanadisai, B. Lonnerdal and S. L. Kelleher, *J. Biol. Chem.*, 2006, **281**, 39699–39707.
31. N. Itsumura, Y. Kibihara, K. Fukue, A. Miyata, K. Fukushima, R. Tamagawa-Mineoka, N. Katoh, Y. Nishito, R. Ishida, H. Narita, H. Kodama and T. Kambe, *Pediatr. Res.*, 2016, **80**, 586–594.
32. S. Schmitt, S. Kury, M. Giraud, B. Dreno, M. Kharfi and S. Bezieau, *Hum. Mutat.*, 2009, **30**, 926–933.

33. S. Nagamatsu, Y. Nishito, H. Yuasa, N. Yamamoto, T. Komori, T. Suzuki, H. Yasui and T. Kambe, *Sci. Rep.*, 2022, **12**, 7334.
34. V. Yuzbasiyan-Gurkan and E. Bartlett, *Genomics*, 2006, **88**, 521–526.
35. J. Dufner-Beattie, B. P. Weaver, J. Geiser, M. Bilgen, M. Larson, W. Xu and G. K. Andrews, *Hum. Mol. Genet.*, 2007, **16**, 1391–1399.
36. J. Geiser, K. J. Venken, R. C. De Lisle and G. K. Andrews, *PLoS Genet.*, 2012, **8**, e1002766.
37. F. Wang, B. E. Kim, J. Dufner-Beattie, M. J. Petris, G. Andrews and D. J. Eide, *Hum. Mol. Genet.*, 2004, **13**, 563–571.
38. B. E. Kim, F. Wang, J. Dufner-Beattie, G. K. Andrews, D. J. Eide and M. J. Petris, *J. Biol. Chem.*, 2004, **279**, 4523–4530.
39. Y. Jiang, K. MacRenaris, T. V. O'Halloran and J. Hu, *J. Biol. Chem.*, 2024, **300**, 107661.
40. X. Mao, B. E. Kim, F. Wang, D. J. Eide and M. J. Petris, *J. Biol. Chem.*, 2007, **282**, 6992–7000.
41. A. Hashimoto, M. Nakagawa, N. Tsujimura, S. Miyazaki, K. Kizu, T. Goto, Y. Komatsu, A. Matsunaga, H. Shirakawa, H. Narita, T. Kambe and M. Komai, *Am. J. Physiol.: Regul., Integr. Comp. Physiol.*, 2016, **310**, R459–R468.
42. T. Kambe and G. K. Andrews, *Mol. Cell. Biol.*, 2009, **29**, 129–139.
43. T. Zhang, D. Sui and J. Hu, *Nat. Commun.*, 2016, **7**, 11979.
44. T. Zhang, E. Kuliyev, D. Sui and J. Hu, *Biochem. J.*, 2019, **476**, 1791–1803.
45. T. Zhang, J. Liu, M. Fellner, C. Zhang, D. Sui and J. Hu, *Sci. Adv.*, 2017, **3**, e1700344.
46. M. E. Ahern, E. M. Bafaro, A. Cowan and R. E. Dempski, *Biochemistry*, 2019, **58**, 1705–1708.
47. T. Zhang, D. Sui, C. Zhang, L. Cole and J. Hu, *FASEB J.*, 2020, **34**, 237–247.
48. E. M. Bafaro, S. Antala, T. V. Nguyen, S. P. Dzul, B. Doyon, T. L. Stemmler and R. E. Dempski, *Metallomics*, 2015, **7**, 1319–1330.
49. C. Zhang, D. Sui, T. Zhang and J. Hu, *Cell Rep.*, 2020, **31**, 107582.
50. C. Pang, J. Chai, P. Zhu, J. Shanklin and Q. Liu, *Nat. Commun.*, 2023, **14**, 3404.
51. H. Chun, T. Korolnek, C. J. Lee, H. J. Coyne 3rd, D. R. Winge, B. E. Kim and M. J. Petris, *J. Biol. Chem.*, 2019, **294**, 2815–2826.
52. A. Wiuf, J. H. Steffen, E. R. Becares, C. Gronberg, D. R. Mahato, S. G. F. Rasmussen, M. Andersson, T. Croll, K. Gotfryd and P. Gourdon, *Sci. Adv.*, 2022, **8**, eabn4331.
53. Y. Zhang, Y. Jiang, K. Gao, D. Sui, P. Yu, M. Su, G. W. Wei and J. Hu, *Nat. Commun.*, 2023, **14**, 385.
54. T. Kambe and T. Wagatsuma, *Biophys. Rev.*, 2023, **4**, 041302.
55. L. A. Gaither and D. J. Eide, *J. Biol. Chem.*, 2000, **275**, 5560–5564.
56. E. Hoch, M. Levy, M. Hershfinkel and I. Sekler, *Int. J. Mol. Sci.*, 2020, **21**, 734.
57. R. D. Palmiter and S. D. Findley, *EMBO J.*, 1995, **14**, 639–649.
58. M. Tsuda, K. Imaizumi, T. Katayama, K. Kitagawa, A. Wanaka, M. Tohyama and T. Takagi, *J. Neurosci.*, 1997, **17**, 6678–6684.

59. T. Jirakulaporn and A. J. Muslin, *J. Biol. Chem.*, 2004, **279**, 27807–27815.
60. C. J. Stocks, J. B. von Pein, J. E. B. Curson, J. Rae, M. D. Phan, D. Foo, N. J. Bokil, T. Kambe, K. M. Peters, R. G. Parton, M. A. Schembri, R. Kapetanovic and M. J. Sweet, *J. Leukocyte Biol.*, 2021, **109**, 287–297.
61. R. F. Krall, A. Moutal, M. B. Phillips, H. Asraf, J. W. Johnson, R. Khanna, M. Hershfinkel, E. Aizenman and T. Tzounopoulos, *Sci. Adv.*, 2020, **6**, eabb1515.
62. J. Rege, S. Bandulik, K. Nanba, C. Kosmann, A. R. Blinder, A. Plain, P. Vats, C. Kumar-Sinha, A. M. Lerario, T. Else, Y. Yamazaki, F. Satoh, H. Sasano, T. J. Giordano, T. A. Williams, M. Reincke, A. F. Turcu, A. M. Udager, R. Warth and W. E. Rainey, *Nat. Genet.*, 2023, **55**, 1623–1631.
63. S. J. Langmade, R. Ravindra, P. J. Daniels and G. K. Andrews, *J. Biol. Chem.*, 2000, **275**, 34803–34809.
64. Y. Li, J. Ma, R. Wang, Y. Luo, S. Zheng and X. Wang, *Cell Metab.*, 2024, **36**, 2118–2129e2116.
65. S. Sun, E. Xie, S. Xu, S. Ji, S. Wang, J. Shen, R. Wang, X. Shen, Y. Su, Z. Song, X. Wu, J. Zhou, Z. Cai, X. Li, Y. Zhang, J. Min and F. Wang, *Adv. Sci.*, 2024, e2406421.
66. Z. Wang, X. Li and B. Zhou, *Biochem. Biophys. Res. Commun.*, 2020, **533**, 1004–1011.
67. R. D. Palmiter, *Proc. Natl. Acad. Sci. U. S. A.*, 2004, **101**, 4918–4923.
68. C. Sun, B. He, Y. Gao, X. Wang, X. Liu and L. Sun, *Sci. Adv.*, 2024, **10**, eadk5128.
69. Y. Long, Z. Zhu, Z. Zhou, C. Yang, Y. Chao, Y. Wang, Q. Zhou, M. W. Wang and Q. Qu, *EMBO Rep.*, 2024, **25**, 5006–5025.
70. H. B. Bui, S. Watanabe, N. Nomura, K. Liu, T. Uemura, M. Inoue, A. Tsutsumi, H. Fujita, K. Kinoshita, Y. Kato, S. Iwata, M. Kikkawa and K. Inaba, *Nat. Commun.*, 2023, **14**, 4770.
71. J. Xue, T. Xie, W. Zeng, Y. Jiang and X.-C. Bai, *eLife*, 2020, **9**, e58823.
72. S. Gupta, J. Chai, J. Cheng, R. D'Mello, M. R. Chance and D. Fu, *Nature*, 2014, **512**, 101–104.
73. N. Coudray, S. Valvo, M. Hu, R. Lasala, C. Kim, M. Vink, M. Zhou, D. Provasi, M. Filizola, J. Tao, J. Fang, P. A. Penczek, I. Ubarretxena-Belandia and D. L. Stokes, *Proc. Natl. Acad. Sci. U. S. A.*, 2013, **110**, 2140–2145.
74. M. Lopez-Redondo, S. Fan, A. Koide, S. Koide, O. Beckstein and D. L. Stokes, *J. Gen. Physiol.*, 2021, **153**, e202112873.
75. E. Shusterman, O. Beharier, L. Shiri, R. Zarivach, Y. Etzion, C. R. Campbell, I. H. Lee, K. Okabayashi, A. Dinudom, D. I. Cook, A. Katz and A. Moran, *Metallomics*, 2014, **6**, 1656–1663.
76. N. Gottesman, H. Asraf, M. Bogdanovic, I. Sekler, T. Tzounopoulos, E. Aizenman and M. Hershfinkel, *Cell Calcium*, 2022, **101**, 102505.
77. M. Levy, N. Elkoshi, S. Barber-Zucker, E. Hoch, R. Zarivach, M. Hershfinkel and I. Sekler, *J. Biol. Chem.*, 2019, **294**, 5879–5889.
78. W. Maret and Y. Li, *Chem. Rev.*, 2009, **109**, 4682–4707.
79. T. Kimura and T. Kambe, *Int. J. Mol. Sci.*, 2016, **17**, 336.

80. N. H. McCormick and S. L. Kelleher, *Am. J. Physiol.: Cell Physiol.*, 2012, **303**, C291–C297.
81. L. C. Erway and A. Grider Jr, *J. Hered.*, 1984, **75**, 480–484.
82. C. P. Kirschke and L. Huang, *J. Biol. Chem.*, 2003, **278**, 4096–4102.
83. L. Huang, Y. Y. Yu, C. P. Kirschke, E. R. Gertz and K. K. Lloyd, *J. Biol. Chem.*, 2007, **282**, 37053–37063.
84. F. Wang, B. E. Kim, M. J. Petris and D. J. Eide, *J. Biol. Chem.*, 2004, **279**, 51433–51441.
85. G. J. Guthrie, T. B. Aydemir, C. Troche, A. B. Martin, S. M. Chang and R. J. Cousins, *Am. J. Physiol.: Gastrointest. Liver Physiol.*, 2015, **308**, G171–G178.
86. T. B. Aydemir, M. H. Kim, J. Kim, L. M. Colon-Perez, G. Banan, T. H. Mareci, M. Febo and R. J. Cousins, *J. Neurosci.*, 2017, **37**, 5996–6006.
87. J. Dufner-Beattie, Y. M. Kuo, J. Gitschier and G. K. Andrews, *J. Biol. Chem.*, 2004, **279**, 49082–49090.
88. J. Geiser, R. C. De Lisle and G. K. Andrews, *PLoS One*, 2013, **8**, e82149.
89. A. Weiss, C. C. Murdoch, K. A. Edmonds, M. R. Jordan, A. J. Monteith, Y. R. Perera, A. M. Rodriguez Nassif, A. M. Petoletti, W. N. Beavers, M. J. Munneke, S. L. Drury, E. S. Krystofiak, K. Thalluri, H. Wu, A. R. S. Kruse, R. D. DiMarchi, R. M. Caprioli, J. M. Spraggins, W. J. Chazin, D. P. Giedroc and E. P. Skaar, *Cell*, 2022, **185**, 2148–2163e2127.
90. M. Pasquini, N. Grosjean, K. K. Hixson, C. D. Nicora, E. F. Yee, M. Lipton, I. K. Blaby, J. D. Haley and C. E. Blaby-Haas, *Cell Rep.*, 2022, **39**, 110834.
91. Y. Nishito, Y. Kamimura, S. Nagamatsu, N. Yamamoto, H. Yasui and T. Kambe, *FASEB J.*, 2024, **38**, e23605.
92. S. E. Smith and E. J. Larson, *J. Biol. Chem.*, 1946, **163**, 29–38.
93. H. N. Hoffman 2nd, R. L. Phyliky and C. R. Fleming, *Gastroenterology*, 1988, **94**, 508–512.
94. A. Krezel and W. Maret, *Int. J. Mol. Sci.*, 2017, **18**, 1237.

CHAPTER 6

Zinc Homeostasis Ensures Protein Homeostasis in the Early Secretory Pathway

YUTA AMAGAI[a] AND KENJI INABA*[a,b]

[a] Medical Institute of Bioregulation, Kyushu University, Fukuoka, 812-8582, Japan; [b] Core Research for Evolutional Science and Technology (CREST), Japan Agency for Medical Research and Development (AMED), Chiyoda-ku, Tokyo, Japan
*Email: kenji.inaba@bioreg.kyushu-u.ac.jp

6.1 Introduction

Zinc is an essential trace element involved in numerous biological processes. A recent comprehensive mass spectrometry analysis revealed that more than 4800 human proteins contain zinc-associated cysteine residues.[1] Since this analysis focused specifically on cysteine residues, the total number of zinc-binding proteins, including those with zinc-associated histidine, aspartate, or glutamate residues, must be even higher. This suggests that zinc ions (Zn^{2+}) potentially regulate the functions of these proteins, and highlights the physiological importance of Zn^{2+} in maintaining the overall health and activities of living organisms. In other words, disruption to zinc homeostasis can lead to a variety of diseases.[2]

The human genome incudes two types of zinc transporter membrane protein families: zinc transporters (ZnTs) and Zrt-/Irt-like proteins (ZIPs). ZnT proteins are members of the SLC30 family and generally transport Zn^{2+} from the cytosol to the extracellular space or intracellular organelles. ZIP proteins are SLC39 family members and transport Zn^{2+} in the opposite

Coordination Chemistry Perspectives No. 2
Zinc in Biology: Molecular Structures, Cellular Processes and Living Systems
Edited by Hiroyuki Yasui and Taiho Kambe

Published by the Royal Society of Chemistry, www.rsc.org

direction to ZnTs. 10 ZnTs and 14 ZIPs have been identified in the human genome. Each zinc transporter differs in tissue distribution and subcellular localization to maintain zinc homeostasis in organs, cells, and organelles.[2,3]

The endoplasmic reticulum (ER) and Golgi apparatus (GA) constitute the early secretory pathway (ESP). Typically, the ER forms a tubular network structure throughout the cytoplasm, while the GA forms a ribbon-like structure with multiple cisternae near the centrosome, and is anchored to microtubules in mammalian cells (Figure 6.1A). Newly synthesized secretory and membrane proteins enter the ER lumen, where they fold into their native structures. Once properly folded, these proteins are transported to the GA *via* vesicular transport mechanisms. In the GA they undergo further modifications, including processing of glycans, during maturation. The resultant mature proteins are then transported to their final destinations, where they fulfill their unique functions.[4] Approximately one-third of the

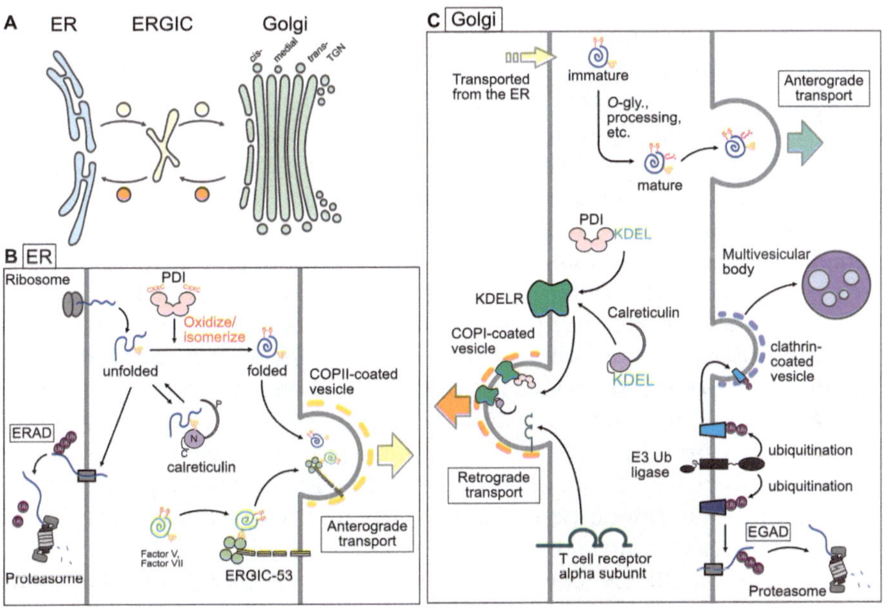

Figure 6.1 Overview of protein quality control in the ER and Golgi apparatus (GA). (A) Schematic illustrations of the ER, ER-Golgi intermediate compartment (ERGIC), and GA. (B) Newly synthesized secretory proteins are inserted into the ER, where PDI enzymes assist their folding by introducing disulfide bonds. Properly folded proteins exit the ER *via* COPII-coated vesicles, while unfolded proteins are managed by ER quality control systems such as the calreticulin cycle and ER-associated degradation (ERAD). Proteins including Factor V and VII are actively sorted into COPII vesicles by ERGIC-53. (C) After further modifications in the GA, secretory proteins are transported to their destinations. Misfolded proteins and ER chaperones are retrogradely transported *via* COPI vesicles, and ubiquitinated proteins are degraded by lysosomes or proteasomes.

human proteome passes through the ESP, including a considerable number of zinc-associated proteins.

This chapter summarizes our current understanding of how ZnTs and ZIPs maintain zinc homeostasis in the mammalian ESP, and how Zn^{2+} contribute to protein quality control (PQC) in this compartment.

6.2 The Mammalian Early Secretory Pathway and Protein Quality Control

Accumulation of misfolded proteins leads to cellular stress, potentially causing neurodegenerative diseases such as Parkinson's and Alzheimer's.[5] To prevent the production and accumulation of misfolded proteins, cells have evolved PQC systems.

The ER has established a highly specialized system for PQC (Figure 6.1B). Most secretory and membrane proteins contain intra- and/or intermolecular disulfide bonds formed by oxidizing thiol groups of pairs of cysteine residues that stabilize their native structures. However, non-native disulfide bonds induce protein misfolding. Disulfide bond formation is generally catalyzed in the ER by protein disulfide isomerase (PDI) family members. Some PDIs function as disulfide isomerases or reductases to facilitate the formation of native protein structures. More than 20 members of the PDI family with different structures and functions act on a variety of substrate proteins in the mammalian ER. PDIA1, a representative PDI, is conserved from yeast to human and is usually referred to simply as PDI or canonical PDI. PDIs with Cys–X–X–Cys (CXXC) motifs in thioredoxin-like (Trx) domains introduce disulfide bonds into substrates through thiol-disulfide exchange reactions. The reduced enzyme active sites are then oxidized by an upstream oxidase, ER oxidoreductin 1α (Ero1α), to drive the redox cycle of PDI-mediated oxidative protein folding. In the human genome, several oxidases of PDI family members have been identified including Ero1α, Ero1β, peroxiredoxin 4 (Prdx4), and glutathione peroxidase 7 (GPX7). These enzymes form diverse redox networks within the ER for PQC.[6]

Furthermore, the ER contains various kinds of molecular chaperones, such as calnexin, calreticulin, and binding immunoglobulin protein (BiP). Calnexin and calreticulin are lectin-like proteins that bind to unfolded *N*-glycosylated proteins to prevent their aggregation in the ER.[7] BiP is a molecular chaperone located in the ER that belongs to the heat shock protein 70 (Hsp70) family. Like other Hsp70 members, BiP hydrolyzes ATP to ADP during its chaperone cycle to control substrate binding and release.[8]

Although the ER has a well-developed system to facilitate protein folding, misfolded proteins still occur. N-linked glycans on misfolded proteins are trimmed to Man7A by ER degradation-enhancing alpha-mannosidase-like protein 1 (EDEM1), and aberrantly formed disulfide bonds are reduced by ERdj5, a PDI family member.[9] Subsequently, ATP-bound BiP recognizes and binds the misfolded proteins, which are retrogradely translocated into the cytosol through a dislocon channel and eventually degraded by a ubiquitin-proteasome system. This degradation process is known as ER-associated degradation (ERAD).[10–12]

On the other hand, correctly folded proteins in the ER are sorted into coat protein complex II (COPII)-coated vesicles or vesicular tubular compartments budding from the ER.[13,14] The vesicles fuse with each other to form the ER-Golgi intermediate compartment (ERGIC).[15] To achieve effective cargo transport, several cargo proteins are actively sorted into these compartments. In line with this, coagulation factor V and factor VIII proteins are captured by the multiple coagulation factor deficiency 2 (MCFD2)-ERGIC-53 complex and likely transported to the ER exit site.[16,17] The GA is composed of multiple flat membrane sacs, which are classified into *cis*, medial, and *trans* cisternae. Cargo proteins enter the GA *via* vesicle transport machineries at the *cis* cisternae, then move through the medial and *trans* cisternae, during which they undergo various modifications for maturation. Mature proteins are then sorted and delivered to their destinations from the *trans*-Golgi network (TGN).[18]

The GA harbors many enzymes for protein modifications, including *O*-glycosylation and protease-catalyzed processing (Figure 6.1C). Advanced cell imaging studies employing super-resolution microscopy have revealed that each Golgi-resident enzyme localizes to a specific subregion of the GA, which is referred to as a 'zone'.[19–23] Cargo proteins traverse these zones and are sequentially modified.[22]

Like the ER, the GA also includes several PQC systems.[24,25] A variety of ubiquitin E3 ligases are located on the Golgi membrane, where they ubiquitinate substrate proteins. The ubiquitinated proteins are degraded either by lysosomal proteases or the proteasome. For lysosomal degradation, ubiquitinated proteins are sorted at the TGN *via* clathrin and Golgi-localized, gamma-adaptin ear domain homology, ARF-binding protein (GGA) clathrin adaptor proteins, then transported to the lysosome *via* vesicle transport.[25,26] Proteasomal degradation of ubiquitinated Golgi proteins has been observed in budding yeast. This system is referred to as endosome and Golgi-associated degradation (EGAD).[27] It is also known that unassembled T cell receptor alpha subunits are captured by BiP and KDEL receptor (KDELR) in the GA, and retrogradely transported to the ER *via* COPI-coated vesicles. KDELR recognizes and binds to the C-terminal Lys-Asp-Glu-Leu (KDEL) sequence of BiP. This binding induces allosteric conformational change in KDELR, exposing its di-lysine motif to the cytosol, which is in turn recognized by COPI coatomer proteins.[28,29] KDELR can recognize other C-terminal KDEL-like sequences such as RDEL, KEEL, and HDEL, which are commonly found at the C-termini of PDI family members and other ER-resident chaperone proteins.[30] The affinity between the KDEL sequence and KDELR is stronger under acidic conditions,[28,31] suggesting that their interaction should be enhanced in the GA, whereas the neutral pH of the ER may promote their dissociation. Thus, the Golgi-to-ER transport system regulates the trafficking of both immature secretory proteins and ER-resident proteins, and thereby plays a key role for PQC in the ESP.

6.3 Zinc-regulated PQC in the ESP

Considering that many proteins are thought to be zinc-associated, and that approximately one-third of the human proteome enters the ESP, it is likely that the functions of large numbers of secretory and/or membrane proteins are regulated by binding Zn^{2+}.

6.3.1 Zinc-regulated PQC in the ER

ZIP7 is a zinc transporter localized to the ER membrane. Cellular treatment with NVS-ZP7-4, an inhibitor of ZIP7, increases the labile Zn^{2+} concentration in the ER, indicating that ZIP7 mobilizes Zn^{2+} from the ER lumen to the cytosol.[32] Notably, downregulation of ZIP7 expression through knockdown or knockout, or its inhibition with NVS-ZP7-4, causes ER stress responses in several types of cells, including mouse mesenchymal stem cells and leukemia cell lines,[32–34] suggesting that Zn^{2+} regulate the PQC system in the ER.

Calreticulin and calnexin function as ER-resident lectin chaperones and interact with $GlcMan_9GlcNAc_2$-modified unfolded proteins to prevent their aggregation. Calreticulin consists of globular N-domain, P-domain, and C-domain with a C-terminal KDEL sequence. Calnexin shares N- and P-domain structural features with calreticulin, but has a transmembrane domain at the C-terminus that anchors it to the ER membrane. The N- and P-domains of calreticulin have binding sites for N-glycans and PDI family members including ERp57 and ERp29, respectively. While calreticulin is known to bind multiple calcium ions (Ca^{2+}) *via* its P- and C-domains, the N-domain contains Zn^{2+}-binding sites (Figure 6.2A). Studies using purified proteins revealed that calreticulin can prevent aggregation of citrate synthase and malate dehydrogenase *in vitro*, and Zn^{2+} enhance this chaperone activity.[35] Thermal stability and chymotrypsin digestion analyses also revealed that calreticulin contains a domain whose structure is destabilized or more loosely packed upon binding Zn^{2+}, whereas Ca^{2+} ions stabilize the overall structure of calreticulin.[34,36,37]

PDI also binds to Zn^{2+}, forming dimers or oligomers *in vitro*.[33,37] Further biochemical studies revealed that PDI with reduced cysteines at its active sites (reduced PDI) has higher affinity for Zn^{2+} than PDI with disulfide bonds at the sites (oxidized PDI).[38] Addition of Zn^{2+} alters the thermal stability of reduced PDI, indicating Zn^{2+}-dependent conformational changes in reduced PDI.[38] Thus, cysteine residues at the active sites of PDI are likely Zn^{2+}-binding sites. In line with this, the reductase activity of PDI against insulin and the artificial substrate di-eosin-glutathione disulfide (Di-E-GSSG) is impaired in the presence of Zn^{2+}, with a 50% effective

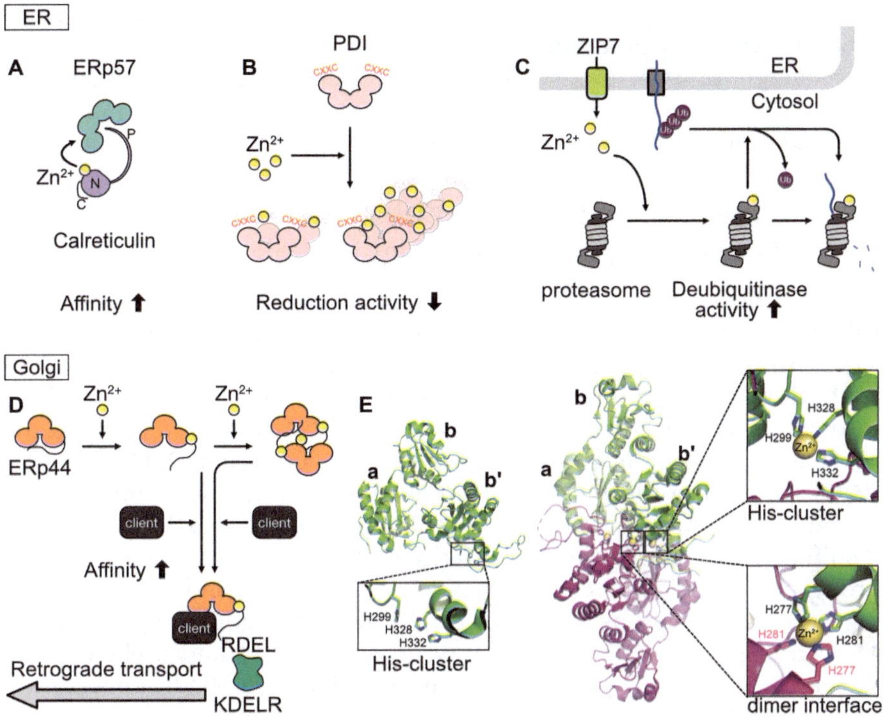

Figure 6.2 The zinc-regulated protein quality control systems in the ESP. (A) Zn^{2+} binding to calreticulin increases its affinity for ERp57. (B) Zn^{2+} promotes dimer/oligomer formation of PDI and inhibits its reduction activity. (C) ZIP7-mediated Zn^{2+} elevation in the cytosol enhances the deubiquitinase activity of proteasomes, supporting ERAD. (D) In the GA, ERp44 binds a Zn^{2+} ion at its His-cluster domain, and Zn^{2+}-bound ERp44 molecules form homodimers with His-277 and His-281 at the dimer interface. Zn^{2+}-bound ERp44 has a stronger affinity for client proteins, ensuring efficient retrieval. (E) Crystal structures of ERp44 in Zn^{2+}-unbound and -bound forms. PBDID 5GU6 (left) and 5XWM (right).

concentration (EC_{50}) value of 2.2 μM (Figure 6.2B).[33,38,40] These observations suggest a close link between the redox activity of PDI and Zn^{2+}.[39]

Zn^{2+} controls the PQC system of the ER not only through redox regulation, but also through enhanced proteasomal degradation.[40] A sophisticated screening system utilizing border cell migration in the *Drosophila* egg chamber revealed that ZIP7-silencing leads to incomplete migration of border cells with accumulation of unfolded protein response markers.[40] This suggests that ER stress impairs the motility of these cells. Exogenous expression of *Drosophila* ZIP7 restores the accumulation of ubiquitinated proteins and mitigates misfolded protein-induced ER stress. Interestingly, the study also showed that the activity of deubiquitinase Rpn11, a functional partner of the proteasome, is enhanced upon binding Zn^{2+}, promoting

ERAD (Figure 6.2C).[40] Furthermore, accumulation of polyubiquitinated proteins caused by proteasomal inhibition is mitigated by treatment with a zinc ionophore or by exogenous expression of human ZIP7 in HeLa cells. Thus, ZIP7 transports Zn^{2+} from the ER lumen to the cytosol, facilitating ERAD and promoting protein homeostasis in the ER.

6.3.2 ERp44: A Zinc-regulated Chaperone Protein Functioning at the ER–Golgi Interface

Ero1α oxidizes PDI in the ER, indicating that Ero1α is an ER-localized oxidase.[6,41,42] However, Ero1α lacks an ER retention signal, such as the KDEL-like sequence, at the C-terminus. ERp44, a member of the PDI family consisting of 377 amino acids, binds to Ero1α and retains it in the ER through its own C-terminal RDEL sequence.[43,44] ERp44 consists of three Trx-like domains (**a**, **b**, and **b′**) followed by a C-terminal tail (C-tail) with an RDEL sequence at the end. Unlike other typical PDI family members, ERp44 contains a Cys–Arg–Phe–Ser motif at the active site of the **a** domain, forming disulfide-linked complexes with various client proteins including Ero1α, Prdx4, and ER-associated aminopeptidase 1 (ERAP1), and immature oligomeric proteins such as immunoglobulin M (IgM), adiponectin, and serotonin transporter.[43,45–52] The affinity between ERp44 and a client protein increases under acidic pH, and pH neutralization in the GA by knockdown of the Golgi pH regulator (GPHR)[53] impairs the client retrieval activity of ERp44.[54] These results suggest that ERp44 captures its clients in the weakly acidic GA and retrogradely transports them to the ER in concert with KDELR.

Crystal structure analyses of ERp44 revealed that the three Trx-like domains form a clover leaf-like structure, while the C-tail forms hydrogen bonds with the **a** domain, including Cys-29 in the Cys–Arg–Phe–Ser motif, and with the **b′** domain.[55,56] Although the C-terminal RDEL sequence was invisible in the electron density map, presumably due to its high flexibility, the client-binding surface of ERp44 is likely masked by the C-tail due to the hydrogen bond between Cys-29 and Thr-369, a residue located near the C-terminus. Structural analysis and molecular dynamics simulations of ERp44 at neutral and acidic pH revealed that the flexibility of the C-tail increases and the number of hydrogen bonds between the C-tail and the **a** and **b′** domains decreases at acidic pH.[56] This partly explains the molecular mechanism by which acidic conditions enhance interactions between ERp44 and client proteins.

More importantly, ERp44 is also regulated by Zn^{2+} (Figure 6.2D). The high-resolution crystal structure of ERp44 shows that His-299, His-328, and His-332 are located in close proximity near the base of the C-tail, forming a His cluster.[56] Biochemical analyses revealed that ERp44 binds 1.5 molar equivalents of Zn^{2+} with apparent K_d values of 135–295 nM at pH 7.2–6.2. Furthermore, Zn^{2+} induce dimerization of ERp44 and increase the affinity between ERp44 and Ero1α six-fold. A cell biological approach showed that zinc chelation by *N*,*N*,*N′*,*N′*-tetrakis(2-pyridylmethyl)ethylenediamine (TPEN) causes accumulation of ERp44 in the GA and allows some for it to be secreted from cells.[57] Zinc replenishment by treatment with zinc pyrithione (ZPT), a Zn^{2+} ionophore, relocates ERp44 to the ER.

Crystal structures of the Zn^{2+}-bound ERp44 dimer revealed that Zn^{2+} bind to the His cluster and to a central site at the dimer interface consisting of His-277 and His-281 from each protomer (Figure 6.2E). Binding of Zn^{2+} at the His cluster significantly changes the conformation of the C-tail, fully exposing Cys-29 to the solvent and increasing the affinity for client proteins. Calorimetric and size-exclusion chromatography analyses in combination with systematic His mutations revealed that Zn^{2+} binds to the His cluster first, and another Zn^{2+} binds to the His-277–His-281 pair to form a ERp44 homodimer. The C-terminal RDEL sequence is buried inside the dimer interface, likely preventing the Golgi-to-ER retrograde transport of ERp44. In this context, the ERp44 H277/281A mutant is constitutively monomeric and localizes in the Golgi to a lesser extent.[57,58] On the other hand, ERp44 with Ala mutations at the His cluster is unable to bind Zn^{2+} and accumulates predominantly in the GA. These observations suggest that ERp44 acquires Zn^{2+} in the Golgi to be relocated to the ER. The His cluster and H277/288 pair are highly conserved among ERp44 homologs from insects to humans, suggesting that Zn^{2+}-mediated regulation of ERp44 is evolutionarily and functionally significant.

6.4 Zinc Regulation in the ESP

As discussed above, Zn^{2+} play critical roles in the PQC of the ESP, highlighting the physiological importance of Zn^{2} homeostasis within the ESP. Multiple ZnT/ZIP transporters localized in specific compartments contribute to Zn^{2+} regulation and dynamics across various cell types and organelles. Among these transporters, six kinds of ZnT members, ZnT4, ZnT5, ZnT6, ZnT7, ZnT9, and ZnT10, and four kinds of ZIP members, ZIP7, ZIP9, ZIP11, and ZIP13, are reported to localize in the ESP.[2] Compared with other organelles including lysosomes and mitochondria, the ESP (especially the GA) harbors a greater number of ZnT/ZIP transporters, suggesting that Zn^{2+} is dynamically and tightly regulated in the ESP.

6.4.1 Localization of ZnT in the ESP

Immunofluorescence studies showed that endogenous ZnT4 and ZnT6 in normal rat kidney (NRK) cell lines exhibit a ribbon-like distribution pattern near the nucleus, a typical feature of GA localization.[59] Another study showed that both endogenous and exogenously expressed ZnT4 localize at the GA and plasma membrane in the HV11 mouse mammary gland-derived epithelial cell line.[60] Excess supplementation of Zn^{2+} in the culture medium relocated ZnT4 to the plasma membrane, disrupting its Golgi-like localization.[59,60] Similarly, exogenously expressed chicken ZnT4-HA and human ZnT4-FLAG displayed Golgi localization in DT40 cells.[61,62] Overexpression of

ZnT4-HA in HC11 cells lowered the activity of galactosyltransferases localized at the TGN.[60] Galactosyltransferases require Zn^{2+} for their enzymatic activity, but the activity can be impaired by excess Zn^{2+} *in vitro*.[63] These observations suggest that overexpressed ZnT4 primarily localizes at the TGN and imports excessive Zn^{2+} into this compartment. However, endogenous ZnT4 was mainly localized at cytoplasmic granules in the PMC42 human breast cancer epithelial cell line.[64] ZnT4-FLAG exogenously expressed in HeLa Kyoto cells, a subline derived from HeLa cells, was detected at early endosomes, recycling endosomes, late endosomes, lysosomes, and the plasma membrane (PM), with less ZnT4 localized at the GA.[65] Thus, localization of ZnT4 is likely affected by cell type, expression levels, and zinc abundance.

ZnT5 and ZnT6 form a heterodimer (see Chapter 9 for more detail).[66,67] Endogenous and exogenously expressed ZnT6 are observed at the GA in several cell types.[59,65,67] Similar to ZnT4, Golgi localization of ZnT6 is disrupted by treatment with excess Zn^{2+} in NRK cells.[59] Since ZnT6 localization at the GA depends on the expression of ZnT5,[67] their localizations are expected to be identical to each other.

Like ZnT5/ZnT6, endogenous and exogenously expressed ZnT7 are also observed at the GA in Chinese hamster ovary (CHO) and HeLa Kyoto cells.[65,68] Immunoelectron microscopy analysis also reported that ZnT7 is localized at the *cis*-side of GA in mouse choroid epithelial cells.[69] The GA is composed of multiple stacks anchored by microtubules nucleated from the γ-tubulin ring complex (γTuRC), which is connected to the GA *via* A-Kinase Anchoring Protein 450 (AKAP450) and Golgi matrix protein 130 (GM130).[70,71] Treatment with nocodazole, a microtubule polymerization inhibitor, disassembles the Golgi structure into mini-stacks, which retain the native-like multi-cisternal structure and cargo transport functions of the Golgi.[19–21] Super-resolution microscopy analysis in combination with nocodazole treatment revealed that while ZnT6 localizes at the *cis*- and medial Golgi, ZnT7 was primarily localized immediately upstream of the GM130-labeled *cis*-Golgi, hereafter referred to as pre-*cis*-Golgi in this chapter. Thus, Golgi-resident ZnTs are likely localized at distinct and specific Golgi cisternae.

Some studies reported that ZnT9 localizes in mitochondria and plays important roles in mitochondrial homeostasis by exporting Zn^{2+} to the cytosol.[72,73] However, exogenously expressed dTAG-tagged ZnT9 in HAP1 cells and GFP-tagged ZnT9 in SH-SY5Y human neuroblastoma cells are reported to localize at the ER.[74,75] Compared with those of other ZnTs, the cellular functions of ZnT9 in the ESP are poorly understood.

FLAG-tagged ZnT10 has also been shown to localize at the GA in SH-SY5Y cells, and excess Zn^{2+} treatment relocated it to the PM, as observed for ZnT4 and ZnT6.[76] This suggests that ZnT10 transports cytoplasmic Zn^{2+} into the GA and/or extracellular space, and plays a role in cellular Zn^{2+} homeostasis.[76] Another study demonstrated that ZnT10-myc is localized at recycling endosomes in HEK293 cells.[77] Additionally, ZnT10 was reported to localize at the PM, where it likely conducts manganese efflux.[78] These findings suggest that ZnT10 alters intracellular localization and function depending on cell types and extracellular conditions.

6.4.2 Localization of ZIP in the ESP

Exogenously expressed ZIP7 localizes to the ER[33,34,79] and is closely associated with protein homeostasis in the ER, as discussed above. ZIP7 primarily resides on the ER membrane where it releases Zn^{2+} into the cytosol.

Exogenously expressed ZIP9-HA has been shown to localize at the GA in HeLa cells.[80] ZIP9 is thought to be essential for protein kinase B and MAP kinase signaling, potentially by regulating cytosolic Zn^{2+} homeostasis.[80,81] More recently, mass spectrometry analysis using GA fractions isolated from HEK293 cells combined with microscopy analysis conformed that ZIP9 is a Golgi-resident protein.[82] Thus, ZIP9 is likely responsible for transporting Zn^{2+} from the GA to the cytosol.

ZIP11 was found to be localized to the GA in mammary epithelial cells.[83] Expression of ZIP11 is elevated in lactating tissues, and a possible function in zinc supplementation to milk has been suggested.[83] However, another study reported that ZIP11 localizes at the nucleus in mouse stomach and colon tissues,[84] and more recent studies suggest that ZIP11 localizes at perinuclear vesicles in HeLa cells, but not in the ESP.[85,86] Similarly, although colocalization of endogenous ZIP13 with the GA in mouse dermal fibroblast cells was reported,[87,88] another study observed that ZIP13 localized to some vesicles in HeLa cells, with no colocalization at either the GA or the ER.[89] Thus, subcellular localization of ZIP11 and ZIP13 remains contentious and requires further investigation.

6.4.3 Golgi-resident ZnTs Regulate Zinc Homeostasis in the GA

The different localizations of ZnT4, ZnT5/6, and ZnT7 at the Golgi cisternae suggest their distinct roles in Zn^{2+} homeostasis within the GA. Newly developed organelle-targetable chemical zinc probes, ZnDA-1H, -2H, and -3H, enable the measurement of labile Zn^{2+} concentration ($[Zn^{2+}]$) in a specific organelle (see Chapter 1 for more details). Analyses using these probes revealed that labile $[Zn^{2+}]$ was 14 pM at the ER and 23 nM at the Golgi in HeLa cells.[90,91] Using several types of HaloTag-fused Golgi marker proteins and ZnDA-1H, each Golgi stack was found to contain different labile $[Zn^{2+}]$ in HeLa Kyoto cells; pre-*cis*-Golgi and *cis*-, medial, and *trans*-Golgi regions contain approximately 100 nM and 60–80 nM of labile Zn^{2+}, respectively (Figure 6.3A).[65]

Importantly, a series of knockdown experiments suggest that ZnT4, ZnT5/6, and ZnT7 govern Zn^{2+} uptake into the medial/*trans*-Golgi, *cis*-/medial Golgi, and pre-*cis*-/*cis*-Golgi subregions, respectively.[65] These results are consistent with their subcellular localizations revealed by super-resolution microscopy,[65]

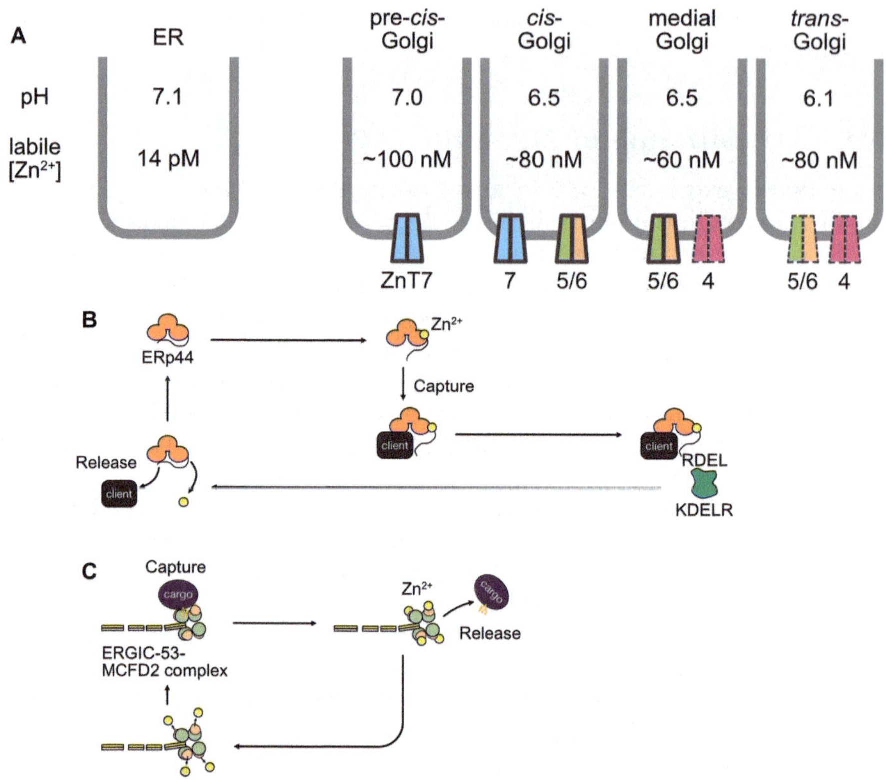

Figure 6.3 ZnT-mediated zinc homeostasis in the ESP, and proposed models of zinc-dependent ERp44 and ERGIC-53 regulations. (A) The GA contains higher concentrations of labile [Zn^{2+}] than the ER. Interestingly, Zn^{2+} levels vary across the Golgi cisternae. Zn^{2+} homeostasis is maintained by Golgi-resident ZnTs. (B) ERp44 acquires Zn^{2+} at the pre-*cis*-Golgi and is transported back to the ER *via* KDELR from the more acidic distal Golgi. (C) ERGIC-53 captures cargo in the ER and releases it in the GA upon Zn^{2+} binding to MCFD2, sustaining the recycling of ERGIC-53. This model explains how ERGIC-53 facilitates efficient cargo transport.

indicating that each ZnT regulates labile [Zn^{2+}] homeostasis in the specific organelle subregion where it is localized.

6.5 Zinc in the ESP may Regulate ESP Cargo Trafficking

Considering that the K_d value of ERp44 for Zn^{2+} is 134 nM at pH 7.2,[57] ERp44 most likely releases Zn^{2+} in the ER and acquires Zn^{2+} primarily at the pre-*cis*-Golgi subregion. Consistently, ZnT7-silencing caused ERp44 to accumulate predominantly in the GA, similar to TPEN treatment, whereas ZnT4- or ZnT5/6-silencing did not. However, the client retention activity of

ERp44 is impaired by ZnT4- or ZnT5/6-silencing, suggesting that Zn^{2+} homeostasis in the acidic distal Golgi region ensures the appropriate function of ERp44, presumably because the conditions facilitate the formation of the client-ERp44-KDELR ternary complex (Figure 6.3B).[65]

The retention using the selective hooks (RUSH) system allows visualization of synchronized ER-to-Golgi transport *via* live-cell imaging.[92] RUSH assays clearly demonstrated ERp44 trafficking from the ER to the GA, with the amount of ERp44 in the GA area reaching a plateau, suggesting that anterograde and retrograde transports reached an equilibrium. However, ZnT7-silencing delayed the time taken to reach the plateau, suggesting that retrograde transport of ERp44 was inhibited under this condition. Importantly, anterograde transport of the GFP-fused transmembrane domain of sialyltransferase, a positive control cargo that is insensitive to Zn^{2+}, was unaffected by knockdown of ZnT4, ZnT5+6, or ZnT7, indicating that perturbation in Golgi Zn^{2+} homeostasis does not inhibit ER-to-Golgi cargo trafficking *per se*.

In another study, ERGIC-53, a representative cargo receptor that promotes the ER-to-Golgi transport of secretory proteins, accumulated in the GA when cells were treated with TPEN.[93] Recent biophysical analysis using isothermal titration calorimetry revealed that MCFD2, a partner protein of ERGIC-53 for glycoprotein transport, binds up to two molar equivalents of Zn^{2+} with micromolar affinity. Cryo-electron microscopy (cryo-EM) analyses further revealed the tetrameric structure of full-length ERGIC-53 in complex with MCFD2. Because Zn^{2+}-bound MCFD2 undergoes conformational changes to mask its substrate-binding site, Zn^{2+} may facilitate the cargo release. As the labile $[Zn^{2+}]$ is much higher in the GA than in the ER, ERGIC-53 binds its substrates in the ER and releases them in the GA, ensuring efficient trafficking of folded glycoproteins (Figure 6.3C).[94]

6.6 Concluding Remarks

This chapter reviews the mechanisms of PQC and ZnT/ZIP-mediated Zn^{2+} homeostasis in the ESP, and the interplay between these two. As the GA contains the highest concentration of labile Zn^{2+} among the organelles investigated to date, it is clear that Zn^{2+}-associated reactions are a feature of this organelle. In particular, redox reactions involving Cys residues are important events in the ESP. Given the frequent involvement of Cys residues in Zn^{2+} binding, it would be interesting to explore whether and how Zn^{2+} participate in redox-relevant reactions and redox homeostasis in the ESP.

As described above, the ESP harbors multiple ZnT and ZIP family members. Although the localizations and functions of these transporters are gradually being elucidated, the picture remains incomplete because the localizations and functions of ZnT and ZIP members vary depending on cell types, cellular conditions, and culture environments. The mechanisms by which these transporters sense Zn^{2+} to regulate their localization, expression, and activity require further investigation. Future studies with advanced techniques including genome-editing, super-resolution microscopy, and cryo-electron tomography will provide further insights into the functions and mechanisms of ZnTs and ZIPs with higher spatiotemporal resolution.

The Zn^{2+}-dependent regulations of proteins and cellular functions are expected to be an area of intensifying research. Among the various aspects of this field, focusing on PQC is likely to make significant contributions to medical science.

Acknowledgements

This work was supported by funding from JST (JP19K16065, JP20H05494, and JP22H04797) to Y. A., Asahi Glass Foundation to Y. A., AMED-CREST (21gm1410006h0001) to K. I., and JSPS KAKENHI (18H03978, 21H04758, and 24H00561) to K. I.

References

1. N. Burger, M. J. Mittenbühler, H. Xiao, S. Shin, S. M. Wei, E. K. Henze, S. Schindler, S. Mehravar, D. M. Wood, J. J. Petrocelli, Y. Sun, H.-G. Sprenger, P. Latorre-Muro, A. L. Smythers, L. H. M. Bozi, N. Darabedian, Y. Zhu, H. S. Seo, S. Dhe-Paganon, J. Che and E. T. Chouchani, *Cell*, 2025, **188**, 832–850.
2. T. Kambe, T. Tsuji, A. Hashimoto and N. Itsumura, *Physiol. Rev.*, 2015, **95**, 749–784.
3. L. Huang and S. Tepaamorndech, *Mol. Aspects Med.*, 2013, **34**, 548–560.
4. M. A. De Matteis and A. Luini, *Nat. Rev. Mol. Cell Biol.*, 2008, **9**, 273–284.
5. P. Sweeney, H. Park, M. Baumann, J. Dunlop, J. Frydman, R. Kopito, A. McCampbell, G. Leblanc, A. Venkateswaran, A. Nurmi and R. Hodgson, *Transl. Neurodegener.*, 2017, **6**, 6.
6. M. Okumura, H. Kadokura and K. Inaba, *Free Radical Biol. Med.*, 2015, **83**, 314–322.
7. M. Michalak, *J. Cell. Mol. Med.*, 2024, **28**, e17839.
8. K. F. R. Pobre, G. J. Poet and L. M. Hendershot, *J. Biol. Chem.*, 2019, **294**, 2098–2108.
9. R. Ushioda, J. Hoseki, K. Araki, G. Jansen, D. Y. Thomas and K. Nagata, *Science*, 2008, **321**, 569–572.
10. M. Hagiwara, K. Maegawa, M. Suzuki, R. Ushioda, K. Araki, Y. Matsumoto, J. Hoseki, K. Nagata and K. Inaba, *Mol. Cell*, 2011, **41**, 432–444.
11. K. Maegawa, S. Watanabe, K. Noi, M. Okumura, Y. Amagai, M. Inoue, R. Ushioda, K. Nagata, T. Ogura and K. Inaba, *Structure*, 2017, **25**, 846–857.e4.
12. X. Cai, S. Ito, K. Noi, M. Inoue, R. Ushioda, Y. Kato, K. Nagata and K. Inaba, *J. Biol. Chem.*, 2023, **299**, 105274.

13. A. V. Weigel, C. L. Chang, G. Shtengel, C. S. Xu, D. P. Hoffman, M. Freeman, N. Iyer, J. Aaron, S. Khuon, J. Bogovic, W. Qiu, H. F. Hess and J. Lippincott-Schwartz, *Cell*, 2021, **184**, 2412–2429.e16.
14. O. Shomron, I. Nevo-Yassaf, T. Aviad, Y. Yaffe, E. E. Zahavi, A. Dukhovny, E. Perlson, I. Brodsky, A. Yeheskel, M. Pasmanik-Chor, A. Mironov, G. V. Beznoussenko, A. A. Mironov, E. H. Sklan, G. H. Patterson, Y. Yonemura, M. Sannai, C. Kaether and K. Hirschberg, *J. Cell Biol.*, 2021, **220**, e201907224.
15. C. Appenzeller-Herzog and H. P. Hauri, *J. Cell Sci.*, 2006, **119**, 2173–2183.
16. W. C. Nichols, U. Seligsohn, A. Zivelin, V. H. Terry, C. E. Hertel, M. A. Wheatley, M. J. Moussalli, H.-P. Hauri, N. Ciavarella, R. J. Kaufman and D. Ginsburg, *Cell*, 1998, **93**, 61–70.
17. C. Zheng and B. Zhang, *Semin. Thromb. Hemostasis*, 2013, **39**, 613–620.
18. B. S. Glick and A. Nakano, *Annu. Rev. Cell Dev. Biol.*, 2009, **25**, 113–132.
19. H. C. Tie, D. Mahajan, B. Chen, L. Cheng, A. M. J. VanDongen and L. Lu, *Mol. Biol. Cell*, 2016, **27**, 848–861.
20. H. C. Tie, A. Ludwig, S. Sandin and L. Lu, *eLife*, 2018, 7, 1–26.
21. H. C. Tie, D. Mahajan and L. Lu, *J. Cell Biol.*, 2022, **221**, e202109114.
22. Y. Shimizu, J. Takagi, E. Ito, Y. Ito, K. Ebine, Y. Komatsu, Y. Goto, M. Sato, K. Toyooka, T. Ueda, K. Kurokawa, T. Uemura and A. Nakano, *Nat. Commun.*, 2021, **12**, 1901.
23. A. Harada, M. Kunii, K. Kurokawa, T. Sumi, S. Kanda, Y. Zhang, S. Nadanaka, K. M. Hirosawa, K. Tokunaga, T. Tojima, M. Taniguchi, K. Moriwaki, S. I. Yoshimura, M. Yamamoto-Hino, S. Goto, T. Katagiri, S. Kume, M. Hayashi-Nishino, M. Nakano, E. Miyoshi, K. G. N. Suzuki, H. Kitagawa and A. Nakano, *Nat. Commun.*, 2024, **15**, 4514.
24. Z. Sun and J. L. Brodsky, *J. Cell Biol.*, 2019, **218**, 3171–3187.
25. L. Buzuk and D. Hellerschmied, *Front. Mol. Biosci.*, 2023, **10**, 1197921.
26. F. Reggiori and H. R. B. Pelham, *Nat. Cell Biol.*, 2002, **4**, 117–123.
27. O. Schmidt, Y. Weyer, V. Baumann, M. A. Widerin, S. Eising, M. Angelova, A. Schleiffer, L. Kremser, H. Lindner, M. Peter, F. Fröhlich and D. Teis, *EMBO J.*, 2019, **38**, 1–23.
28. P. Bräuer, J. L. Parker, A. Gerondopoulos, I. Zimmermann, M. A. Seeger, F. A. Barr and S. Newstead, *Science*, 2019, **363**, 1103–1107.
29. A. Gerondopoulos, P. Bräuer, T. Sobajima, Z. Wu, J. L. Parker, P. Biggin, F. A. Barr and S. Newstead, *eLife*, 2021, **10**, 1–29.
30. I. Raykhel, H. Alanen, K. Salo, J. Jurvansuu, D. N. Van, M. Latva-Ranta and L. Ruddock, *J. Cell Biol.*, 2007, **179**, 1193–1204.
31. D. W. Wilson, M. J. Lewis and H. R. Pelham, *J. Biol. Chem.*, 1993, **268**, 7465–7468.
32. E. Nolin, S. Gans, L. Llamas, S. Bandyopadhyay, S. M. Brittain, P. Bernasconi-Elias, K. P. Carter, J. J. Loureiro, J. R. Thomas, M. Schirle, Y. Yang, N. Guo, G. Roma, S. Schuierer, M. Beibel, A. Lindeman, F. Sigoillot, A. Chen, K. X. Xie, S. Ho, J. Reece-Hoyes, W. A. Weihofen, K. Tyskiewicz, D. Hoepfner, R. I. McDonald, N. Guthrie, A. Dogra, H. Guo, J. Shao, J. Ding, S. M. Canham, G. Boynton, E. L. George,

Z. B. Kang, C. Antczak, J. A. Porter, O. Wallace, J. A. Tallarico, A. E. Palmer, J. L. Jenkins, R. K. Jain, S. M. Bushell and C. J. Fryer, *Nat. Chem. Biol.*, 2019, **15**, 179–188.
33. B. H. Bin, J. Bhin, J. Seo, S. Y. Kim, E. Lee, K. Park, D. H. Choi, T. Takagishi, T. Hara, D. Hwang, H. Koseki, Y. Asada, S. Shimoda, K. Mishima and T. Fukada, *J. Invest. Dermatol.*, 2017, **137**, 1682–1691.
34. W. Ohashi, S. Kimura, T. Iwanaga, Y. Furusawa, T. Irié, H. Izumi, T. Watanabe, A. Hijikata, T. Hara, O. Ohara, H. Koseki, T. Sato, S. Robine, H. Mori, Y. Hattori, H. Watarai, K. Mishima, H. Ohno, K. Hase and T. Fukada, *PLoS Genet.*, 2016, **12**, e1006349.
35. Y. Saito, Y. Ihara, M. R. Leach, M. F. Cohen-Doyle and D. B. Williams, *Calreticulin functions in vitro as a molecular chaperone for both glycosylated and non-glycosylated proteins*, 1999, vol. 18.
36. Z. Li, W. F. Stafford and M. Bouvier, *Biochemistry*, 2001, **40**, 11193–11201.
37. A. Solovyov and H. F. Gilbert, *Protein Sci.*, 2004, **13**, 1902–1907.
38. A. I. S. Moretti, V. E. Baksheeva, A. Y. Roman, T. C. De Bessa, F. Devred, H. Kovacic and P. O. Tsvetkov, *Int. J. Mol. Sci.*, 2024, **25**, 2095.
39. W. Maret, *Antioxid. Redox Signaling*, 2006, **8**, 1419–1441.
40. X. Guo, M. Mutch, A. Y. Torres, M. Nano, N. Rauth, J. Harwood, D. McDonald, Z. Chen, C. Montell, W. Dai and D. J. Montell, *Dev. Cell*, 2024, **59**, 1655–1667.e6.
41. A. Mezghrani, A. Fassio, A. Benham, T. Simmen, I. Braakman and R. Sitia, *EMBO J.*, 2001, **20**, 6288–6296.
42. Y. Sato and K. Inaba, *FEBS J.*, 2012, **279**, 2262–2271.
43. T. Anelli, M. Alessio, A. Mezghrani, T. Simmen, F. Talamo, A. Bachi and R. Sitia, *EMBO J.*, 2002, **21**, 835–844.
44. T. Tempio and T. Anelli, *J. Cell Sci.*, 2020, **133**, jcs240366.
45. T. Kakihana, K. Araki, S. Vavassori, S. I. Iemura, M. Cortini, C. Fagioli, T. Natsume, R. Sitia and K. Nagata, *J. Biol. Chem.*, 2013, **288**, 29586–29594.
46. T. Anelli, M. Alessio, A. Bachi, L. Bergamelli, G. Bertoli, S. Camerini, A. Mezghrani, E. Ruffato, T. Simmen and R. Sitia, *EMBO J.*, 2003, **22**, 5015–5022.
47. T. Anelli, S. Ceppi, L. Bergamelli, M. Cortini, S. Masciarelli, C. Valetti and R. Sitia, *EMBO J.*, 2007, **26**, 4177–4188.
48. C. Hisatsune, E. Ebisui, M. Usui, N. Ogawa, A. Suzuki, N. Mataga, H. Takahashi-Iwanaga and K. Mikoshiba, *Mol. Cell*, 2015, **58**, 1015–1027.
49. Z. V. Wang, T. D. Schraw, J.-Y. Kim, T. Khan, M. W. Rajala, A. Follenzi and P. E. Scherer, *Mol. Cell. Biol.*, 2007, **27**, 3716–3731.
50. L. Hampe, M. Radjainia, C. Xu, P. W. R. Harris, G. Bashiri, D. C. Goldstone, M. A. Brimble, Y. Wang and A. K. Mitra, *J. Biol. Chem.*, 2015, **290**, 18111–18123.
51. S. Freyaldenhoven, Y. Li, A. M. Kocabas, E. Ziu, S. Ucer, R. Ramanagoudr-Bhojappa, G. P. Miller and F. Kilic, *J. Biol. Chem.*, 2012, **287**, 17801–17811.

52. C. Giannone, M. R. Chelazzi, A. Orsi, T. Anelli, T. Nguyen, J. Buchner and R. Sitia, *EMBO J.*, 2022, **41**, e108518.
53. Y. Maeda, T. Ide, M. Koike, Y. Uchiyama and T. Kinoshita, *Nat. Cell Biol.*, 2008, **10**, 1135–1145.
54. S. Vavassori, M. Cortini, S. Masui, S. Sannino, T. Anelli, I. R. Caserta, C. Fagioli, M. F. Mossuto, A. Fornili, E. vanAnken, M. Degano, K. Inaba and R. Sitia, *Mol. Cell*, 2013, **50**, 783–792.
55. K. Yang, D. F. Li, X. Wang, J. Liang, R. Sitia, C. chen Wang and X. Wang, *Structure*, 2016, **24**, 1755–1765.
56. S. Watanabe, M. Harayama, S. Kanemura, R. Sitia and K. Inaba, *Proc. Natl. Acad. Sci. U. S. A.*, 2017, **114**, E3224–E3232.
57. S. Watanabe, Y. Amagai, S. Sannino, T. Tempio, T. Anelli, M. Harayama, S. Masui, I. Sorrentino, M. Yamada, R. Sitia and K. Inaba, *Nat. Commun.*, 2019, **10**, 603.
58. S. Sannino, T. Anelli, M. Cortini, S. Masui, M. Degano, C. Fagioli, K. Inaba and R. Sitia, *J. Cell Sci.*, 2014, **127**, 4260–4269.
59. L. Huang, C. P. Kirschke and J. Gitschier, *J. Biol. Chem.*, 2002, **277**, 26389–26395.
60. N. H. McCormick and S. L. Kelleher, *Am. J. Physiol.: Cell Physiol.*, 2012, **303**, 291–297.
61. A. Fukunaka, Y. Kurokawa, F. Teranishi, I. Sekler, K. Oda, M. L. Ackland, V. Faundez, M. Hiromura, S. Masuda, M. Nagao, S. Enomoto and T. Kambe, *J. Biol. Chem.*, 2011, **286**, 16363–16373.
62. T. Tsuji, Y. Kurokawa, J. Chiche, J. Pouysségur, H. Sato, H. Fukuzawa, M. Nagao and T. Kambe, *J. Biol. Chem.*, 2017, **292**, 2159–2173.
63. E. A. Permyakov, I. L. Reyzer and L. J. Berliner, *Effects of Zn(II) on Galactosyltransferase Activity*, 1993, vol. 12.
64. A. A. Michalczyk, J. Allen, R. C. Blomeley and M. L. Ackland, *Biochem. J.*, 2002, **364**, 105–113.
65. Y. Amagai, M. Yamada, T. Kowada, T. Watanabe, Y. Du, R. Liu, S. Naramoto, S. Watanabe, J. Kyozuka, T. Anelli, T. Tempio, R. Sitia, S. Mizukami and K. Inaba, *Nat. Commun.*, 2023, **14**, 2683.
66. A. Fukunaka, T. Suzuki, Y. Kurokawa, T. Yamazaki, N. Fujiwara, K. Ishihara, H. Migaki, K. Okumura, S. Masuda, Y. Yamaguchi-Iwai, M. Nagao and T. Kambe, *J. Biol. Chem.*, 2009, **284**, 30798–30806.
67. E. Suzuki, N. Ogawa, T. Takeda, Y. Nishito, Y. Tanaka, T. Fujiwara, M. Matsunaga, S. Ueda, N. Kubo, T. Tsuji, A. Fukunaka, T. Yamazaki, K. M. Taylor, Y. Ogra and T. Kambe, *J. Biol. Chem.*, 2020, **295**, 5669–5684.
68. C. P. Kirschke and L. Huang, *J. Biol. Chem.*, 2003, **278**, 4096–4102.
69. Z.-H. Chi, X. Wang, Z.-Y. Wang, H.-L. Gao, A. Dahlstrom and L. Huang, *NeuroReport*, 2006, **17**, 1807–1811.
70. K. Chabin-Brion, J. Me Marceiller, F. Perez, C. Settegrana, A. Drechou, G. Ve Durand and C. Poü, *The golgi complex is a microtubule-organizing organelle*, 2001, vol. 12.
71. A. Akhmanova and L. C. Kapitein, *Nat. Rev. Mol. Cell Biol.*, 2022, **23**, 541–558.

72. H. Deng, X. Qiao, T. Xie, W. Fu, H. Li, Y. Zhao, M. Guo, Y. Feng, L. Chen, Y. Zhao, L. Miao, C. Chen, K. Shen and X. Wang, *Proc. Natl. Acad. Sci. U. S. A.*, 2021, **118**, e2023909118.
73. A. Kowalczyk, O. Gbadamosi, K. Kolor, J. Sosa, L. Andrzejczuk, G. Gibson, C. S. Croix, M. Chikina, E. Aizenman, N. Clark and K. Kiselyov, *Biochem. J.*, 2021, **478**, 3205–3220.
74. A. Bensimon, M. D. Pizzagalli, F. Kartnig, V. Dvorak, P. Essletzbichler, G. E. Winter and G. Superti-Furga, *Cell Chem. Biol.*, 2020, **27**, 728–739.e9.
75. Y. Perez, Z. Shorer, K. Liani-Leibson, P. Chabosseau, R. Kadir, M. Volodarsky, D. Halperin, S. Barber-Zucker, H. Shalev, R. Schreiber, L. Gradstein, E. Gurevich, R. Zarivach, G. A. Rutter, D. Landau and O. S. Birk, *Brain*, 2017, **140**, 928–939.
76. H. J. Bosomworth, J. K. Thornton, L. J. Coneyworth, D. Ford and R. A. Valentine, *Metallomics*, 2012, **4**, 771–779.
77. N. Patrushev, B. Seidel-Rogol and G. Salazar, *PLoS One*, 2012, 7, e33211.
78. C. E. Zogzas, M. Aschner and S. Mukhopadhyay, *J. Biol. Chem.*, 2016, **291**, 15940–15957.
79. C. Anzilotti, D. J. Swan, B. Boisson, M. Deobagkar-Lele, C. Oliveira, P. Chabosseau, K. R. Engelhardt, X. Xu, R. Chen, L. Alvarez, R. Berlinguer-Palmini, K. R. Bull, E. Cawthorne, A. P. Cribbs, T. L. Crockford, T. S. Dang, A. Fearn, E. J. Fenech, S. J. de Jong, B. C. Lagerholm, C. S. Ma, D. Sims, B. van den Berg, Y. Xu, A. J. Cant, G. Kleiner, T. R. Leahy, M. T. de la Morena, J. M. Puck, R. S. Shapiro, M. van der Burg, J. R. Chapman, J. C. Christianson, B. Davies, J. A. McGrath, S. Przyborski, M. Santibanez Koref, S. G. Tangye, A. Werner, G. A. Rutter, S. Padilla-Parra, J. L. Casanova, R. J. Cornall, M. E. Conley and S. Hambleton, *Nat. Immunol.*, 2019, **20**, 350–361.
80. W. Matsuura, T. Yamazaki, Y. I. Yuko, S. Masuda, M. Nagao, G. K. Andrews and T. Kambe, *Biosci., Biotechnol., Biochem.*, 2009, **73**, 1142–1148.
81. M. Taniguchi, A. Fukunaka, M. Hagihara, K. Watanabe, S. Kamino, T. Kambe, S. Enomoto and M. Hiromura, *PLoS One*, 2013, **8**, e58022.
82. R. Fasimoye, W. Dong, R. S. Nirujogi, E. S. Rawat, M. Iguchi, K. Nyame, E. Bagnoli, A. R. Prescott, D. R. Alessi, M. Abu-Remaileh and T. K. Phung, *Proc. Natl. Acad. Sci. U. S. A.*, 2023, **120**, e2219953120.
83. S. L. Kelleher, V. Velasquez, T. P. Croxford, N. H. McCormick, V. Lopez and J. Macdavid, *J. Cell. Physiol.*, 2012, **227**, 1761–1770.
84. R. J. Cousins, A. B. Martin, T. B. Aydemir, G. J. Guthrie, D. A. Samuelson and S. M. Chang, *J. Nutr.*, 2013, **143**, 1882–1888.
85. E. Y. Kim, O. Verdejo-Torres, K. Diaz-Rodriguez, F. Hasanain, L. Caromile and T. Padilla-Benavides, *Metallomics*, 2024, **16**, mfae006.
86. M. Olea-Flores, J. Kan, A. Carlson, S. A. Syed, C. McCann, V. Mondal, C. Szady, H. M. Ricker, A. McQueen, J. G. Navea, L. A. Caromile and T. Padilla-Benavides, *Front. Cell Dev. Biol.*, 2022, **10**, 895433.
87. T. Fukada, N. Civic, T. Furuichi, S. Shimoda, K. Mishima, H. Higashiyama, Y. Idaira, Y. Asada, H. Kitamura, S. Yamasaki, S. Hojyo, M. Nakayama, O. Ohara, H. Koseki, H. G. dos Santos, L. Bonafe, R. Ha-Vinh, A. Zankl,

S. Unger, M. E. Kraenzlin, J. S. Beckmann, I. Saito, C. Rivolta, S. Ikegawa, A. Superti-Furga and T. Hirano, *PLoS One*, 2008, **3**, e3642.

88. B. H. Bin, T. Fukada, T. Hosaka, S. Yamasaki, W. Ohashi, S. Hojyo, T. Miyai, K. Nishida, S. Yokoyama and T. Hirano, *J. Biol. Chem.*, 2011, **286**, 40255–40265.
89. J. Jeong, J. M. Walker, F. Wang, J. G. Park, A. E. Palmer, C. Giunta, M. Rohrbach, B. Steinmann and D. J. Eide, *Proc. Natl. Acad. Sci. U. S. A.*, 2012, **109**, E3530.
90. T. Kowada, T. Watanabe, Y. Amagai, R. Liu, M. Yamada, H. Takahashi, T. Matsui, K. Inaba and S. Mizukami, *Cell Chem. Biol.*, 2020, **27**, 1521–1531.e8.
91. R. Liu, T. Kowada, Y. Du, Y. Amagai, T. Matsui, K. Inaba and S. Mizukami, *ACS Sens.*, 2022, **7**, 748–757.
92. G. Boncompain, S. Divoux, N. Gareil, H. De Forges, A. Lescure, L. Latreche, V. Mercanti, F. Jollivet, G. Raposo and F. Perez, *Nat. Methods*, 2012, **9**, 493–498.
93. T. Tempio, A. Orsi, D. Sicari, C. Valetti, E. D. Yoboue, T. Anelli and R. Sitia, *iScience*, 2021, **24**, 102244.
94. S. Watanabe, Y. Kise, K. Yonezawa, M. Inoue, N. Shimizu, O. Nureki and K. Inaba, *Nat. Commun.*, 2024, **15**, 2404.

CHAPTER 7

Regulation of Pluripotent Stem Cell Pluripotency and Differentiation Through Zinc Mobilization

N. SHIRAKI* AND S. KUME*

Institute of Science Tokyo, School of Life Science and Technology, 4259-B-25 Nagatsuta-cho, Midori-ku, Yokohama, Kanagawa 226-8501, Japan
*Emails: shiraki@life.isct.ac.jp; skume@life.isct.ac.jp

7.1 Introduction

Zinc, an essential trace element, is crucial for various biological processes, including cell division, DNA synthesis, enzyme activation, protein stabilization, and protein production. It also plays a significant role in embryonic development and stem cell function. This chapter explores the important role of zinc in maintaining stem cell pluripotency and differentiation. The discussion begins with examining the role of zinc in the development of bovine and sea urchin embryos, providing a foundation to highlight its significance in various types of stem cells. By exploring these subjects in depth, we aim to understand zinc's multifaceted effect on stem cell biology and underscore its potential applications in directed differentiation, thereby contributing to the advancement of stem cell research and regenerative medicine.

Coordination Chemistry Perspectives No. 2
Zinc in Biology: Molecular Structures, Cellular Processes and Living Systems
Edited by Hiroyuki Yasui and Taiho Kambe

Published by the Royal Society of Chemistry, www.rsc.org

One investigation focused on zinc treatment during *in vitro* culture of bovine embryos by culturing with 0, 2, 20, or 40 μM $ZnSO_4$. The percentage of blastocytes was not affected by 2 or 20 μM zinc but reduced by 40 μM zinc. Only 2 μM zinc supplementation significantly increased the inner cell mass (ICM), a cluster of cells within a blastocyst that gives rise to embryonic stem cells (ESCs), and total cell numbers in day eight blastocysts. This improvement in ICM cell numbers is particularly important, as larger ICM sizes are associated with better pregnancy outcomes in cattle.[1] This strongly suggests that zinc is vital for cell proliferation of ESCs and that there is an optimal concentration for cell growth.

Interestingly, the effects of zinc on embryonic development are not limited to mammals. Sea urchin embryos were exposed to various concentrations (0, 1, 10, 100, and 1000 μM) of $ZnCl_2$ during different stages of development, and their progress was observed. These studies on sea urchin development have shown that high concentrations of zinc (100 μM) inhibit the endoderm and skeletal tissue development, while not affecting ectodermal tufts formation.[2] Metallothioneins are cysteine-rich proteins that bind and regulate zinc. They act as zinc buffers, storing excess zinc and releasing it when needed, thus playing a crucial role in maintaining cellular zinc homeostasis and protecting against zinc toxicity.[3] Metallothionein mRNA is present in large quantities in the ectoderm of sea urchin embryos but is hardly detectable in the mesoderm and endoderm tissue fractions. Treatment with 750 μM $ZnSO_4$ induces the accumulation of metallothionein mRNA in the mesoderm and endoderm but not in the ectoderm, which already contains a large amount of metallothionein mRNA.[4]

These findings indicate that zinc is crucial in early embryonic development across species, but its effects are concentration-dependent. At optimal levels, zinc supports vital developmental processes, while high concentrations of zinc can be toxic, emphasizing the importance of maintaining a delicate balance. Notably, the impact of zinc varies among different germ layers, with the mesoderm and endoderm showing greater sensitivity to zinc treatment than the ectoderm.

7.2 Zinc in Specific Stem Cell Types

Zinc plays a crucial role in stem cell biology, influencing pluripotency, self-renewal, and differentiation across different types of stem cells. Recent studies have uncovered the significance of zinc in mesenchymal stem cells (MSCs), hematopoietic stem cells (HSCs), and pluripotent stem cells (PSCs), including ESCs and induced pluripotent stem cells (iPSCs). The following subsections describe the role of zinc in these stem cell populations.

7.2.1 Mesenchymal Stem Cells (MSCs)

Mesenchymal stem cells (MSCs) are multipotent adult stem cells in several tissues, such as adipose tissue, bone marrow, and umbilical cord. MSCs can differentiate into osteocytes, chondrocytes, and adipocytes. MSCs have potential in regenerative medicine due to their anti-inflammatory and tissue repair properties. Zinc enhances the proliferation, differentiation, and therapeutic potential of MSCs. The following three paragraphs outline the role of zinc in these MSCs.

One study explored the potential of zinc supplementation to inhibit apoptosis and enhance the regenerative capacity of rat MSCs in ischemic tissue.[5] In a mouse ischemic hindlimb model, zinc-treated bone marrow-derived MSCs (BM-MSCs) significantly improved cell viability and survival and increased angiogenesis. *In vitro*, 10 μM zinc was found to inhibit H_2O_2-induced apoptosis and promote BM-MSC proliferation through the upregulation of the phosphorylation level of the Akt and increased vascular endothelial growth factor (VEGF) secretion. When administered to a mouse model of hindlimb ischemia, BM-MSCs (5×10^6 cells per mouse) that had been pre-treated with 10 μM zinc for 24 hours and then transplanted, showed significantly improved blood flow and cell viability compared to untreated BM-MSCs, as confirmed by laser doppler imaging and histological analysis.

Human adipose-derived MSCs (AD-MSCs) also benefit from zinc supplementation, particularly in neuronal differentiation.[6] Zinc supplementation (30 μM and 100 μM $ZnCl_2$) promotes AD-MSC proliferation *via* upregulation of ERK1/2 activity at the undifferentiated stage. Then, 10 μM $ZnCl_2$ treatment enhances neuronal differentiation by facilitating neurite outgrowth through the inactivation of RhoA signaling at the differentiation period. Zinc-treated AD-MSCs exhibited upregulation of key neuronal markers such as *MAP2* and *NESTIN*, while zinc chelation using CaEDTA impaired cell proliferation and neurite extension. These findings suggest zinc's potential role in the applications of neuroregenerative medicine using AD-MSCs.

In human umbilical cord-derived MSCs (UC-MSCs), zinc enhanced proliferation, migration, and self-renewal.[7] Lower concentrations (20 μM) of $ZnCl_2$ promoted cell growth and wound healing, while higher concentrations (250 μM and 500 μM) of $ZnCl_2$ showed concentration-dependent cytotoxicity. Zinc treatment improved wound healing, cell migration, adhesion, and colony-forming efficiency. Additionally, 20 μM of $ZnCl_2$ upregulated key genes involved in the cell cycle (*CDC20, CDK1, CCNA2, CDCA2*), cell proliferation (*TGFβ1, GDF5, HIF1α*), cell migration (*CXCR4, VEGF-A, V-CAM1*), and self-renewal (*POU5F1, SOX2, NANOG*). Protein expression of Lin28, a pluripotency marker, was also increased in cells treated with zinc. As shown above, the upregulation of key genes contributing to cell proliferation and pluripotency underscores zinc's essential role in maintaining UC-MSCs' viability and functionality, emphasizing its potential in cell therapies.

Together, these studies demonstrate the broad utility of zinc in cell proliferation and differentiation of MSCs across different sources and treatment conditions, providing a simple and effective tool for improving MSC-based cell therapy in regenerative medicine.

7.2.2 Hematopoietic Stem Cells (HSCs)

Hematopoietic stem cells (HSCs) are also multipotent adult stem cells responsible for the formation of all blood cells, including red blood cells, white blood cells, and platelets. They carry out this role through a process called hematopoiesis. HSCs are mainly found in the bone marrow and are capable of self-renewal and differentiation into various blood cell lineages. Their role is closely linked to angiogenesis, the formation of new blood vessels, and these processes often coincide during development and tissue repair.

Two recent studies using zebrafish embryos focusing on zinc transporters have revealed a previously unknown role of zinc in hematopoiesis and angiogenesis.[8,9] Zinc homeostasis is tightly regulated by two major transporter families, *SLC30* (ZnT) and *SLC39* (Zrt- and Irt-like protein, ZIP), and the zinc-binding proteins metallothionein.[3] The *SLC30* family mainly contributes to the efflux of zinc, preventing toxicity by transporting excess zinc outside the cell or into intracellular organelles. This family has ten members, numbered *SLC30A1*(ZnT1)–*SLC30A10*(Znt10). On the other hand, the *SLC39* family promotes the influx of zinc, and the cell takes up zinc from the extracellular environment. This family comprises 14 members, *SLC39A1*(ZIP1)–*SLC39A14*(ZIP14). In addition, metallothionein act as intracellular zinc storage proteins, binding to zinc and releasing it as needed to maintain the cell. This delicate balance between zinc transport and storage is essential for many biological processes, including cell signaling, growth, and differentiation. It is becoming increasingly clear that disorders of zinc metabolism involving these transporters and metallothioneins are associated with various diseases, such as Acrodermatitis enteropathica, Alzheimer's disease, and Parkinson's disease.

Xuyan He and colleagues reported, using zebrafish and mice, that *slc39a10* (ZIP10) is a significant zinc transporter in hematopoiesis.[8] Functional screening was performed using zebrafish embryos to investigate the role of zinc in embryonic hematopoiesis. The transgenic (cmyb:eGFP) zebrafish were exposed to 100 μM zinc chelator *N*,*N*,*N*,*N*-tetrakis(2-pyridylmethyl) ethylenediamine (TPEN) at 60–72 hours post fertilization (hpf) in the absence or presence of 100 μM $ZnCl_4$. cmyb+ HSCs were sorted using a flow cytometer, and mRNA levels were examined using quantitative PCR. *Slc39a10* was upregulated following TPEN treatment, and this upregulation was reduced by zinc supplementation. To investigate the phenotype in detail, morpholino-induced *slc39a10* knocking down (KD) and CRISPR/Ca9-based *slc39a10* knocked out (KO) zebrafish embryos were used, *slc39a10* conditional KO mice (*Slc39a10*$^{fl/fl}$; Vav-Cre+) expressing Cre recombinase specifically in hematopoietic cells were also used. Both *slc39a10* mutant zebrafish and *Slc39a10*-deficient mice show more severe hematopoietic impairment than animals lacking transferrin receptor 1, which supplies iron for fetal hematopoiesis by promoting the uptake of ferrous iron transferrin,[10] highlighting zinc's crucial role in early HSCs. The loss of *slc39a10* leads to zinc deficiency in fetal HSCs, causing upregulation of ROS levels, DNA damage, apoptosis, and cell cycle arrest in the G_1 phase. 100 μM $ZnSO_4$ supplementation restores colony formation in *slc39a10*-deficient mouse HSCs, and inhibiting necroptosis partially rescues hematopoiesis. These findings demonstrate that *slc39a10* protects against zinc deficiency-induced cell death, supporting its role in HSC development and its potential as a therapeutic target for anemia and zinc deficiency-related disorders.[8]

Another study identified that treating zebrafish with 1 mM $ZnSO_4$ significantly and selectively increased the mRNA expression level of the *slc39a5* gene, which encodes ZIP5 but did not affect other *slc39* family genes.[9] *Slc39a5* at the basolateral side of intestinal enterocytes removes excess zinc from the circulation. Knocking down *slc39a5* using morpholino causes red blood cells to accumulate in the caudal vein plexus, delayed venous sprouting, and reduced cardiac blood flow, resulting in cardiac ischemia. In addition, *slc39a5* KD/KO in zebrafish embryos resulted in decreased cell proliferation and delayed migration, accompanied by increased zinc levels measured with the fluorescent zinc probe FluoZin-3 AM. Moreover, high-dose zinc (1 mM) treatment in wild-type embryos impaired caudal vein plexus formation, mimicking the phenotype in *slc39a5* morphants. Finally, zinc chelation by TPEN in *slc39a5* morphants restored angiogenesis. These findings highlight the critical role of *slc39a5* in regulating zinc homeostasis to ensure appropriate endothelial sprouting and migration during venous angiogenesis. This research provides the first evidence that zinc plays a key regulatory role in vascular development, particularly through the activity of *slc39a5*.

These studies have revealed that the zinc transporters *slc39a10* and *slc39a5* play essential roles in hematopoiesis and angiogenesis. *Slc39a10* is essential for maintaining zinc levels in HSCs, preventing cell death, DNA damage, and cell cycle arrest caused by zinc deficiency. On the other hand, *slc39a5* regulates zinc homeostasis in endothelial cells, thereby supporting appropriate angiogenesis. These findings emphasize the important regulatory role of zinc in both HSC development and vascularization.

7.2.3 Pluripotent Stem Cells (PSCs)

PSCs exist in two states: naive and primed. Naive PSCs, exemplified by mouse ESCs/iPSCs, represent a more primitive state resembling the pre-implantation blastocyst. They form dome-shaped colonies and depend on leukemia inhibitory factor (LIF)/signal transducer and transcription 3 (STAT3) signaling activator. Primed PSCs, typical of human ESCs/iPSCs, correspond to a later developmental stage. They form flattened colonies and rely on fibroblast growth factors 2 (FGF2)/activin signaling. Recent research has shown that zinc affects the regulation of pluripotency and differentiation of these PSCs. The following two lower subsections explain reports on zinc in mouse ESCs (naïve type PSCs) and human ESCs (primed type PSCs).

7.2.3.1 Mouse Embryonic Stem Cells (ESCs)

One study demonstrated that zinc could transiently maintain pluripotency in mouse ESCs even under conditions typically favorable for cell differentiation, such as LIF removal, retinoic acid treatment, and embryoid bodies formation that enables non-fate-selective differentiation by forming three-dimensional spheroids using ultra-low adhesion plates.[11] Treatment with 2 μM $ZnCl_2$ increased the expression of pluripotency-related genes (*Pou5f1*, *Sox2*, and *Nanog*) at 48 hours of culture compared to the untreated group in all three differentiation conditions described above. However, when the culture time was extended to 72, 96, and 120 hours in LIF removal culture, the mouse ESCs treated with 2 μM $ZnCl_2$ began to show a tendency to differentiate, and both mRNA and protein expression levels of pluripotency markers (OCT4 and NANOG) gradually decreased. Therefore, these data suggest that zinc only temporally maintains pluripotency in mouse ESCs. Mechanistically, 2 μM $ZnCl_2$ was shown to activate the STAT3 pathway by promoting STAT3 phosphorylation, which plays an important role in an undifferentiated state. Significantly, inhibition of the STAT3 pathway using a chemical initiator AG490 neutralized the effect of $ZnCl_2$, further emphasizing zinc's role as a key regulator of STAT3 signaling in maintaining mouse ESCs. On the other hand, it was also shown that 2 μM $ZnCl_2$ does not affect the pluripotency or cell proliferation of mouse ESCs in the presence of LIF.

In 2019, three years after the above report, other researchers also identified zinc as an essential factor for mouse ESC self-renewal.[12] First, to determine zinc concentrations that are toxic for mouse ESCs, cell viability was analyzed using concentrations of $ZnCl_2$ from 40 to 240 μM. $ZnCl_2$ maintained mouse ESC viability up to 140 μM, but 160 μM or higher $ZnCl_2$ showed toxicity. Under conditions without LIF, adding zinc contributed to partially maintaining pluripotency. Specifically, when cultured for 30 days without LIF, adding 100 μM $ZnCl_2$ maintained the expression of *Pou5f1* and *Klf4*, although it did not prevent the decrease in *Nanog* expression. Zinc transporter *Slc39a7*(ZIP7) is expressed in various organelles and cytoplasmic vesicles related to zinc storage and promotes zinc release from intracellular storage, such as the endoplasmic reticulum and Golgi, into the cytoplasm. It is also widely known that the activation of *Slc39a7* is closely related to the phosphorylation of AKT,[13] a primary protein kinase regulator of the self-renewal pathway of mouse ESCs.[14] To clarify the roles of *Slc39a7* and AKT in ESCs, experiments using the PI3K/AKT inhibitor LY-294002 or *Slc39a7* KD were carried out. The study further demonstrated that zinc activates *via* *Slc39a7*(ZIP7) and AKT signaling pathways, which are known to support stem cell self-renewal. Furthermore, when 100 μM $ZnCl_2$ was combined with LIF, the effect of LIF was enhanced, and the activity of both the zinc-derived AKT and LIF-derived STAT3 pathways increased. This combination suppressed differentiation markers while promoting the expression of significant pluripotency regulators.

The findings emphasize the importance of zinc in regulating pluripotency in mouse ESCs (naïve type PSCs). Its ability to modulate essential signaling pathways, such as STAT3 and AKT, positions zinc as a promising factor for maintaining the undifferentiated state of mouse ESCs. Interestingly, low concentrations (2 μM) of $ZnCl_2$ treatment do not have an additive effect on LIF treatment,[11] while another study demonstrates that high concentrations (100 μM) of $ZnCl_2$ enhance LIF signaling.[12]

7.2.3.2 Human Embryonic Stem Cells (ESCs)

Intracellular zinc can be broadly divided into two forms: labile zinc (mobile reactive zinc), which is free or undergoes low-affinity binding to small molecules, and protein-bound zinc (Immobile zinc), which is firmly bound to zinc-related protein, serving as either a structural component or a co-factor.[3] The following two papers on the role of zinc in human ESCs also explore this specific type of zinc.

In 2010, Janet L. Wolford and colleagues used fluorescence X-ray microprobe (XFM) analysis to map the metal distribution in human ESCs at high resolution and analyze the metal content in the cells, providing evidence that the loss of pluripotency indicated by decreased OCT4 expression is directly correlated with increased zinc in the nucleus.[15] After long-term culture under standard growth conditions, human ESCs spontaneously differentiate along the colony boundaries. Examination of OCT4 signals showed a gradual increase in ploidy from the colony boundaries toward the center. On the other hand, zinc content was higher at the colony's edge and lower toward the center. In the five-day retinoic acid treatment, the OCT4 level decreased in the colony's center, opposite to the spontaneous differentiation described above. Still, the zinc content was high in the center of the colony, and the tendency for zinc to be high in areas with low OCT4 was the same as in spontaneous differentiation. Although the zinc content of 3-day differentiated endoderm cells was lower than that of undifferentiated cells, the zinc level of the cells treated with Activin increased compared to the untreated cells.

Another interesting report on zinc based on single-cell analysis of endoderm differentiation was made about human PSCs in 2018.[16] The differentiation of human PSCs into definitive endoderm (DE) is an essential first step in generating cells that generate organs such as the lung, liver, stomach, intestine, and pancreas. During this *in vitro* differentiation process, some cells fail to differentiate while maintaining the expression of pluripotency factors such as OCT4, encoded by the *POU5F1* gene, resulting in a heterogeneous population. Single-cell RNA sequencing analysis of cells harvested during the four-day differentiation process using both Activin A and CHIR99021 revealed high metallothionein gene expression in the remaining OCT4-positive cells that failed to differentiate into DE. Using XFM and multi-isotope mass spectrometry, the authors found that high intracellular zinc levels corresponded to persistent expression of OCT4 and failure to differentiate into DE. Although XFM analysis would measure total zinc content (labile and protein-bound type of zinc), fluorescent zinc indicators such as Fluozin-3 exclusively bind labile zinc. Among the four-day endoderm differentiated cells, Fluozin-3 did not reveal any zinc difference, indicating that the different zinc levels demonstrated by XFM are due to protein-bound zinc. The endoderm differentiation protocol in this report is based on Advanced RPMI 1640, which contains 3 μM of zinc. Therefore, the presence of 3 μM zinc in the differentiation medium may cause heterogeneous cell reactions that lead to the arrest of some cells during DE differentiation.

Wolford *et al.* reported an inverse correlation between OCT4 expression and zinc content in transiently differentiated cells during long-term culture under undifferentiated maintenance conditions.[15] On the other hand, Lu *et al.* reported higher zinc content in the remaining OCT4-positive cells under 4-day endoderm differentiation conditions.[16] This difference might be attributed to the varying culture conditions for maintaining undifferentiated cells and endoderm differentiation, but the details of this difference were unknown. The detailed role of zinc in maintaining the undifferentiated state of human PSCs has yet to be elucidated, and its role in the differentiation of human PSCs into more mature cells, such as the pancreas, has been largely unexplored.

7.3 Role of Zinc in Methionine Metabolism and Pluripotency in Human PSCs

PSCs have unique characteristics that set them apart from somatic cells. These cells can replicate indefinitely, are pluripotent, and differentiate into all cell types. They also possess a distinct transcriptional circuitry that helps maintain their pluripotent state. PSCs are in a specific epigenetic state that allows for rapid cell fate determination. Additionally, histone methylation plays a dynamic role in regulating the pluripotency and plasticity of PSCs. In this section, we would like to outline the role of zinc in human PSCs based on the results of our research. During our research into the role of nutritional factors in PSCs, we first discovered that methionine, a type of essential amino acid, is critical for the survival of human PSCs.[17,18] Furthermore, in our research aimed at elucidating the molecular mechanisms that promote differentiation following methionine deprivation, we discovered an interaction between methionine metabolism and zinc dynamics. Our findings revealed that zinc plays an important role in maintaining the pluripotency of human PSCs.[19]

In the following subsections, we outline the role of methionine in human PSCs (Section 7.3.1) and then explore how methionine regulates zinc dynamics (Section 7.3.2). We also introduce the function of zinc in PSCs (Section 7.3.3) and elucidate how zinc regulates methionine metabolism (Section 7.3.4). Lastly, we will explain how zinc is involved in the differentiation into the endoderm and pancreatic β cells (Section 7.3.5).

7.3.1 Methionine Contributes to Pluripotency in Human PSCs

There have been many studies on the role of amino acid metabolism in PSCs, and it was shown in 2009 that threonine catabolism is essential in mouse ESCs.[20] Mouse ESCs express high levels of threonine dehydrogenase (Tdh) and Tdh catabolism threonine to glycine, which supports the folate cycle through the glycine cleavage system (GCS) (see Figure 7.1). When culture medium deprived of any of the 20 amino acids was applied to mouse ESCs, they were found to be highly dependent on one essential amino acid: threonine. Shyh-Chang, Ng, and his colleagues also reported that depletion of threonine from the culture medium or *Tdh* KD from mouse PSCs suppresses folate metabolism derived from threonine, inhibiting the methionine cycle and resulting in a decrease in *S*-adenosylmethionine (SAM) accumulation. This leads to a decline in trimethylation of histone H3 lysine 4 (H3K4me3), a reduction in proliferation, and an increase in differentiation.[21]

We then examined the importance of methionine metabolism and revealed that it is crucial in maintaining human PSCs.[17] Human PSCs require SAM to maintain their pluripotency, like mouse PSCs. In human PSCs, the enzyme TDH, which contributes to the supply of methyl groups to the folate cycle and methionine metabolic cycle in mice, is a non-functional gene. In human PSCs, methionine is the source of SAM (see Figure 7.1). Methionine deprivation decreases intracellular SAM, making human PSCs more susceptible to differentiation. Furthermore, since undifferentiated iPSCs

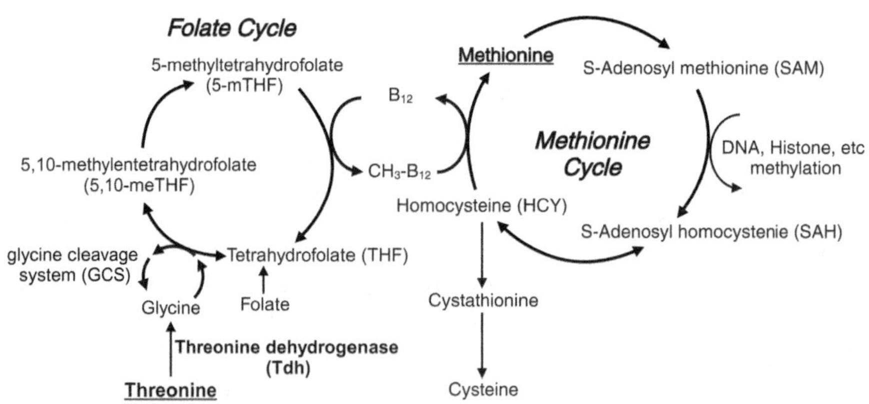

Figure 7.1 One-carbon metabolism includes the folate cycle, the methionine cycle, and the glycine cleavage system. *S*-Adenosylmethionine (SAM), which functions as a methyl donor when DNA and histones are methylated, is supplied by one-carbon metabolism. In mouse PSCs, threonine is used as a methyl group donor, while in human PSC cells, methionine is used. Created in BioRender. Shiraki, N. (2024) BioRender.com/a69q258.

consume methionine at a much higher rate than differentiated cells, we discovered that by culturing them in a methionine-free medium for 48 hours, while undifferentiated cells and endodermal cells are mixed during the endoderm differentiation process, it is possible to selectively eliminate the remaining undifferentiated cells and promote hepatic cell differentiation.

Our research then delved into the impact of methionine on PSCs in detail, using RNA-sequencing (RNAseq) and ChIP-sequencing to evaluate the epigenetic modifications.[18] We found that short-term (5 hours) methionine deprivation in PSCs causes rapid metabolic changes, including a decrease in SAM, followed by reversible epigenetic modifications. Notably, H3K4me3 was markedly decreased compared to other histone modifications. Methionine deficiency significantly reduced the expression of pluripotency genes such as *NANOG* and genes involved in the TGFβ and cholesterol biosynthesis processes, correlated with the downregulation of H3K4me3. Furthermore, during endoderm differentiation, methionine deprivation increases the expression of various endoderm-specific genes, such as *GATA6*. This causes the loss of H3K27me3 in many endodermal genes, switching from a bivalent (H3K4me3+/(H3K27me3+) to a monovalent (H3K4me3) state.

7.3.2 Methionine Regulates Zinc Dynamics in Human PSCs

To search for genes whose expression changes due to short-term methionine deprivation, we used human PSCs cultured for 5 hours in a complete medium or methionine-deprived medium for gene profiling.[19] We found that the expression of *SLC30A1*, which encodes the zinc efflux transporter ZNT1 present in undifferentiated iPSCs, increased significantly. Furthermore, this increase was more pronounced with methionine restriction than other essential amino acid deprivation and was not observed in other zinc transporters.

Apart from the above methionine-related research, we also previously used high-performance liquid chromatography (HPLC) coupled to inductively coupled plasma mass spectrometry (HPLC–ICP–MS) to examine the intracellular labile and protein-bound zinc content during cell differentiation from human iPSCs to pancreatic cells.[22] Undifferentiated iPSCs contained the highest content of protein-bound zinc, which decreased in endoderm and pancreatic endocrine cells. From the results of this HPLC–ICP–MS analysis, we confirmed the contribution of zinc to the pluripotency of human PSCs when we determined that methionine metabolism regulates zinc dynamics. So, we measured the concentrations of labile and protein-bound zinc, copper, and iron in human iPSCs cultured for 5 hours in a methionine-deprived medium. As a result, we found that only the protein-bound zinc decreased by 5-hour methionine deprivation. In addition, human cervical cancer cells (HeLa), human hepatoma cells (HepG2), and iPSC-derived definitive endoderm cells did not exhibit a decrease in protein-bound zinc triggered by methionine restriction. These findings suggest that methionine

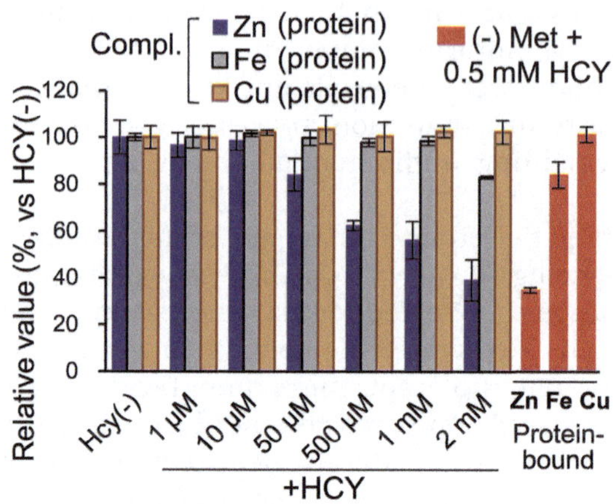

Figure 7.2 Concentration-dependent homocysteine treatment reduced protein-bound zinc. Methionine-deprived human iPSC lysates required a lower homocysteine to reduce protein-bound Zn. Adapted from ref. 19 with permission from Elsevier, Copyright 2022.

plays a role in regulating zinc dynamics, specifically in the undifferentiated human PSCs.

The following is a discussion of the relationship between methionine metabolism and zinc. Proteins bind to zinc through the sulfur in the sulfhydryl group of cysteine. Homocysteine (HCY) contains a sulfhydryl group and can replace zinc from its binding proteins.[23] The excretion of HCY into the extracellular space stops when methionine is deprived of the medium,[17] and we hypothesized that a temporary increase in HCY occurs in human PSCs and that this competitively displaces zinc from binding proteins under methionine deprivation. As expected, HCY added excess to human iPSC-extracts, reducing protein-bound zinc (see Figure 7.2). These results suggest that methionine metabolism regulates the level of protein-bound zinc *via* HCY.

7.3.3 Zinc Contributes to Maintaining the Pluripotency of Human PSCs

To investigate the role of zinc in human PSCs, we first made a zinc-free medium by treating the complete medium with Chelex-100, a chelating resin with a high affinity for metal ions. Before and after Chelex-100 treatment, the amount of zinc, copper, iron, magnesium, and calcium in the medium was measured, and the amount removed by Chelex-100 was supplemented (except zinc) to make a zinc-free medium. Although this chelating method is useful for eliminating zinc from the medium, some inconveniences, such as measuring the metal concentration when preparing the medium, required establishing a method for preparing a culture medium without zinc.

Only the F12 medium contains a high amount of zinc, approximately 3 μM, among the various basal media. If F12 itself or a medium containing

F12 (such as IMDM/F12 and DMEM/F12) is not used, there will possibly be no contribution of zinc from the basic culture medium.[24] Zinc is also in fetal bovine serum (FBS), commonly used in cell culture experiments. One report states that zinc is present in bovine serum at a concentration of 0.8–0.9 $\mu g\,mL^{-1}$ (12–14 μM).[11] When using 10% FBS, the concentration of zinc in the standard culture medium is thought to be approximately 1.2–1.4 μM. In serum-free cultures, insulin is used as a growth signal, but insulin solutions typically contain zinc, as zinc plays a crucial role in insulin hexamer formation and stability.[25] If researchers use insulin or supplements containing insulin (such as Insulin-Transferrin-Selenium: ITS), it is essential to check the zinc content beforehand. If there is a risk of zinc contamination, consider using IGF (insulin-like growth factors) instead of insulin. In the case of experiments to understand the role of zinc in cell cultures with undisclosed supplement contents and those concentrations, it is also crucial to measure the zinc in the culture environment.

Our research has led to the development of a novel custom medium for culturing human iPSCs, a significant advancement in the field. This medium allows precise control of zinc concentration. We made a zinc-free basal medium, substituting insulin with IGF1 as a growth factor, and cultured human iPSCs in a medium containing several zinc concentrations (0, 0.1, 0.3, 0.5, 1, 3, 5, and 10 μM). Intracellular zinc was reduced by culturing with zinc-free media ((−)Zn) for three days (see Figure 7.3A). We found that cell proliferation is directly influenced by zinc concentration, with just 3 μM $ZnSO_4$ sufficient to maintain undifferentiated human iPSCs, and zinc at concentrations below ten μM did not exhibit cytotoxicity (see Figure 7.3B–D). Notably, culturing in a zinc-free medium (Zn0) for several days led to a decrease in the expression of pluripotency-related genes, such as *NANOG*, and a significant increase in the expression of differentiation marker genes, such as *GATA4* (see Figure 7.3E and F). The above results confirmed that zinc is essential for the proliferation and maintenance of undifferentiated human iPSCs and that removing zinc results in proliferation suppression and loss of pluripotency.

Interestingly, alkaline phosphatase (ALP), a zinc-dependent enzyme that requires zinc for its catalytic activity, was significantly inhibited in the zinc-free culture (see Figure 7.3G). ALP is widely used as a marker for stem cells, particularly in identifying and characterizing PSCs. ALP activity is high in undifferentiated states and decreases upon differentiation, making it a valuable tool for assessing stemness and monitoring cellular differentiation processes. Our research underscores the need for caution when evaluating the impact of intracellular zinc depletion on stem cell properties. While ALP is a helpful marker indicating undifferentiated stem cells, its activity depends on zinc. This dependency introduces potential issues when studying the effects of zinc depletion in culture medium, chelation treatment, and inhibition of zinc transport by KD/KO of transporters. Our research results highlight the complexity of this field of study.

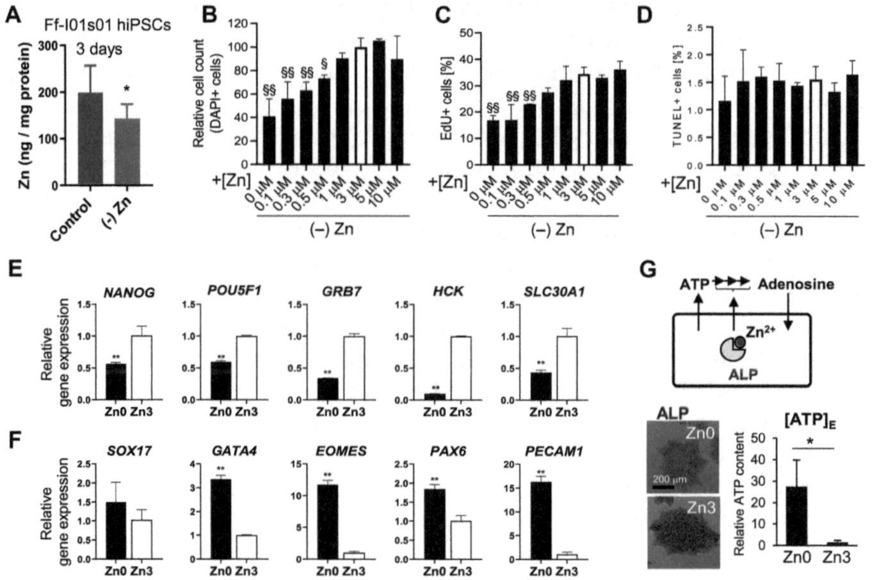

Figure 7.3 Culturing human iPSCs under zinc-free conditions decreased proliferation, pluripotency, and increased differentiation marker expression, as when human iPSCs were cultured under methionine-deprived conditions. Human iPSCs were cultured under a custom-developed zinc-free medium (IGF1-based medium). (A) A decreased protein-bound zinc level after culturing in a zinc-deprived medium. (B) Cell number. (C) 5-Ethynyl-2′-deoxyuridine (EdU) incorporation assay. (D) Apoptosis (TUNEL assay). (E) Gene expressions of pluripotency and proliferation marker genes and SLC30A1. (F) Differentiation marker gene expressions. (G) Alkaline phosphatase staining (lower left) and extracellular ATP contents (lower right). Data are expressed as means ± SDs, $N = 3$. Differences between groups were analyzed by one-way ANOVA Dunnett's multiple comparison test. Adapted from ref. 19 with permission from Elsevier, Copyright 2022.

7.3.4 Zinc Regulates Methionine Metabolism in Human PSCs

Having investigated the effect of methionine metabolism on zinc dynamics, we next evaluated the effect of zinc restriction ((−)Zn) on methionine metabolism. RNAseq analysis revealed that zinc deficiency downregulates *DNMT3B*, suggesting that human iPSCs sense extracellular zinc levels (see Figure 7.4).

Next, we investigated the changes in intracellular and medium-based methionine-related metabolites between zinc-free ((−)Zn) and 3 μM zinc treatments (control). In zinc-free conditions, intracellular adenosine levels were significantly reduced due to decreased adenosine nucleotide metabolism caused by decreased zinc-dependent ALP activity (see Figure 7.4). In addition, 5-methyltetrahydrofolate homocysteine methyltransferase (MTR) are also zinc-dependent enzymes that catalyze the reaction that converts Hcy

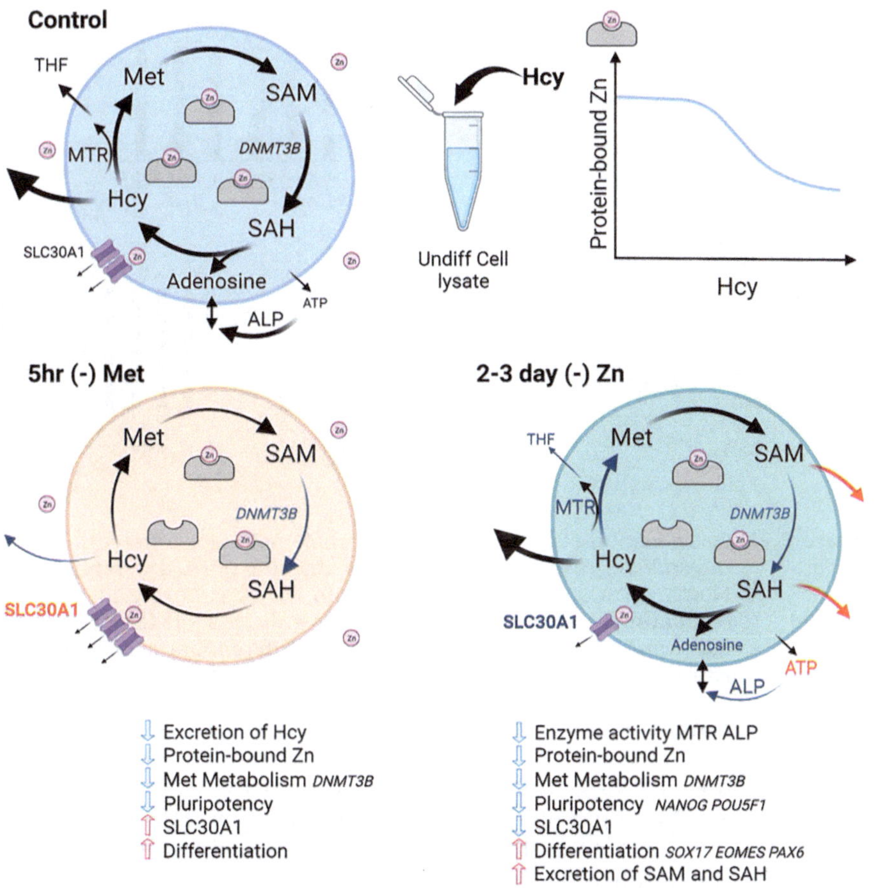

Figure 7.4 Methionine or zinc deprivation regulates cell metabolism and gene expression in human iPSCs. Adapted from ref. 19 with permission from Elsevier, Copyright 2022.

to methionine. In the absence of zinc, a decrease in tetrahydrofolate (THF) in the culture medium was observed due to inhibition of MTR activity, indicating that Hcy was accumulating. Human PSCs take up methionine in the culture medium, metabolize it to SAM and SAH, and maintain the amount of these metabolites in the cell. These SAM and SAH were significantly excreted into the extracellular space under (−)Zn conditions. These results suggest that intracellular Hcy accumulates in zinc-free conditions, causing a "traffic jam" (decrease in flow) in methionine metabolism. The above results suggest that the metabolism of zinc and methionine is interdependent.

7.3.5 Zinc in Endoderm and Pancreatic Differentiation

The role of zinc during endoderm differentiation was investigated using various human iPSCs. Endoderm differentiation under zinc-free conditions

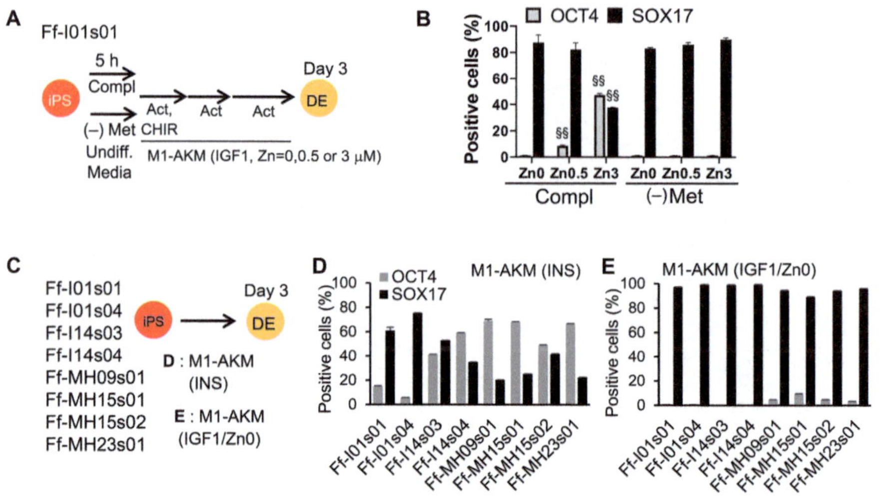

Figure 7.5 Optimization of Zn concentration supplements for differentiation into the endoderm. (A) and (C) Schematic drawing of DE differentiation (A). Human iPSCs were differentiated into DE by culturing in a medium containing 0, 0.5, or 3 μM zinc, with or without short-term methionine deprivation pretreatment. (D) and (E) Differentiation into DE under zinc-containing culture medium (Insulin base media) (D) or zinc-free culture medium (IGF1 base media) (E). Adapted from ref. 19 with permission from Elsevier, Copyright 2022.

resulted in low OCT3/4– and high SOX17+ cells being detected. Zinc concentrations of 0.5 or 3 μM increased OCT3/4 positivity and decreased SOX17 positivity (see Figure 7.5A and B). There was a variation between cell lines in the standard insulin-based culture medium with enough zinc concentration. A considerable number of OCT4-positive undifferentiated cells remained in all cell lines. On the other hand, when cultured in an IGF1-based zinc-free medium during differentiation, the OCT4-positive cells almost disappeared, and many cells differentiated into SOX17-positive definitive endoderm cells (see Figure 7.5C–E). These results suggest that endoderm differentiation is promoted in human iPSCs cultured in a zinc-free medium.

Furthermore, to investigate the role of zinc in pancreatic differentiation, we compared the effects of 3 μM zinc treatment with the differentiation of pancreatic progenitor cells into endocrine progenitor cells. The results revealed that cells cultured in a zinc-containing medium at this stage exhibited higher differentiation into PDX1+NKX6.1+ endocrine progenitor cells compared to cells cultured without zinc. Interestingly, we confirmed that IGF1 does not sufficiently support cell proliferation during the differentiation process into endocrine precursor cells. To summarize, we determined that the best conditions for endodermal differentiation were a zinc-free medium supplemented with IGF1. In contrast, the pancreatic differentiation was optimal in an insulin-based medium containing 3 μM zinc. Finally, we generated functional pancreatic β cells with glucose-responsive insulin secretion using a short 5-hour methionine deprival pretreatment, following zinc-deficient culture medium during the endodermal differentiation stage.[26]

7.4 Conclusion

Zinc is essential in stem cell biology, affecting pluripotency, self-renewal, and differentiation in various stem cell types. In mesenchymal stem cells, zinc enhances the potential for proliferation and differentiation. It is also essential for the proliferation of hematopoietic stem cells and angiogenesis. In pluripotent stem cells, zinc contributes to pluripotency, and the best condition for initial endoderm differentiation is the absence of zinc. Furthermore, functional pancreatic beta cells were generated by combining a methionine-free and a zinc-free culture medium. Zinc contributes to the maintenance of many types of stem cells, so by accurately utilizing these properties, we can contribute to research on regenerative medicine.

Abbreviations

AD-MSCs	Adipose-derived mesenchymal stem cells
ALP	Alkaline phosphatase
BM-MSCs	Bone marrow-derived mesenchymal stem cells
DE	Definitive endoderm
ERK1/2	Extracellular signal-regulated kinases 1/2
ESCs	Embryonic stem cells
FGF2	Fibroblast growth factor 2
FBS	Fetal bovine serum
H3K4me3	Trimethylated histone H3 lysine 4
H3K27me3	Trimethylated histone H3 lysine 27
HSCs	Hematopoietic stem cells
ICM	Inner cell mass
IGF	Insulin-like growth factors
ITS	Insulin-transferrin-selenium
iPSCs	Induced pluripotent stem cells
KO	Knocked out
KD	Knocked down
LIF	Leukemia inhibitory factor
MSCs	Mesenchymal stem cells
PSCs	Pluripotent stem cells
RNAseq	RNA sequencing
ROS	Reactive oxygen species
SAM	*S*-Adenosylmethionine
SAH	*S*-Adenosylhomocysteine
SLC30A	Solute carrier family 30 member A (zinc transporter family)

SLC39A	Solute carrier family 39 member A (zinc transporter family)
STAT3	Signal transducer and activator of transcription 3
UC-MSCs	Umbilical cord-derived mesenchymal stem cells

Acknowledgements

This work was supported by Grants-in-Aid from the Ministry of Education, Culture, Sports, Science and Technology (MEXT) Japan (#18H02154 and 23H03304 to NS), and by the Japanese Insulin Dependent Diabetes Mellitus (IDDM) Network Foundation.

References

1. L. K. Wooldridge, M. E. Nardi and A. D. Ealy, *J. Anim. Sci.*, 2019, **97**, 4946–4950.
2. H. Timourian, *J. Exp. Zool.*, 1968, **169**, 121–132.
3. B. Chen, P. Yu, W. N. Chan, F. Xie, Y. Zhang and L. Liang, *et al.*, *Signal Transduction Targeted Ther.*, 2024, **9**, 6.
4. M. Nemer, E. C. Travaglini, E. Rondinelli and J. D'Alonzo, *Dev. Biol.*, 1984, **102**, 471–482.
5. D. Zhang, Y. Li, T. Zhu, F. Zhang, Z. Yang and D. Miao, *Cytotherapy*, 2011, **13**, 156–164.
6. M. Y. Moon, H. J. Kim, B. Y. Choi, M. Sohn, T. N. Chung and S. W. Suh, *Stem Cells Int.*, 2018, **2018**, 5736535.
7. I. Sahibdad, S. Khalid, G. R. Chaudhry, A. Salim, S. Begum and I. Khan, *World J. Stem Cells*, 2023, **15**, 751–767.
8. X. He, C. Ge, J. Xia, Z. Xia, L. Zhao and S. Huang, *et al.*, *Adv. Sci.*, 2023, **10**, e2205345.
9. Z. Xia, X. Bi, J. Lian, W. Dai, X. He, L. Zhao, J. Min and F. Wang, *Open Biol.*, 2020, **10**, 200281.
10. S. Wang, X. He, Q. Wu, L. Jiang, L. Chen and Y. Yu, *et al.*, *Haematologica*, 2020, **105**, 2071–2082.
11. J. Hu, Z. Yang, J. Wang, J. Yu, J. Guo, S. Liu, C. Qian, L. Song, Y. Wu and J. Cheng, *PLoS One*, 2016, **11**, e0148994.
12. H. Mnatsakanyan, I. S. R. Sabater, M. Salmeron-Sanchez and P. Rico, *Front. Cell Dev. Biol.*, 2019, 7, 180.
13. K. M. Taylor, P. Vichova, N. Jordan, S. Hiscox, R. Hendley and R. I. Nicholson, *Endocrinology*, 2008, **149**, 4912–4920.
14. S. Watanabe, H. Umehara, K. Murayama, M. Okabe, T. Kimura and T. Nakano, *Oncogene*, 2006, **25**, 2697–2707.
15. J. L. Wolford, Y. Chishti, Q. Jin, J. Ward, L. Chen, S. Vogt and L. Finney, *PLoS One*, 2010, **5**, e12308.
16. J. Lu, A. Baccei, E. Lummertz da Rocha, C. Guillermier, S. McManus, L. A. Finney, C. Zhang, M. L. Steinhauser, H. Li and P. H. Lerou, *Stem Cell Res.*, 2018, **28**, 48–55.

17. N. Shiraki, Y. Shiraki, T. Tsuyama, F. Obata, M. Miura, G. Nagae, H. Aburatani, K. Kume, F. Endo and S. Kume, *Cell Metab.*, 2014, **19**, 780–794.
18. H. Ozawa, A. Kambe, K. Hibi, S. Murakami, A. Oikawa and T. Handa, *et al.*, *Stem Cells*, 2023, **41**, 271–286.
19. E. Z. Sim, T. Enomoto, N. Shiraki, N. Furuta, S. Kashio and T. Kambe, *et al.*, *Cell Rep.*, 2022, **40**, 111120.
20. J. Wang, P. Alexander, L. Wu, R. Hammer, O. Cleaver and S. L. McKnight, *Science*, 2009, **325**, 435–439.
21. N. Shyh-Chang, J. W. Locasale, C. A. Lyssiotis, Y. Zheng, R. Y. Teo and S. Ratanasirintrawoot, *et al.*, *Science*, 2013, **339**, 222–226.
22. A. Arakawa, N. Shiraki, T. Tsuyama, S. Kume, D. Iwahata and N. Yamada, *J. Anal. Bioanal. Tech.*, 2016, **7**, 295.
23. T. Kambe, *Biochemistry*, 2020, **59**, 74–79.
24. D. Jayme, T. Watanabe and T. Shimada, *Cytotechnology*, 1997, **23**, 95–101.
25. K. Pounot, G. W. Grime, A. Longo, M. Zamponi, D. Noferini and V. Cristiglio, *et al.*, *Biophys. J.*, 2021, **120**, 886–898.
26. E. Z. Sim, T. Enomoto, N. Shiraki and S. Kume, *STAR Protoc.*, 2023, **4**, 102183.

CHAPTER 8

Importance of Zinc in Epigenetic Control

TAKAO FUJISAWA*[a,b] AND HIDENORI ICHIJO[b]

[a] Laboratory of Cell Signaling, Graduate School of Pharmaceutical Sciences, The University of Tokyo, Tokyo, Japan; [b] Cell Signaling and Stress Responses Laboratory, Advanced Research Departments, Institute of Science Tokyo, Tokyo, Japan
*Email: fujisawa@mol.f.u-tokyo.ac.jp

8.1 Introduction

Zinc, an indispensable micronutrient, plays a pivotal role in diverse biological functions.[1] Its role extends to the regulation of gene expression, where it functions both as a structural component in various proteins and as a key modulator of molecular pathways governing chromatin architecture and transcriptional activity. In recent years, increasing attention has been directed toward the impact of zinc on epigenetic mechanisms—processes that influence gene expression without altering the DNA sequence itself.

Epigenetic regulation, including modifications such as DNA methylation, histone modifications and non-coding RNA-mediated regulation, is crucial for maintaining cellular homeostasis and responding to environmental changes.[2] The involvement of zinc in these processes is increasingly recognized, as it can affect both the establishment and maintenance of epigenetic marks. Zinc deficiency in particular, has been associated with dysregulation of these epigenetic mechanisms, contributing to the development and progression of various diseases.

Understanding how zinc modulates epigenetic pathways provides valuable insights into its broader implications for medical research. In this chapter,

Coordination Chemistry Perspectives No. 2
Zinc in Biology: Molecular Structures, Cellular Processes and Living Systems
Edited by Hiroyuki Yasui and Taiho Kambe

Published by the Royal Society of Chemistry, www.rsc.org

we explore the current knowledge regarding the involvement of zinc in gene regulation, with a focus on its impact on epigenetic modifications.

8.2 DNA Methylation

DNA methylation is a versatile and essential epigenetic mechanism that plays fundamental roles in gene regulation. DNA methylation typically occurs at cytosine residues within CpG dinucleotides.[3] The addition of methyl groups is facilitated by DNA methyltransferases, which are enzymes responsible for transferring methyl groups to DNA. Conversely, DNA demethylation, the process of removing these methyl groups, is carried out by DNA demethylases, which help reverse methylation, thereby allowing gene expression to be activated when necessary. Methylated DNA represses gene transcription by obstructing the binding of transcriptional machinery and recruiting proteins that inhibit gene expression.

8.2.1 DNA Methyltransferases

DNA methyltransferases (DNMTs) are enzymes that catalyze the addition of methyl groups to DNA, primarily at cytosine bases within CpG dinucleotides.[4] There are several types of DNMTs in mammals, including DNMT1, DNMT3A, DNMT3B, DNMT3L and DNMT3C. DNMTs maintain DNA methylation patterns during cell division and contribute to the maintenance of epigenetic modifications that can influence gene expression and genomic stability.

8.2.1.1 CXXC Domain in DNMT1

DNMT1 catalyzes the transfer of methyl groups to CpG sites in DNA, ensuring the faithful replication of epigenetic patterns. DNMT1 contains several zinc-binding domains, including the CXXC domain, a zinc-binding motif within the Replication Foci Targeting Sequence (RFTS) domain, and additional zinc-binding sites in the catalytic domain, all of which play critical roles in maintaining the enzyme's structure and function.[5] In mouse DNMT1, the CXXC domain contains eight cysteine residues arranged in two clusters—one with six cysteines and the other with two cysteines.[6,7] These clusters coordinate two zinc ions, which are essential for maintaining the domain's structural integrity. Notably, structural analyses have revealed that the CXXC domain plays an autoinhibitory role by binding to unmethylated CpG sites, while zinc ions in other domains are required for the activity.[7] When bound to unmethylated DNA, the CXXC domain positions an autoinhibitory linker between the DNA and DNMT1's active site, thereby preventing *de novo* methylation. Recent studies have reported that zinc signaling from the zinc transporter ZIP13 regulates the activity of DNMT1.[8]

8.2.1.2 ADD Domain in DNMT3L

DNMT3L is a catalytically inactive regulatory factor of DNMTs. DNMT3L plays a supportive role by enhancing the activity of other DNMTs, specifically DNMT3A and DNMT3B, which are responsible for *de novo* DNA methylation.[9] DNMT3L contains the ATRX–DNMT3–DNMT3L (ADD) domain, which is a zinc-binding domain. The ADD domain is a cysteine-rich region that consists of a C2C2-type zinc finger and a closely located domain of an imperfect plant homeodomain (PHD)-type zinc finger. Crystallographic studies of human DNMT3L showed that the tail of histone H3 bound to the ADD domain.[10] Upon binding to the unmethylated histone tail, DNMT3L facilitates the recruitment or activation of DNMT3A, a DNA methyltransferase, and this recruitment is essential for catalyzing the addition of methyl groups to DNA.

8.2.1.3 Others

Zinc plays a crucial role in regulating DNMT1 expression.[11] Zinc treatment enhances DNMT1 transcription by increasing the binding of MTF-1 to the DNMT1 promoter. This process is counteracted by the nuclear receptor small heterodimer partner (SHP), which represses MTF-1 activity. When zinc levels are sufficient, MTF-1 promotes DNMT1 expression, whereas SHP inhibits it. In SHP-knockout models, DNMT1 expression is significantly upregulated, indicating that SHP serves as a repressor in the presence of zinc. Furthermore, another study has shown that offspring of mice overfed with zinc during pregnancy exhibit significantly higher levels of fetal liver DNA methylation compared to offspring of mice fed a zinc-adequate diet.[12] Thus, maternal zinc overfeeding may influence DNA methylation patterns in offspring, potentially through increased DNMT1 expression.

8.2.2 DNA Demethylases

DNA demethylation is a critical regulatory mechanism that governs gene expression, serving as a counterpart to DNA methylation. The process involves the active removal of methyl groups from cytosine residues in DNA, primarily mediated by the ten-eleven translocation (TET) family of enzymes. These include TET1, TET2, and TET3, which catalyze the oxidation of 5-methylcytosine (5mC) into 5-hydroxymethylcytosine (5hmC), and further oxidative derivatives.[13] Subsequent steps in the DNA repair pathways restore unmodified cytosine, completing the demethylation process.

8.2.2.1 CXXC Domain in TET

TET enzymes catalyze the oxidation of 5mC to 5hmC, which serves as a crucial intermediate in the active DNA demethylation process. Among the TET family proteins, TET1 and TET3 contain a CXXC domain, which can bind with DNA directly.[14] *In vitro* binding assay showed that TET1 exhibits a preference for binding unmethylated CpG sites over methylated ones.[15] It also exhibits a strong affinity for CpG-enriched promoters, as evidenced by chromatin immunoprecipitation sequencing (ChIP-seq) studies, to regulate the gene expression in mouse embryonic stem cells.[16,17] Similarly, the CXXC domain in TET3 facilitates its binding to regions in CpG rich sites, with a significant portion of these bindings occurring at gene promoters to regulate key genes involved in eye and neural development in Xenopus.[18]

8.2.3 DNA Methylation Readers

Proteins that recognize methylated DNA play essential roles in gene regulation by recruiting chromatin-modifying enzymes, maintaining DNA methylation patterns, and participating in processes like genomic imprinting.[19] Multiple types of proteins, including Methyl-CpG Binding Domain (MBD) proteins, SET and RING Associated (SRA) domain proteins, and zinc finger proteins, can recognize methylated DNA, although reports on the involvement of zinc are limited.

8.2.3.1 Zinc Finger Domain in ZFP57

Zinc finger protein 57 (ZFP57) is a transcription factor with structural features essential for DNA binding and methylation recognition. It contains several zinc finger motifs, two of which are directly involved in recognizing DNA sequences. The structure of its DNA-binding domain, in complex with methylated DNA, has been resolved, providing key insights into its recognition mechanism.[20] The first zinc finger interacts with the 5′ segment (TGC) of the recognition sequence, while the second binds to the methylated 3′ segment (CGC). ZFP57 plays a vital role in regulating genomic imprinting by preserving DNA methylation imprints across various regions in early mouse embryos and embryonic stem cells.[21] It achieves this by recruiting a co-repressor complex composed of KAP1 (also known as TRIM28 or TIF1β), DNMTs, and UHRF1. This complex ensures the maintenance of DNA methylation and repressive histone modifications at specific sites. Mutations in the zinc finger domains of human ZFP57 have been linked to cases of transient neonatal diabetes, which results from hypomethylation at multiple imprinted loci, highlighting the critical role of zinc in ZFP57's function.[20]

8.2.3.2 Zinc Finger Domain in Kaiso

Kaiso, a member of the BTB/POZ domain protein family, is essential for maintaining DNA methylation homeostasis.[22] Kaiso contains three C2H2 zinc finger domains that are crucial for its DNA-binding function. All three zinc finger domains, along with adjacent N-terminal and C-terminal regions, are required for optimal DNA binding.[23] The zinc finger domains of Kaiso exhibit remarkable structural flexibility, enabling the recognition of both methylated DNA and specific DNA sequences through a combination of direct

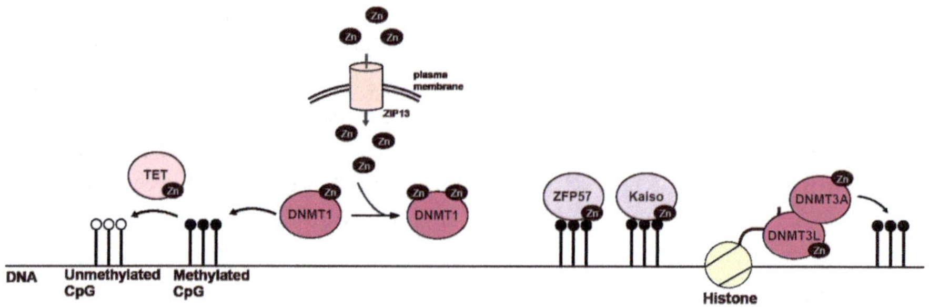

Figure 8.1 Roles of zinc in DNA methylation.

base interactions and conformational adjustments. This dual specificity allows Kaiso to function both as a methyl-CpG binding protein and a sequence-specific transcriptional regulator. Studies utilizing Kaiso-deficient cells have shown that the absence of Kaiso leads to mild but statistically significant genome-wide hypermethylation.[22,24] Additionally, Kaiso has been shown to co-immunoprecipitate with DNMT3a and DNMT3b, underscoring its critical role in regulating DNA methylation homeostasis (Figure 8.1).

8.3 Histone Methylation

Histone methylation is a key epigenetic modification that regulates chromatin structure and gene expression.[25] This process involves the addition of methyl groups to lysine or arginine residues on histone proteins, catalyzed by histone methyltransferases. It is reversible through the action of demethylases, allowing dynamic control of gene regulation. Depending on the specific site and context of the methylation, this modification can either activate or repress gene transcription. Histone methylation patterns are recognized by reader proteins, such as the PHD domain, which mediate downstream effects on chromatin organization.

8.3.1 Histone Methyltransferases

Histone methyltransferases are enzymes that catalyze the addition of methyl groups to specific amino acids on histone tails, thereby modulating gene expression. HMTs are generally categorized into two main groups: lysine methyltransferases and arginine methyltransferases.[26] Lysine methyltransferases include enzymes with a SET domain and those without it. In contrast, arginine methyltransferases (PRMTs) specifically methylate arginine residues. Each type of HMT exhibits distinct specificity for different histone residues, influencing diverse epigenetic outcomes.

8.3.1.1 Zinc Finger Domain in CXXC1

The COMPlex ASsociated with SET1 (COMPASS) is a histone H3 lysine 4 (H3K4) methyltransferase complex crucial for gene transcription. While yeast contains a single COMPASS complex, mammals have evolved at least six distinct versions, each defined by different catalytic subunits.[27] One of the key subunits, CXXC1, is essential for COMPASS activity and contains a monovalent zinc finger domain.[28] This zinc finger domain in CXXC1 specifically interacts with the RbBP5 subunit of COMPASS. This interaction is vital for stimulating the complex's methyltransferase activity, facilitating efficient H3K4 methylation. Disruption of this interaction significantly reduces the complex's enzymatic activity. Another structural analysis revealed that the zinc finger domain in CXXC1 binds to unmethylated CpG islands.[29] This mechanism would contribute to facilitating the recruitment of the COMPASS complex, leading to H3K4 methylation at specific genomic loci. Additionally, a study demonstrated that a deletion mutant of the zinc finger domain in CXXC1 was unable to methylate H3K4, highlighting the critical role of zinc in CXXC1's function as a histone methyltransferase.[30]

8.3.1.2 Pre-SET Domain in SUV39H1

SUV39H1 is a histone methyltransferase that catalyzes the trimethylation of histone H3 at lysine 9 (H3K9me3).[31] This modification is important for the formation of heterochromatin and gene repression. SUV39H1 contains a SET domain, which is responsible for its methyltransferase activity. The SET domain is flanked by two additional domains: the pre-SET domain at the N-terminal and the post-SET domain at the C-terminal.[31] The pre-SET domain contains nine conserved cysteine residues that coordinate three zinc ions, forming a triangular cluster. This structure is critical for stabilizing the overall protein architecture. Mutation in the zinc-binding cysteine residues leads to the impaired catalytic function, suggesting the importance of zinc for SUV39H1's activity.[32]

8.3.2 Histone Demethylases

Histone demethylation, a critical epigenetic modification, plays a pivotal role in regulating gene expression, influencing a wide range of physiological and pathological processes.[33] This process involves the removal of methyl groups from specific lysine residues on histone proteins, leading to alterations in chromatin structure and accessibility to transcription factors. These changes, in turn, modulate gene expression by facilitating or inhibiting the binding of transcriptional machinery to DNA. This mechanism is primarily mediated by two major classes of demethylases: the lysine-specific demethylases (KDMs) and the Jumonji C (JmjC) domain-containing demethylases.[34] These enzymes remove methyl groups from specific lysine residues on histones, thereby altering chromatin structure and accessibility to transcription factors, which in turn regulates gene expression.

8.3.2.1 Zinc Finger Domain in KMD1B

KMD1B, also referred to as LSD2, is a flavin-dependent histone demethylase that exhibits structural homology with KDM1A (LSD1). Both enzymes catalyze the demethylation of mono- and di-methylated lysine residues within histone tails, particularly H3K4 and H3K9.[34] However, KMD1B is distinguished by the presence of a zinc finger domain at its N-terminus. KMD1B possesses two distinct zinc finger domains within its N-terminal region. These domains are characterized as a C4H2C2-type zinc finger and a CW-type zinc finger, which serve multiple functions.[35] They are indispensable for the enzyme's demethylase activity, as evidenced by an *in vitro* demethylase assay using zinc finger mutants.[36] Moreover, the zinc finger domains exhibit an unexpected role in binding flavin adenine dinucleotide (FAD), a cofactor typically associated with the oxidase domain in other proteins. Thus, the zinc finger domain of KMD1B is a unique structural feature that plays critical roles in the enzyme's demethylase activity, cofactor binding, and overall regulation.

8.3.2.2 Zinc Finger Domain in KDM2A and KDM2B

KDM2A and KDM2B are homologous lysine demethylases that share a conserved N-terminal JmjC domain.[34] This domain catalyzes the demethylation of mono- and di-methylated histone H3 lysine 36 (H3K36me1/2). Despite their structural similarity, KDM2A and KDM2B exhibit distinct functional roles. KDM2B possesses a broader substrate specificity, capable of demethylating both H3K36me1/2 and H3K4me3.[37] This enzyme has been implicated in gene silencing and chromatin compaction through its interaction with polycomb repressive complex 1 (PRC1).[38] Conversely, KDM2A is primarily involved in the repression of pericentromeric heterochromatin, mediated by its interaction with HP1α.[39] The CXXC-type zinc finger domain within KDM2A and KDM2B plays a crucial role in their functional differentiation. KDM2A's CXXC domain exhibits a specific affinity for unmethylated CpG dinucleotides, leading to the depletion of H3K36me2 from these regions.[40] Methylation of CpG sites inhibits this binding, enabling KDM2A to discriminate between methylated and unmethylated DNA segments. In contrast, KDM2B preferentially targets polycomb-repressed CpG islands.[38] Wild-type KDM2B demonstrates a specific localization to CpG islands, whereas a zinc finger mutant fails to bind any class of CpG islands. This observation suggests that KDM2B can target CpG islands independently of PRC1 components, potentially functioning as a targeting module for PRC1 and contributing to polycomb-mediated repression.

8.3.2.3 Zinc Finger Domain in KDM5C

The KDM5 family, comprising four members—KDM5A, KDM5B, KDM5C, and KDM5D—are histone lysine demethylases that play crucial roles in gene regulation.[41] As members of the JmjC domain-containing family of histone demethylases, KDM5 proteins specifically remove di- and tri-methyl marks from histone H3 lysine 4 (H3K4me2/3), epigenetic modifications typically associated with active transcription. These enzymes possess a C5HC2 zinc finger domain. Previous research has shown that mutation in this zinc finger leads to altered gene expression patterns, suggesting that the zinc finger domain is essential for KDM5C to function as a histone demethylase for gene regulation.[42]

8.3.2.4 Zinc Finger Domain in KDM6A and KDM6B

KDM6 demethylases are members of the JmjC domain-containing protein family, which comprises three members: KDM6A, KDM6B, and KDM6C.[43] These enzymes specifically remove di- and tri-methylation marks from histone H3 lysine 27 (H3K27me2/3). By removing these repressive marks, KDM6 demethylases generally activate gene expression. All three KDM6 members possess a GATA-like zinc finger domain at their C-terminus. Structural analysis has revealed that the zinc finger domain in KDM6A and KDM6B interacts with the histone H3 tail, which accounts for the substrate specificity that enables KDM6 to discriminate H3K27me2/3 from other methylated lysines with distinct functions, including the near-cognate H3K9me2/3, in histones.[44,45]

8.3.3 Histone Methylation Readers

Proteins that recognize methylated histones play a pivotal role in interpreting the histone code, a complex system of epigenetic modifications that regulate gene expression.[46] These reader proteins possess specialized domains that enable them to bind selectively to specific methylated lysine or arginine residues on histone tails. Common binding domains include chromodomains, tudor domains, PWWP domains, and PHD zinc fingers.[46] These domains exhibit the capacity to differentiate between various methylation states (mono-, di-, or tri-methylation) and the adjacent amino acid sequence. By binding to specific methylated histone residues, reader proteins can recruit additional proteins to chromatin, leading to the formation of regulatory complexes that can either activate or repress transcription.

8.3.3.1 PHD Domain in BPTF

BPTF is the largest subunit of the nucleosome remodeling factor (NURF) complex.[47] NURF catalyzes the sliding of nucleosomes on DNA. BPTF possesses two PHD finger domains, which contain zinc ions for structural integrity.[48] One PHD finger domain is located at the N-terminal region, and the other is at the C-terminal region. The PHD finger domain located at the C-terminal region specifically binds to trimethylated histone H3 lysine 4 (H3K4me3), a histone modification associated with active gene promoters.[48] In addition, the bromodomain in BPTF interacts with acetylated lysines, particularly H4K16ac.[49] By recognizing both histone modifications, BPTF can interpret multiple chromatin states simultaneously, enhancing its specificity for particular chromatin states. This allows NURF to be recruited to transcriptionally active regions and remodel chromatin.

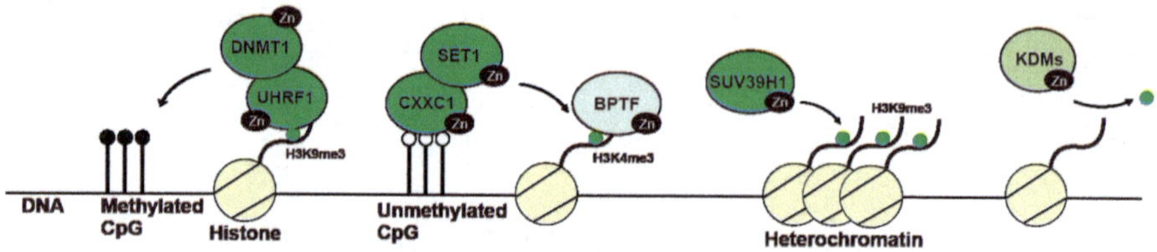

Figure 8.2 Roles of zinc in histone methylation.

8.3.3.2 PHD Domain in UHRF1

UHRF1 is a pivotal protein in epigenetic regulation, particularly in the maintenance of DNA methylation patterns.[50] UHRF1 is a multi-domain protein composed of several critical structural elements, including the SRA domain, RING domain, PHD domain, Tandem Tudor Domain (TTD), and Ubiquitin-like (UBL) domain. The PHD finger of UHRF1 is stabilized by three zinc ions, forming a rod-like structure that consists of a short α-helix, a double-stranded antiparallel β-sheet, and three loops.[51] In collaboration with the adjacent TTD, the PHD domain specifically recognizes the H3K9me3 histone modification, thereby directing the UHRF1 complex to relevant genomic loci. UHRF1 also interacts directly with DNMT1 through its SRA domain.[52] This interaction not only recruits DNMT1 but also activates its DNA methyltransferase activity. These coordinated mechanisms are essential for the preservation of DNA methylation patterns (Figure 8.2).

8.4 Histone Acetylation

Histone acetylation is a critical post-translational modification that modulates chromatin structure and gene expression.[53] This process involves the addition of acetyl groups to lysine residues on histone tails by histone acetyltransferases (HATs), leading to a more relaxed chromatin conformation, which promotes transcriptional activation. In contrast, histone deacetylases (HDACs) remove these acetyl groups, restoring chromatin to a more condensed state, thereby repressing gene expression. Acetylated histones are recognized by reader proteins, such as bromodomain-containing proteins, which further influence chromatin dynamics and transcriptional outcomes. The balance between histone acetylation and deacetylation is essential for physiological processes. Dysregulation of these mechanisms is associated with various pathologies, highlighting the importance of histone acetylation in both normal and disease states.

8.4.1 Histone Acetyltransferases

Histone acetyltransferases (HATs) are a family of enzymes that catalyze the acetylation of lysine residues on histone proteins, a critical process in chromatin remodeling and gene expression regulation.[54] The conserved catalytic core of HATs is essential for their enzymatic activity, while their domain architecture exhibits diversity, influencing their functional specificity. For example, the GNAT family of HATs, including GCN5 and PCAF, possess a bromodomain at the C-terminus, enabling them to recognize acetylated histone tails.[55] The p300/CBP family is a large multi-domain protein, allowing for diverse interactions with chromatin and transcriptional regulators.[56] MYST family HATs, such as MOZ and KAT7, typically feature a conserved MYST domain, which includes an acetyl-CoA binding site and a unique C2HC-type zinc finger.[57] Zinc-binding domains present in certain HATs have been implicated in their enzymatic activity and interactions with other proteins.

8.4.1.1 Zinc Finger Domain in p300

p300, a multidomain protein, is composed of several distinct domains, including a HAT domain, Bromodomain, ZZ domain, Nuclear Receptor Interaction (RID) domain, and CREB and MYB Interaction domain.[58] Each of these domains contributes significantly to p300's diverse functions, which encompass gene transcription, chromatin remodeling, and protein–protein interactions. The ZZ domain, a zinc finger motif characterized by its ability to bind two zinc ions, is crucial for maintaining the structural integrity and functional activity of p300. Recent research has unveiled a novel role for the ZZ domain in p300 as a histone H3 reader.[59] This interaction facilitates selective acetylation of histone H3 lysine residues 27 and 18 by the p300 HAT domain in conjunction with the p300 bromodomain.

8.4.1.2 MYST Domain in MOF

MOF, also known as KAT8, is a member of the MYST family.[60] These enzymes possess a unique C2HC-type zinc finger domain adjacent to their HAT domains. The zinc finger domain of MOF serves multiple functions. Mutations within this domain result in a reduction of MOF's HAT activity, underscoring its critical role in the enzyme's catalytic function.[61] Moreover, the zinc finger domain facilitates MOF's interaction with chromatin. *In vitro* binding study has demonstrated that the zinc finger domain interacts with the histone H4 tail, a key determinant of MOF's nucleosome binding affinity to acetylate H4K16.[61]

8.4.1.3 MYST Domain in KAT7

KAT7, also known as HBO1 or MYST2, is a member of the MYST family.[60] Mutations in KAT7 that disrupt zinc coordination are reported to significantly impair KAT7's function. For example, a C371A mutation in the MYST domain results in a marked reduction in zinc coordination and loss of HAT activity, highlighting the critical role of zinc.[62] Moreover, under conditions of zinc deficiency, KAT7 exhibits decreased HAT activity, leading to reduced acetylation of histone H3 at lysine 14 (H3K14ac) in cultured cells. This epigenetic modification alteration results in the transcriptional upregulation of ZIP10, a plasma membrane-localized zinc transporter responsible for importing zinc from extracellular sources to maintain cellular zinc homeostasis.[62]

8.4.1.4 Others

Zinc transporters play a critical role in maintaining intracellular zinc homeostasis. Recent investigations have revealed that perturbing zinc homeostasis through knockdown of ZIP10, a zinc transporter, results in a significant reduction of overall HAT activity within the cell.[63] Conversely, knockdown of ZIP6 had no discernible effect on HAT activity. These findings suggest that specific zinc signaling mediated by ZIP10 may exert an epigenetic regulatory influence on histone acetylation.

8.4.2 Histone Deacetylase

HDACs are key enzymes involved in the epigenetic regulation of gene expression.[64] They primarily remove acetyl groups from histone tails, leading to chromatin condensation and transcriptional repression. HDACs are categorized into four major classes based on their sequence homology and their structural and functional characteristics: Class I to IV. Class III HDACs, also known as sirtuins, comprise seven members in mammals (SIRT1-7).[65] Sirtuins require NAD+ as a cofactor for their enzymatic activity. In contrast, the classical HDAC family (Class I, Class IIa, Class IIb, Class IV) utilizes a zinc ion that is indispensable for catalytic activity.[66]

8.4.2.1 Catalytic Domain in the Classical HDAC Family

The classical HDAC family consists of eleven enzymes (HDAC1-11), each categorized into four distinct classes based on their structural and functional characteristics.[66] Class I HDACs (HDAC1, HDAC2, HDAC3, and HDAC8) are primarily nuclear-localized and ubiquitously expressed across various tissues. Class IIa HDACs (HDAC4, HDAC5, HDAC7, and HDAC9) exhibit a dynamic localization pattern, shuttling between the nucleus and cytoplasm. They are preferentially expressed in specific tissues such as the brain, heart, and muscle. Class IIb HDACs (HDAC6 and HDAC10) are predominantly cytoplasmic and often target non-histone proteins, including tubulin. HDAC11 is classified as a Class IV HDAC and is distinguished by its unique sequence homology to other HDAC classes. The catalytic activity of HDACs is dependent on the presence of a zinc ion in their active site.[67] The zinc ion plays a critical role in maintaining the appropriate conformation of the catalytic domain, facilitating substrate binding, and promoting the deacetylation reaction. Specifically, the zinc ion enables a nucleophilic attack of a water molecule on the acetyl group, leading to its removal.

Therefore, zinc is essential for the structural integrity, stability, and enzymatic function of classical HDACs. Treatment of cultured cells with the permeable zinc chelator TPEN leads to the inactivation of HDACs.[68]

8.4.3 Histone Acetylation Readers

Acetylation of lysine residues on histone proteins is a critical epigenetic modification that plays a key role in regulating gene expression. Bromodomain and YEATS domain proteins are two major classes that specifically recognize acetylated lysine residues.[69,70] These domains are modular protein structures that bind to acetylated lysine with high affinity and specificity. Although both bromodomain and YEATS domain proteins recognize acetylated lysine, neither of them coordinates zinc, as they lack zinc-binding motifs. In contrast, the tandem PHD zinc fingers in DFP3b are reported to function as readers of acetylated lysine residues.

8.4.3.1 PHD Domain in DFP3b

DPF3b is a member of the d4 protein family, which also includes DPF1, DPF2, and DPF3a. DPF3b possesses a dual PHD finger domain at its C-terminus. These tandem PHD domains exhibit a characteristic zinc-finger motif composed of a two-stranded antiparallel β-sheet followed by a carboxy-terminal α-helix.[71] This structural arrangement is stabilized by zinc ions coordinated through a specific C4HC3 motif. The tandem PHD finger domain has been demonstrated to preferentially bind to acetylated histones, specifically recognizing acetylation at lysine 14 of histone H3 (H3K14ac).[71] This binding is crucial for its role within the BRG1/BRM-associated factor (BAF) chromatin remodeling complex, which regulates the transcriptional activation and repression of specific genes (Figure 8.3).

8.5 Histone Ubiquitination

Histone ubiquitination is a critical post-translational modification that regulates chromatin dynamics and gene expression.[72] This process involves the covalent attachment of ubiquitin, a 76-amino acid protein, to lysine residues on histones, primarily H2A and H2B. A cascade of enzymes, including E1 ubiquitin-activating enzyme, E2 ubiquitin-conjugating enzyme,

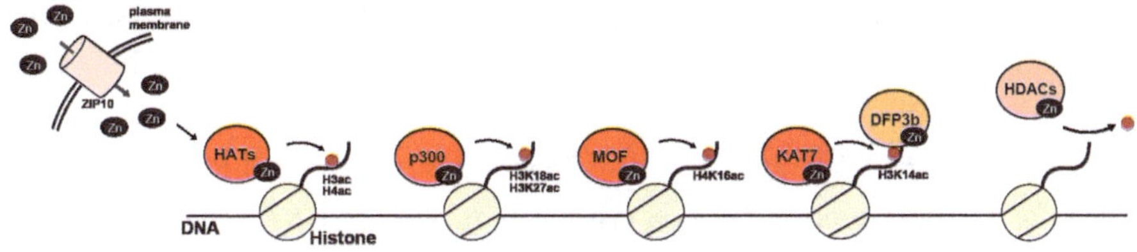

Figure 8.3 Roles of zinc in histone acetylation.

and E3 ubiquitin ligase, catalyze histone ubiquitination. Deubiquitinating enzymes (DUBs) can reverse this process, enabling dynamic regulation of the modification. Histone ubiquitination can occur as monoubiquitination, where a single ubiquitin molecule is attached, or polyubiquitination, where ubiquitin chains form. These modifications can lead to diverse outcomes for chromatin structure. Ubiquitinated histones serve as regulatory signals that are recognized by specialized proteins known as ubiquitin readers. These reader proteins bind to ubiquitinated histones and recruit other chromatin-modifying complexes, ultimately influencing transcriptional activation or repression.

8.5.1 E3 Ligases

E3 ubiquitin ligases play a crucial role in the ubiquitination system by catalyzing the transfer of ubiquitin to substrate proteins. There are over 600 E3 ligase genes in the human genome, representing about 5% of human genes.[73] The major classes are Really Interesting New Gene (RING), Homologous to the E6-AP Carboxyl Terminus (HECT), and RING-in-between-RING (RBR) E3 ligases, with RING being the largest group. Many E3 ligases incorporate zinc ions, which are crucial for their structural integrity and catalytic function.

8.5.1.1 RING Domain in the PRC1 Complex

Polycomb repressive complex 1 (PRC1) is a pivotal epigenetic regulator involved in gene silencing and chromatin modification.[74] It specifically monoubiquitinates histone H2A at lysine 119 (H2AK119ub1). Proteomic and biochemical investigations have elucidated the core constituents of PRC1, comprising the E3 ubiquitin ligase RING1A/B and one of six Polycomb group RING finger (PCGF) proteins. The specific association between RING1A/B and each PCGF protein results in six distinct subtypes of PRC1 complexes. In addition to these core components, various associated proteins contribute to the diversity of PRC1 complexes, each modulating PRC1's functional capacity in unique ways. Several PRC1 components contain zinc. RING1A/B, essential components of PRC1, possess a RING domain, a zinc-binding domain that functions as an E3 ligase.[75] Furthermore, a RING domain-deletion mutant of RING1A failed to interact with Snail, a transcription repressor, emphasizing the critical role of the zinc-binding RING domain as a protein–protein interaction domain for transcriptional repression.[76] BMI1, also known as RNF51, is another protein containing a RING domain in PRC1. This domain is indispensable for the interaction with RING1 to form the PRC1 complex.[77]

8.5.1.2 RING Domain in the BRCA1/BARD1 Complex

The BRCA1/BARD1 complex is a critical regulator of DNA repair and tumor suppression, primarily through its E3 ubiquitin ligase activity targeting histone H2A Lys 125/127/129.[78] The N-terminal RING finger domains of

BRCA1 and BARD1 are indispensable for heterodimer formation and E3 ubiquitin ligase activity.[79] These domains facilitate the recruitment of E2 ubiquitin-conjugating enzymes and the target substrate, thereby promoting efficient ubiquitin transfer. The zinc-coordinating cysteine residue at position 61 in BRCA1 is crucial for the proper folding and stability of the RING finger domain. Mutations in this residue, such as the C61G substitution, can disrupt the zinc-binding site, leading to loss of ubiquitin ligase activity, further emphasizing the significance of zinc in the complex's function.[80]

8.5.2 Histone Deubiquitinases

Histone deubiquitinases (DUBs) are pivotal components of the cellular machinery responsible for regulating histone modifications. These enzymes can be classified into several distinct families, including ubiquitin-specific peptidases (USPs), ovarian tumor proteases (OTUs), JAB1/MPN/Mov34 metalloenzymes (JAMMs), motif interacting with ubiquitin-containing novel DUB family proteases (MINDYs), ubiquitin C-terminal hydroxylases (UCHs), Machado–Josephin domain proteases (MJDs), and zinc finger-containing ubiquitin peptidase 1 (ZUP1).[81] Among these diverse DUB families, JAMM/MPN+ proteases are distinguished by their zinc-dependent metalloprotease activity.

8.5.2.1 Catalytic Domain in BRCC36

BRCC36, a member of the JAMM/MPN+ family of zinc metalloproteases, plays a pivotal role in DNA damage repair.[82] Within the nucleus, it forms the BRCA1-A complex, a multi-protein assembly that is essential for maintaining genomic integrity. This complex is composed of several subunits, including BRCC45/BRE, MERIT40, RAP80, and Abraxas, each contributing to the complex's overall function. This complex exhibits specific catalytic activity, cleaving Lys63-linked ubiquitin chains of histone H2A.[83] The critical importance of zinc for BRCC36's catalytic activity is evident from the complete loss of function when mutations occur at the zinc-coordinating residues, histidine 122 and 124.[84]

8.5.2.2 Catalytic Domain in Rpn11

Rpn11, a member of the JAMM/MPN+ family of zinc-dependent metalloproteases, plays a pivotal role in the DNA damage response pathway. As a component of the 26S proteasome, Rpn11 specifically targets Lys-63-linked polyubiquitin chains on histone H2A.[85] This proteolytic cleavage is essential for facilitating the repair of DNA double-strand breaks (DSBs). Zinc plays a pivotal role in Rpn11's catalytic activity, as mutations within its JAMM motif (H113A and H115A) result in a loss of ubiquitin cleavage at irradiation-induced nuclear foci.[86]

8.5.3 Histone Ubiquitination Readers

Histone ubiquitination readers recognize ubiquitinated histone modifications, which play pivotal roles in diverse cellular processes, including DNA damage repair, transcriptional regulation, and genome stability.[87] Despite the lack of a conserved domain or structural motif among histone ubiquitination readers, some proteins, such as 53BP1, RNF169, and DNMT1, have been identified as capable of recognizing these modifications.

8.5.3.1 RFTS Domain in DMNT1

DNMT1, a DNA methyltransferase, plays a multifaceted role in epigenetic regulation.[88] Beyond its primary function of catalyzing DNA methylation, DNMT1 also acts as a reader for ubiquitinated histone proteins. Structural studies have revealed that the replication foci targeting sequence (RFTS) of DNMT1, containing a zinc-binding motif, is crucial for recognizing histone H3 ubiquitinated at lysine 18 and 23.[89] This specific recognition mechanism allows DNMT1 to be targeted to chromatin regions. Interestingly, the binding of DNMT1 to doubly mono-ubiquitinated H3 is shown to enhance its methyltransferase activity[89] (Figure 8.4).

8.6 miRNA and lncRNA

MicroRNAs (miRNAs) are small, non-coding RNA molecules that have emerged as key regulatory molecules in epigenetics, influencing gene expression without altering the DNA sequence.[90] They are typically 20–25 nucleotides in length and are transcribed from DNA but do not encode proteins. Instead, miRNAs bind to specific messenger RNA (mRNA) molecules, leading to their degradation or preventing their translation into proteins. Long non-coding RNAs (lncRNAs), transcripts exceeding 200 nucleotides in length, are distinguished from smaller non-coding RNAs such as miRNAs.[91] LncRNAs are characterized by their structural diversity, often adopting complex secondary and tertiary structures. These intricate conformations may underlie their diverse biological functions. LncRNAs are known to interact with a variety of biomolecules, including proteins, DNA, and RNA, and are implicated in a wide range of cellular processes. Several

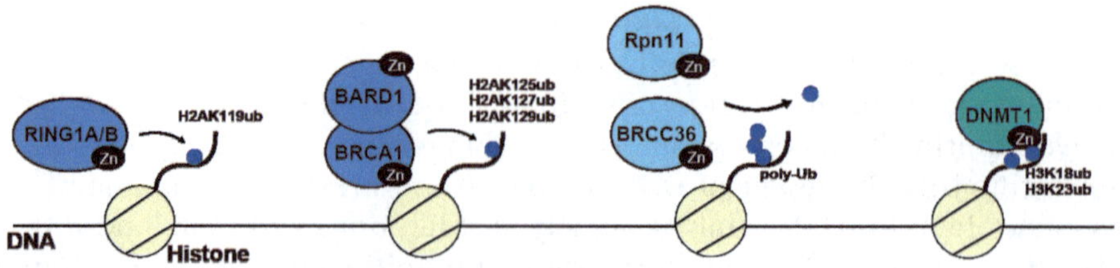

Figure 8.4 Roles of zinc in histone ubiquitination.

studies have shown that the expression level of miRNA or lncRNA was altered when cellular zinc homeostasis is disrupted.

8.6.1 miR-21 and miR-31

A rat model has demonstrated that chronic zinc deficiency can induce an inflammatory gene expression profile that facilitates the development of esophageal squamous cell carcinoma (ESCC).[92] In a separate study, miRNA expression was analyzed in rats fed a zinc-deficient diet, revealing significant upregulation of miR-31 and miR-21.[93] While the precise functions of these miRNAs were not elucidated in this particular investigation, their overexpression was also observed in human ESCC tissues, suggesting their potential pathological significance.

8.6.2 miR-193b

A study investigating the correlation between the cell cycle and zinc revealed that zinc supplementation induces the upregulation of miR-193b.[94] Conversely, zinc deprivation results in the suppression of miR-193b, suggesting a regulatory role for zinc in miR-193b expression. The downregulation of miR-193b under zinc-deficient conditions leads to an elevation in Cyclin D1 and the zinc transporter ZIP5, which are validated targets of miR-193b. The expression of ZIP5 and Cyclin D1 facilitates zinc influx and the progression of the cell cycle.

8.6.3 Braveheart (Bvht)

The long noncoding RNA Braveheart (Bvht) is a critical regulator of cardiovascular development and lineage commitment in mice.[95] Structural characterization of *in vitro* transcribed full-length Bvht (~590 nucleotides) revealed a highly modular architecture, including a 5′ asymmetric G-rich internal loop (AGIL).[96] The AGIL within Bvht interacts with the cellular nucleic acid binding protein CNBP, a protein with seven zinc-finger domains known for its affinity to single-stranded G-rich sequences. This interaction is postulated to play a critical role in the regulation of cardiovascular lineage commitment.

8.6.4 Others

The advent of microarray and next-generation sequencing technologies has enabled comprehensive investigations into the role of miRNA and lncRNA under zinc deficiency. Extensive studies of miRNA expression changes in rice seedlings during zinc deprivation have elucidated the regulatory mechanisms governing gene expression through miRNA modulation in response to zinc deficiency.[97] Moreover, comprehensive analyses of miRNAs in human serum during zinc deficiency have identified a panel of miRNAs as potential

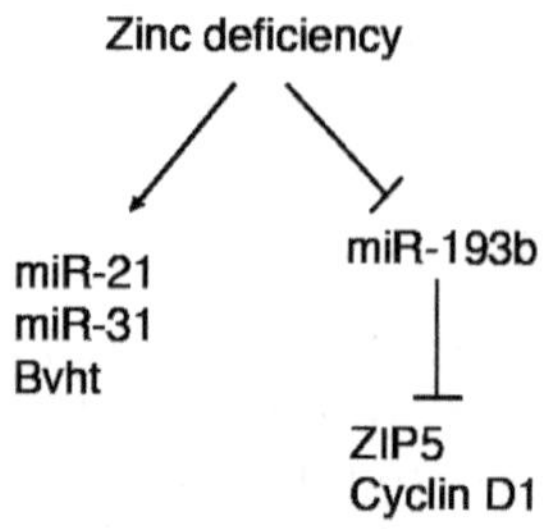

Figure 8.5 Roles of zinc in miRNA and lncRNA regulation.

biomarkers for monitoring zinc status.[98] Another study has reported on the role of lncRNAs. In *Zip14* knockout mice, where zinc homeostasis is perturbed, the expression of several lncRNAs, such as bvhr, was found to be increased[99] (Figure 8.5).

8.7 Conclusion

Epigenetics has profound implications for both physiological and pathological processes. Zinc is a constituent of numerous proteins implicated in epigenetic regulation. Despite its established role in these processes, the specific mechanisms by which zinc influences epigenetic phenomena remain an area of active investigation. Further elucidation of the involvement of zinc in epigenetics could provide valuable insights into the diverse biological events mediated by epigenetic mechanisms.

Acknowledgements

This study was supported by the Japan Agency for Medical Research and Development (AMED) (grant number JP21gm5010001, 23gm0010009s0102, and 23gm1710001s0102 to H. I.), by the Japan Society for the Promotion of Science (JSPS) for the Grants-in-Aid for Scientific Research (KAKENHI; grant numbers JP22K06610 to T. F. and JP21H04760 to H. I.) and the Grant-in-Aid for Scientific Research on Innovative Areas (KAKENHI; grant number JP22H04804 to T. F.), and by the Japan Science and Technology Agency (JST) for Moonshot R&D–MILLENNIA Program (grant number JPMJMS2022-18 to H. I.).

References

1. T. Kambe, T. Tsuji, A. Hashimoto and N. Itsumura, *Physiol. Rev.*, 2015, **95**, 749–784.
2. C. D. Allis and T. Jenuwein, *Nat. Rev. Genet.*, 2016, **17**, 487–500.
3. M. V. C. Greenberg and D. Bourc'his, *Nat. Rev. Mol. Cell Biol.*, 2019, **20**, 590–607.
4. L. D. Moore, T. Le and G. Fan, *Neuropsychopharmacology*, 2013, **38**, 23–38.
5. A. Kikuchi, H. Onoda, K. Yamaguchi, S. Kori, S. Matsuzawa, Y. Chiba, S. Tanimoto, S. Yoshimi, H. Sato, A. Yamagata, M. Shirouzu, N. Adachi,

J. Sharif, H. Koseki, A. Nishiyama, M. Nakanishi, P.-A. Defossez and K. Arita, *Nat. Commun.*, 2022, **13**, 7130.

6. K. Takeshita, I. Suetake, E. Yamashita, M. Suga, H. Narita, A. Nakagawa and S. Tajima, *Proc. Natl. Acad. Sci. U. S. A.*, 2011, **108**, 9055–9059.
7. J. Song, O. Rechkoblit, T. H. Bestor and D. J. Patel, *Science*, 2011, **331**, 1036–1040.
8. M.-G. Lee, M.-A. Choi, S. Chae, M.-A. Kang, H. Jo, J.-M. Baek, K.-R. In, H. Park, H. Heo, D. Jang, S. Brito, S. T. Kim, D.-O. Kim, J.-S. Lee, J.-R. Kim and B.-H. Bin, *Sci. Rep.*, 2019, **9**, 15042.
9. I. Suetake, F. Shinozaki, J. Miyagawa, H. Takeshima and S. Tajima, *J. Biol. Chem.*, 2004, **279**, 27816–27823.
10. S. K. T. Ooi, C. Qiu, E. Bernstein, K. Li, D. Jia, Z. Yang, H. Erdjument-Bromage, P. Tempst, S.-P. Lin, C. D. Allis, X. Cheng and T. H. Bestor, *Nature*, 2007, **448**, 714–717.
11. Y. Zhang, G. K. Andrews and L. Wang, *Nucleic Acids Res.*, 2012, **40**, 4850–4860.
12. G. Davison, J. C. Mathers, D. Ford and R. A. Valentine, *Proc. Nutr. Soc.*, 2010, **69**, E17.
13. X. Wu and Y. Zhang, *Nat. Rev. Genet.*, 2017, **18**, 517–534.
14. C. Xu, K. Liu, M. Lei, A. Yang, Y. Li, T. R. Hughes and J. Min, *Structure*, 2018, **26**, 85–95.e3.
15. H. Zhang, X. Zhang, E. Clark, M. Mulcahey, S. Huang and Y. G. Shi, *Cell Res.*, 2010, **20**, 1390–1393.
16. H. Wu, A. C. D'Alessio, S. Ito, K. Xia, Z. Wang, K. Cui, K. Zhao, Y. E. Sun and Y. Zhang, *Nature*, 2011, **473**, 389–393.
17. K. Williams, J. Christensen, M. T. Pedersen, J. V. Johansen, P. A. C. Cloos, J. Rappsilber and K. Helin, *Nature*, 2011, **473**, 343–348.
18. Y. Xu, C. Xu, A. Kato, W. Tempel, J. G. Abreu, C. Bian, Y. Hu, D. Hu, B. Zhao, T. Cerovina, J. Diao, F. Wu, H. H. He, Q. Cui, E. Clark, C. Ma, A. Barbara, G. J. C. Veenstra, G. Xu, U. B. Kaiser, X. S. Liu, S. P. Sugrue, X. He, J. Min, Y. Kato and Y. G. Shi, *Cell*, 2012, **151**, 1200–1213.
19. H. Zhu, G. Wang and J. Qian, *Nat. Rev. Genet.*, 2016, **17**, 551–565.
20. Y. Liu, H. Toh, H. Sasaki, X. Zhang and X. Cheng, *Genes Dev.*, 2012, **26**, 2374–2379.
21. S. Quenneville, G. Verde, A. Corsinotti, A. Kapopoulou, J. Jakobsson, S. Offner, I. Baglivo, P. V. Pedone, G. Grimaldi, A. Riccio and D. Trono, *Mol. Cell*, 2011, **44**, 361–372.
22. D. Kaplun, A. Starshin, F. Sharko, K. Gainova, G. Filonova, N. Zhigalova, A. Mazur, E. Prokhortchouk and S. Zhenilo, *Int. J. Mol. Sci.*, 2021, **22**, 7587.
23. B. A. Buck-Koehntop, R. L. Stanfield, D. C. Ekiert, M. A. Martinez-Yamout, H. J. Dyson, I. A. Wilson and P. E. Wright, *Proc. Natl. Acad. Sci. U. S. A.*, 2012, **109**, 15229–15234.
24. D. S. Kaplun, R. E. Fok, V. S. Korostina, E. B. Prokhortchouk and S. V. Zhenilo, *Biochemistry*, 2019, **84**, 283–290.
25. E. L. Greer and Y. Shi, *Nat. Rev. Genet.*, 2012, **13**, 343–357.

26. D. Husmann and O. Gozani, *Nat. Struct. Mol. Biol.*, 2019, **26**, 880–889.
27. A. Shilatifard, *Annu. Rev. Biochem.*, 2012, **81**, 65–95.
28. Y. Yang, M. Joshi, Y.-H. Takahashi, Z. Ning, Q. Qu, J. S. Brunzelle, G. Skiniotis, D. Figeys, A. Shilatifard and J.-F. Couture, *Nucleic Acids Res.*, 2020, **48**, 421–431.
29. J. P. Thomson, P. J. Skene, J. Selfridge, T. Clouaire, J. Guy, S. Webb, A. R. W. Kerr, A. Deaton, R. Andrews, K. D. James, D. J. Turner, R. Illingworth and A. Bird, *Nature*, 2010, **464**, 1082–1086.
30. C. M. Tate, J.-H. Lee and D. G. Skalnik, *FEBS J.*, 2010, **277**, 210–223.
31. J. Padeken, S. P. Methot and S. M. Gasser, *Nat. Rev. Mol. Cell Biol.*, 2022, **23**, 623–640.
32. I. Santos-Barriopedro, L. Bosch-Presegué, A. Marazuela-Duque, C. de la Torre, C. Colomer, B. N. Vazquez, T. Fuhrmann, B. Martínez-Pastor, W. Lu, T. Braun, E. Bober, T. Jenuwein, L. Serrano, M. Esteller, Z. Chen, S. Barceló-Batllori, R. Mostoslavsky, L. Espinosa and A. Vaquero, *Nat. Commun.*, 2018, **9**, 101.
33. S. M. Kooistra and K. Helin, *Nat. Rev. Mol. Cell Biol.*, 2012, **13**, 297–311.
34. L. Qu, T. Yin, Y. Zhao, W. Lv, Z. Liu, C. Chen, K. Liu, S. Shan, R. Zhou, X. Li and H. Dong, *Cell Death Discovery*, 2023, **9**, 1–15.
35. Z. Yang, J. Jiang, D. M. Stewart, S. Qi, K. Yamane, J. Li, Y. Zhang and J. Wong, *Cell Res.*, 2010, **20**, 276.
36. Q. Zhang, S. Qi, M. Xu, L. Yu, Y. Tao, Z. Deng, W. Wu, J. Li, Z. Chen and J. Wong, *Cell Res.*, 2013, **23**, 225–241.
37. Y. Tsukada, J. Fang, H. Erdjument-Bromage, M. E. Warren, C. H. Borchers, P. Tempst and Y. Zhang, *Nature*, 2006, **439**, 811–816.
38. A. M. Farcas, N. P. Blackledge, I. Sudbery, H. K. Long, J. F. McGouran, N. R. Rose, S. Lee, D. Sims, A. Cerase, T. W. Sheahan, H. Koseki, N. Brockdorff, C. P. Ponting, B. M. Kessler and R. J. Klose, *eLife*, 2012, **1**, e00205.
39. J. Borgel, M. Tyl, K. Schiller, Z. Pusztai, C. M. Dooley, W. Deng, C. Wooding, R. J. White, T. Warnecke, H. Leonhardt, E. M. Busch-Nentwich and T. Bartke, *Nucleic Acids Res.*, 2017, **45**, 1114–1129.
40. J. C. Zhou, N. P. Blackledge, A. M. Farcas and R. J. Klose, *Mol. Cell. Biol.*, 2012, **32**, 479–489.
41. Y. Ohguchi and H. Ohguchi, *Cancers*, 2022, **14**, 3270.
42. H. M. Belalcazar, E. L. Hendricks, S. Zamurrad, F. L. W. Liebl and J. Secombe, *Cell Rep.*, 2021, **34**, 108753.
43. C. Hua, J. Chen, S. Li, J. Zhou, J. Fu, W. Sun and W. Wang, *Front. Oncol.*, 2021, **11**, 779918.
44. S. E. Jones, L. Olsen and M. Gajhede, *Biochemistry*, 2018, **57**, 585–592.
45. T. Sengoku and S. Yokoyama, *Genes Dev.*, 2011, **25**, 2266–2277.
46. K. Hyun, J. Jeon, K. Park and J. Kim, *Exp. Mol. Med.*, 2017, **49**, e324.
47. D. C. Hargreaves and G. R. Crabtree, *Cell Res.*, 2011, **21**, 396–420.
48. H. Li, S. Ilin, W. Wang, E. M. Duncan, J. Wysocka, C. D. Allis and D. J. Patel, *Nature*, 2006, **442**, 91–95.

49. A. J. Ruthenburg, H. Li, T. A. Milne, S. Dewell, R. K. McGinty, M. Yuen, B. Ueberheide, Y. Dou, T. W. Muir, D. J. Patel and C. D. Allis, *Cell*, 2011, **145**, 692–706.
50. M. Mancini, E. Magnani, F. Macchi and I. M. Bonapace, *Nucleic Acids Res.*, 2021, **49**, 6053–6068.
51. L. Hu, Z. Li, P. Wang, Y. Lin and Y. Xu, *Cell Res.*, 2011, **21**, 1374–1378.
52. A. C. Berkyurek, I. Suetake, K. Arita, K. Takeshita, A. Nakagawa, M. Shirakawa and S. Tajima, *J. Biol. Chem.*, 2014, **289**, 379–386.
53. B. R. Sabari, D. Zhang, C. D. Allis and Y. Zhao, *Nat. Rev. Mol. Cell Biol.*, 2017, **18**, 90–101.
54. R. Marmorstein and M.-M. Zhou, *Cold Spring Harbor Perspect. Biol.*, 2014, **6**, a018762.
55. A. I. M. Salah Ud-Din, A. Tikhomirova and A. Roujeinikova, *Int. J. Mol. Sci.*, 2016, **17**, 1018.
56. B. M. Dancy and P. A. Cole, *Chem. Rev.*, 2015, **115**, 2419–2452.
57. R. T. Utley and J. Côté, in *Protein Complexes that Modify Chromatin*, ed. J. L. Workman, Springer Berlin Heidelberg, Berlin, Heidelberg, 2003, vol. 274, pp. 203–236.
58. B. M. Dancy and P. A. Cole, *Chem. Rev.*, 2015, **115**, 2419–2452.
59. Y. Zhang, Y. Xue, J. Shi, J. Ahn, W. Mi, M. Ali, X. Wang, B. J. Klein, H. Wen, W. Li, X. Shi and T. G. Kutateladze, *Nat. Struct. Mol. Biol.*, 2018, **25**, 841–849.
60. R. T. Utley and J. Côté, in *Protein Complexes that Modify Chromatin*, ed. J. L. Workman, Springer, Berlin, Heidelberg, 2003, pp. 203–236.
61. A. Akhtar and P. B. Becker, *EMBO Rep.*, 2001, **2**, 113–118.
62. T. Fujisawa, S. Takenaka, L. Maekawa, T. Kowada, T. Matsui, S. Mizukami, Y. Kato, M. Suzuki, H. Noma, I. Naguro and H. Ichijo, 2023, DOI: 10.1101/2023.10.18.562865.
63. B.-H. Bin, S.-H. Lee, J. Bhin, T. Irié, S. Kim, J. Seo, K. Mishima, T. R. Lee, D. Hwang, T. Fukada and E.-G. Cho, *Br. J. Dermatol.*, 2019, **180**, 869–880.
64. A. Curcio, R. Rocca, S. Alcaro and A. Artese, *Pharmaceuticals*, 2024, **17**, 620.
65. Q.-J. Wu, T.-N. Zhang, H.-H. Chen, X.-F. Yu, J.-L. Lv, Y.-Y. Liu, Y.-S. Liu, G. Zheng, J.-Q. Zhao, Y.-F. Wei, J.-Y. Guo, F.-H. Liu, Q. Chang, Y.-X. Zhang, C.-G. Liu and Y.-H. Zhao, *Signal Transduction Targeted Ther.*, 2022, 7, 1–74.
66. A. J. M. D. Ruijter, A. H. V. Gennip, H. N. Caron, S. Kemp and A. B. P. V. Kuilenburg, *Biochem. J.*, 2003, **370**, 737–749.
67. P. M. Lombardi, K. E. Cole, D. P. Dowling and D. W. Christianson, *Curr. Opin. Struct. Biol.*, 2011, **21**, 735–743.
68. Y. Zhao, W. Zhong, X. Sun, Z. Song, D. L. Clemens, Y. J. Kang, C. J. McClain and Z. Zhou, *Am. J. Pathol.*, 2011, **179**, 693–702.
69. D. Zhao, Y. Li, X. Xiong, Z. Chen and H. Li, *J. Mol. Biol.*, 2017, **429**, 1994–2002.
70. T. Fujisawa and P. Filippakopoulos, *Nat. Rev. Mol. Cell Biol.*, 2017, **18**, 246–262.

71. L. Zeng, Q. Zhang, S. Li, A. N. Plotnikov, M. J. Walsh and M.-M. Zhou, *Nature*, 2010, **466**, 258–262.
72. J. J. Chen, D. Stermer and J. C. Tanny, *Front. Cell Dev. Biol.*, 2022, **10**, 968398.
73. L. Buetow and D. T. Huang, *Nat. Rev. Mol. Cell Biol.*, 2016, **17**, 626–642.
74. Z. Geng and Z. Gao, *Int. J. Mol. Sci.*, 2020, **21**, 8594.
75. Z. Li, R. Cao, M. Wang, M. P. Myers, Y. Zhang and R.-M. Xu, *J. Biol. Chem.*, 2006, **281**, 20643–20649.
76. J. Chen, H. Xu, X. Zou, J. Wang, Y. Zhu, H. Chen, B. Shen, X. Deng, A. Zhou, Y. E. Chin, F. J. Rauscher III, C. Peng and Z. Hou, *Cancer Res.*, 2014, **74**, 4353–4363.
77. Z. Li, R. Cao, M. Wang, M. P. Myers, Y. Zhang and R.-M. Xu, *J. Biol. Chem.*, 2006, **281**, 20643–20649.
78. S. R. Witus, M. D. Stewart and R. E. Klevit, *Biochem. J.*, 2021, **478**, 3467–3483.
79. P. S. Brzovic, P. Rajagopal, D. W. Hoyt, M.-C. King and R. E. Klevit, *Nat. Struct. Mol. Biol.*, 2001, **8**, 833–837.
80. R. Hashizume, M. Fukuda, I. Maeda, H. Nishikawa, D. Oyake, Y. Yabuki, H. Ogata and T. Ohta, *J. Biol. Chem.*, 2001, **276**, 14537–14540.
81. B. S. Atanassov, E. Koutelou and S. Y. Dent, *FEBS Lett.*, 2011, **585**, 2016–2023.
82. Y. Dong, M.-A. Hakimi, X. Chen, E. Kumaraswamy, N. S. Cooch, A. K. Godwin and R. Shiekhattar, *Mol. Cell*, 2003, **12**, 1087–1099.
83. L. Feng, J. Wang and J. Chen, *J. Biol. Chem.*, 2010, **285**, 30982–30988.
84. H.-M. Ng, L. Wei, L. Lan and M. S. Y. Huen, *J. Biol. Chem.*, 2016, **291**, 16197–16207.
85. M. S. Y. Huen, R. Grant, I. Manke, K. Minn, X. Yu, M. B. Yaffe and J. Chen, *Cell*, 2007, **131**, 901–914.
86. L. R. Butler, R. M. Densham, J. Jia, A. J. Garvin, H. R. Stone, V. Shah, D. Weekes, F. Festy, J. Beesley and J. R. Morris, *EMBO J.*, 2012, **31**, 3918–3934.
87. G. Smeenk and N. Mailand, *Front. Genet.*, 2016, **7**, 122.
88. H. Gowher and A. Jeltsch, *Biochem. Soc. Trans.*, 2018, **46**, 1191–1202.
89. S. Ishiyama, A. Nishiyama, Y. Saeki, K. Moritsugu, D. Morimoto, L. Yamaguchi, N. Arai, R. Matsumura, T. Kawakami, Y. Mishima, H. Hojo, S. Shimamura, F. Ishikawa, S. Tajima, K. Tanaka, M. Ariyoshi, M. Shirakawa, M. Ikeguchi, A. Kidera, I. Suetake, K. Arita and M. Nakanishi, *Mol. Cell*, 2017, **68**, 350–360.e7.
90. R. Shang, S. Lee, G. Senavirathne and E. C. Lai, *Nat. Rev. Genet.*, 2023, **24**, 816–833.
91. J. S. Mattick, P. P. Amaral, P. Carninci, S. Carpenter, H. Y. Chang, L.-L. Chen, R. Chen, C. Dean, M. E. Dinger, K. A. Fitzgerald, T. R. Gingeras, M. Guttman, T. Hirose, M. Huarte, R. Johnson, C. Kanduri, P. Kapranov, J. B. Lawrence, J. T. Lee, J. T. Mendell, T. R. Mercer, K. J. Moore, S. Nakagawa, J. L. Rinn, D. L. Spector, I. Ulitsky, Y. Wan, J. E. Wilusz and M. Wu, *Nat. Rev. Mol. Cell Biol.*, 2023, **24**, 430–447.

92. C. Taccioli, H. Chen, Y. Jiang, X. P. Liu, K. Huang, K. J. Smalley, J. L. Farber, C. M. Croce and L. Y. Fong, *Oncogene*, 2012, **31**, 4550–4558.
93. H. Alder, C. Taccioli, H. Chen, Y. Jiang, K. J. Smalley, P. Fadda, H. G. Ozer, K. Huebner, J. L. Farber, C. M. Croce and L. Y. Y. Fong, *Carcinogenesis*, 2012, **33**, 1736–1744.
94. J. Jin, Y. Guo, X. Dong, J. Liu and Y. He, *Radiother. Oncol.*, 2020, **150**, 104–113.
95. C. A. Klattenhoff, J. C. Scheuermann, L. E. Surface, R. K. Bradley, P. A. Fields, M. L. Steinhauser, H. Ding, V. L. Butty, L. Torrey, S. Haas, R. Abo, M. Tabebordbar, R. T. Lee, C. B. Burge and L. A. Boyer, *Cell*, 2013, **152**, 570–583.
96. Z. Xue, S. Hennelly, B. Doyle, A. A. Gulati, I. V. Novikova, K. Y. Sanbonmatsu and L. A. Boyer, *Mol. Cell*, 2016, **64**, 37–50.
97. H. Zeng, X. Zhang, M. Ding and Y. Zhu, *BMC Plant Biol.*, 2019, **19**, 585.
98. M.-S. Ryu, B. Langkamp-Henken, S.-M. Chang, M. N. Shankar and R. J. Cousins, *Proc. Natl. Acad. Sci. U. S. A.*, 2011, **108**, 20970–20975.
99. F. R. Jimenez-Rondan, C. H. Ruggiero and R. J. Cousins, *Nutrients*, 2022, **14**, 5114.

CHAPTER 9

Role of ZNT5-6 and ZNT7 in Zinc Enzyme Activation

T. KAMBE,* K. MIYAZAKI, M. SHIMOYA AND T. WAGATSUMA

Division of Integrated Life Science, Graduate School of Biostudies, Kyoto University, Kyoto 606-8502, Japan
*Email: kambe.taiho.7z@kyoto-u.ac.jp

9.1 Introduction

Zinc is the second most abundant essential element in humans and plays a crucial role in numerous biological processes under both normal and pathological conditions. This is achieved by its binding to various proteins (metalation), including enzymes, where it acts as an essential cofactor. Zinc-binding motifs in proteins typically involve the coordination of zinc ions to amino acids, including cysteine (Cys), histidine (His), aspartic acid (Asp), and glutamic acid (Glu).[1] In zinc enzymes, zinc facilitates enzymatic functions by coordinating with His, Asp, and Glu residues, as well as substrates and other molecules, such as H_2O, during catalysis at the active site.[2,3] For example, enzymes related to alkaline phosphatase (ALP) share a similar active-site structure, with a bimetallic core consisting of two zinc ions. One zinc ion is directly involved in the enzymatic reaction and is coordinated by one Asp and two His residues, along with the substrate or H_2O, whereas the other zinc ion is coordinated by one His and two Asp residues and one serine (Ser). It is estimated that 40–50% of all enzymes require a metal for proper function,[4,5] the majority of which are zinc-containing enzymes. These enzymes are localized in the cytosol and intracellular organelles, including the early secretory compartments (the endoplasmic reticulum (ER) and Golgi apparatus), endosomes, lysosomes, and extracellular space. Although it is generally

Coordination Chemistry Perspectives No. 2
Zinc in Biology: Molecular Structures, Cellular Processes and Living Systems
Edited by Hiroyuki Yasui and Taiho Kambe

Published by the Royal Society of Chemistry, www.rsc.org

believed that proteins and enzymes spontaneously acquire zinc (zinc metalation) during biosynthesis, the exact molecular mechanisms underlying this process remain unclear. Recent studies, however, have shed light on the sophisticated activation mechanisms of a number of zinc enzymes, mediated by specific zinc transporters across biological membranes.

A proportion of zinc enzymes undergo folding and metalation and thus are activated in the early secretory compartments (although some enzymes need to be further processed (*e.g.*, cleaved) to obtain enzymatic activity) before reaching their final destinations in vertebrate cells. Studies, including ours, have demonstrated that ZNT5-6 and ZNT7 play vital roles in this process.[6,7] Their zinc supply is critical for the activation of zinc enzymes that function in the plasma membrane and in the extracellular space, as reviewed extensively in several articles and book chapters.[8–11] Recently, we reported that they play essential roles in the activation of intracellular organelle-resident zinc enzymes. This chapter provides an overview of our current understanding of the activation of these zinc enzymes by summarizing the contributions of zinc *via* ZNT5-6 and ZNT7 to various biological functions, including *N*-glycosylation, pigmentation, sphingomyelin metabolism, and glycosylphosphatidylinositol (GPI)-anchored glycolipid biosynthesis. Finally, we discuss key issues and unanswered questions related to the activation of zinc enzymes by ZNT5-6 and ZNT7.

9.2 Overview of ZNT5-6 and ZNT7 Functions

Ten ZNT transporters (ZNT1–ZNT10) are found in vertebrates, with each being localized in specific intracellular organelles,[12,13] with the exception of ZNT1 and ZNT10. ZNT5-6 and ZNT7 are found in the early secretory compartments.[14–17] Their essential biological functions were confirmed by the fact that their orthologs are found in eukaryotic cells and play pivotal roles in the maintenance of the secretory pathway, although only one of them is functional in some species (Figure 9.1). For example, *Drosophila melanogaster* possesses only the ZNT7 ortholog, whereas *Saccharomyces cerevisiae, Schizosaccharomyces pombe*, *Caenorhabditis elegans*, and *Arabidopsis thaliana* possess only the ZNT5-6 orthologs.[18–23] ZNT5-6 and ZNT7 have crucial functions in protein quality control at the ER–Golgi interface.[24–26] Similar important quality control functions in the early secretory pathway have been found in *S. cerevisiae* (Msc2p and Zrg17p heterodimers)[18] and *S. pombe* (Cis4 and Zrg17 heterodimers).[19,20] These topics are described in Chapter 6, and this chapter focuses on their functions in zinc enzyme activation.

The essential roles of ZNT5-6 and ZNT7 in zinc enzyme activation were first identified in a study of ALP using chicken cells.[27] ALP activity is substantially reduced in cells lacking ZNT5-6 and ZNT7 functions (both *ZNT5* and *ZNT7* gene-deficient cells, referred to as *Z5Z7*-double knockout (DKO) cells).[28] Loss of ZNT5-6 and ZNT7 functions required for ALP activation is overcome by the expression of ZNT5-6 orthologs in *C. elegans* (Cdf5 and Toc1 heterodimers)

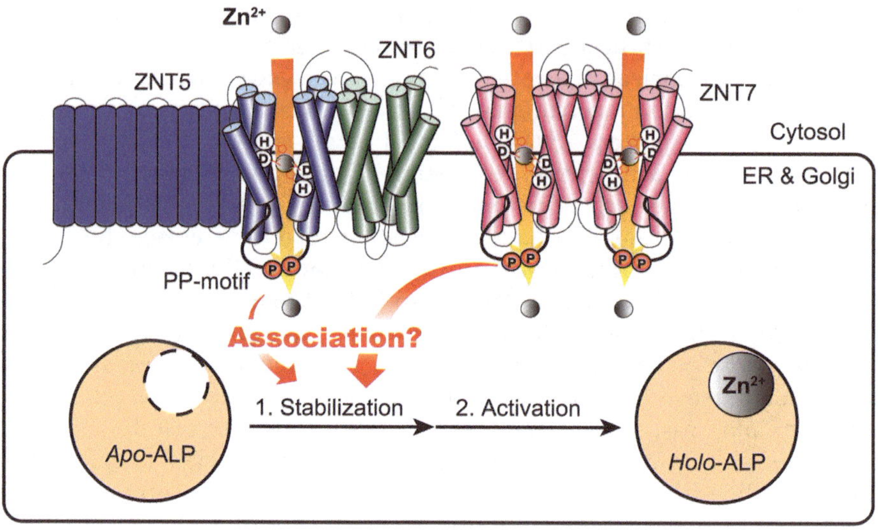

	Heterodimer		Homodimer
H. sapiens	ZNT5	ZNT6	ZNT7
S.cerevisiae	Msc2p	Zrg17p	-
S.pombe	Cis4	Zrg17	-
C.elegans	Cdf5	Toc1	-
A. thaliana	MTP12	MTP5	-
D. melanogaster	-	-	Znt86D

Figure 9.1 Model of the two-step mechanism of ALP activation by ZNT5-6 and ZNT7. In the first step, ALP is stabilized, which may be regulated by the association of the PP motif in the luminal loop of ZNT5 and ZNT7 with ALP. In the second step, ALP is converted from the inactive *apo*-form to the active *holo*-form by acquiring zinc through ZNT5-6 and ZNT7. The motif consisting of two His and two Asp residues (the HDHD motif) forms an intramembranous zinc-binding site essential for the zinc transport activity of ZNT5 and ZNT7, from which zinc is released. ZNT5-6 orthologs are present in various eukaryotes.

and *A. thaliana* (MTP12 and MTP5 heterodimers),[21,22,29] indicating that ZNT5-6 and ZNT7, and their orthologs, are essential for enzyme activation by transporting zinc into the lumen of early secretory compartments. Importantly, ALP activity was not restored by overexpression of ZNT1, ZNT2, ZNT3, ZNT4, or ZNT8.

ALP activation requires conversion of *apo*-ALP (inactive form) to *holo*-ALP (active form) (ALP forms dimers and binds two zinc ions per monomer), in which a two-step mechanism is operative. First, protein stabilization, followed by enzyme activation.[30] Specifically, in cells lacking ZNT5-6 and ZNT7, the ALP protein is degraded mainly by proteasome-mediated degradation,[24,30] which apparently occurs in the Endo-H-sensitive form of ALP before reaching the medial Golgi. The degradation of ALP is inhibited by the overexpression of the zinc-transport incompetent mutant ZNT5-6,[30] suggesting that ZNT5-6 and ZNT7 are involved in the regulation of ALP folding, independent of zinc transport activity. The supply of zinc to ALP by ZNT5-6 and ZNT7 was coupled with the stabilization of ALP protein through escaping the degradation. Although the underlying mechanisms remain unknown, some information has been reported. ALP activation by ZNT5-6 and ZNT7 requires the Pro–Pro (PP)-motif in the loop region of ZNT5-6 or ZNT7 exposed to the lumens of the early secretory compartments (in the case of ZNT7, the luminal loop is between helices III and IV. See Figure 9.1). Mutation of di-proline in the PP-motif in ZNT7 into di-alanine (PP-AA) decreased TNAP activity, although the mutated ZNT7 has zinc transport activity.[22] The PP motif is highly conserved among the ZNT5 and ZNT7 orthologs described above but not among other ZNTs. Considering that the PP-motif is expected to be sterically close to the zinc exit site of ZNT5-6 and ZNT7, a cooperative regulatory mechanism may operate through the direct binding of ALP with ZNT5 or ZNT7 through the PP-motif. However, further investigation is required.

9.3 Zinc Enzymes Located in the Intracellular Organelles

It is theoretically plausible that acquisition of zinc by enzymes localized within their respective organelles is mediated by a co-localizing ZNT protein. However, ZNT5-6 and ZNT7 play particularly significant roles in specific zinc enzymes that function in intracellular organelles. In this section, we provide an overview of the types of zinc enzymes that require zinc supplied by ZNT5-6 and ZNT7 for their biological functions.

9.3.1 Contribution of ZNT5-6 and ZNT7 to *N*-Glycosylation

Glycosylation is essential for various physiological processes, including cell migration, cell–cell adhesion, cell signaling, and cell growth.[31,32] Additionally, alterations in *N*-glycosylation are associated with a wide range of pathological conditions, including cancer metastasis, viral pathogenesis, and chronic inflammatory diseases. Both *N*- and *O*-glycosylation require divalent cations, including calcium and manganese.[33–35] For instance, glycosidases such as ER α-mannosidase I and Golgi α-mannosidase I (members of the glycoside hydrolase family 47 (GH47)) require calcium, whereas glycosyltransferases, such as *N*-acetylglucosaminyltransferases and galactosyltransferases, require manganese as essential cofactors in the biosynthesis of *N*-glycosylation[33–35] (Figure 9.2).

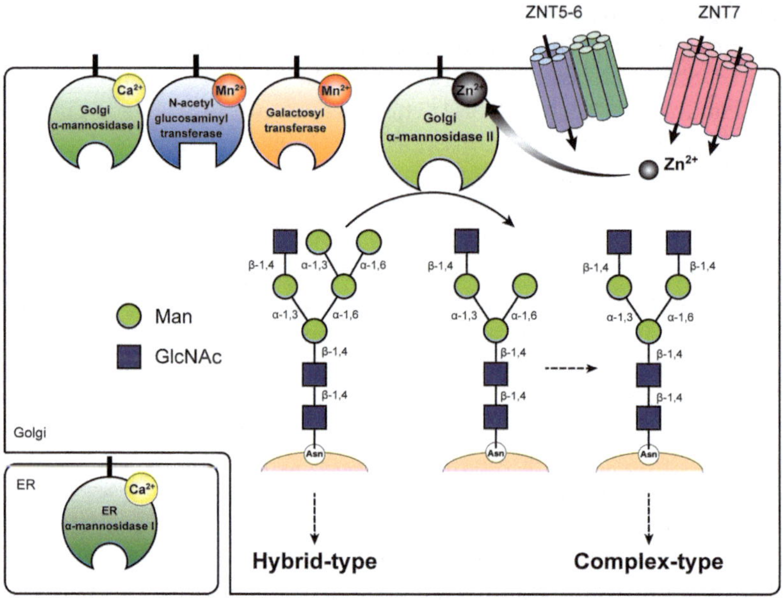

Figure 9.2 Zinc supplied by ZNT5-6 and ZNT7 is required for GMII activity in the Golgi. GMII plays a pivotal role in the biosynthesis of *N*-glycosylation, and loss of its activity reduces levels of complex-type glycans and markedly increases hybrid-type glycans. GMII is activated through the acquisition of zinc, mediated by ZNT5-6 and ZNT7; however, it is still unclear where GMII acquires zinc (ER or Golgi). Zinc activation of GMII is unique, because many other enzymes involved in *N*-glycosylation require calcium and manganese. For example, ER α-mannosidase I and Golgi α-mannosidase I (belonging to the GH47 family) require calcium, while *N*-acetylglucosaminyltransferases and galactosyltransferases require manganese.

Recently, we demonstrated that zinc is crucial for *N*-glycosylation, serving as an essential cofactor for Golgi α-mannosidase II (GMII, MAN2A1, a member of the GH38 family proteins)[36] (Figure 9.2). GMII, a typical type II transmembrane protein, plays a key role in the conversion of hybrid- to complex-type *N*-glycans by catalyzing the removal of both α-1,3-linked and α-1,6-linked mannoses.[37] GMII was reported to contain zinc at its active site in Drosophila in 2001,[38] but how GMII acquires zinc and the importance of this cation in its activity remain unclear. The loss of ZNT5-6 and ZNT7 function causes decreased GMII activity, which is not restored by excess zinc. Thus, *N*-glycosylation in *Z5Z7*-DKO cells resulted in a marked accumulation of hybrid-type glycans, with a concomitant reduction in complex-type glycans, indicating that ZNT5-6 and ZNT7 contribute to protein quality control by regulating GMII activity.

In various tumor cells, the distribution of *N*-linked sugars on the cell surface is altered and correlated with tumor progression, making GMII a potential molecular target for cancer drugs. Many GMII inhibitors, including swainsonine and related compounds, have been administered in clinical trials.[39–41] However, they have not been clinically applied, because they also inhibit the activity of lysosomal α-mannosidase (LAMAN, MAN2B1, a part of the GH38 family), a homolog of GMII, and cause a lysosomal storage disease, α-mannosidosis, resulting from the off-target inhibition.[41–43] Importantly, ZNT5-6 and ZNT7 are essential for GMII activity, but not LAMAN. These findings suggest that ZNT5-6 and ZNT7 are potential targets for cancer therapy. Indeed, the growth of *Z5Z7*-DKO cells from the pancreatic cancer cell line MIA Paca 2 was impaired in a nude mouse xenograft model compared to that of the parental cells. Thus, ZNT5-6 and ZNT7 may be novel chemotherapeutic targets in the future.

9.3.2 Contribution of ZNT5-6 and ZNT7 to Pigmentation

Melanins are the end products of complex multistep transformations of L-tyrosine within specialized organelles called melanosomes in melanocytes, and are the main pigments responsible for skin, hair, and eye color.[44,45] Melanin plays a pivotal role in photoprotection against the harmful effects of ultraviolet light. In melanocytes, two main types of melanin are synthesized: brown-to-black eumelanin and yellow-to-red pheomelanin.[46] Melanin biosynthesis within melanosomes requires three melanogenic enzymes, tyrosinase (TYR), and tyrosinase-related proteins 1 and 2 (TYRP1 and TYRP2), which are classified as type-3 copper proteins.[47] TYR activity requires copper,[48] and TYRs in other organisms, such as mushrooms and fungi, possess two copper atoms at the active site,[49,50] although copper coordination in TYR has not yet been reported in vertebrates. TYR, TYRP1, and TYRP2 are predicted to have homologous structures, and the amino acid residues that coordinate the metal (copper) at the active site are highly conserved among them.[47,51,52]

We recently found that the TYRP1 protein is not expressed in *Z5Z7*-DKO melanoma cells and is degraded *via* the lysosomal degradation pathway in these cells.[53] However, TYR and TYRP2 expression levels were not impaired in *Z5Z7*-DKO cells. *Z5Z7*-DKO melanoma Mewo cells exhibited a lighter brown with decreased melanin content, whereas WT Mewo cells were dark brown in color when observed in cell pellets. This finding in cultured cells is also true *in vivo*; pre-hatched medaka fish (*Oryzias latipes*) deficient in Znt5-6 and Znt7 show lighter brownish melanin, resulting in immature melanosomes with lower melanin content. In contrast, pre-hatched WT medaka showed a black color because of densely packed mature melanosomes with higher melanin content. This color alteration in cells and medaka is consistent with observations reported in other animals lacking TYRP1.[54,55] Therefore, ZNT5-6 and ZNT7 are required for melanin biosynthesis by supplying zinc.

9.3.3 Contribution of ZNT5-6 and ZNT7 to Sphingomyelin Metabolism

Sphingomyelin content is strictly regulated in cells by enzymes of the sphingomyelin metabolic pathways, maintaining a balance between SM synthesis and degradation.[56] One of these enzymes is sphingomyelin phosphodiesterase 1 (SMPD1),[57] also known as acid sphingomyelinase. SMPD1 hydrolyzes sphingomyelin to ceramide and phosphorylcholine. Genetic alterations in the *SMPD1* gene cause toxic accumulation of sphingomyelin in lysosomes, leading to a progressive, life-threatening disorder called Niemann–Pick disease.[58] SMPD1 is localized in lysosomes and the extracellular space, with these two forms of enzyme being splicing variants.[59] SMPD1 contains two zinc ions in its active site.

It was previously thought that the active site in lysosomal SMPD1 was metalated within the lumen of endosomes/lysosomes, whereas that in secretory SMPD1 was metalated in the extracellular milieu.[60,61] However, we recently demonstrated that SMPD1 is activated by metalation with zinc through ZNT5-6 and ZNT7,[62] indicating that zinc metalation of SMPD1 occurs in early secretory compartments. In *Z5Z7*-DKO cells, the ratio of ceramide to sphingomyelin content significantly decreased in terms of dominant molecular species and sphingomyelin content increased in terms of minor species. Transmission electron microscopy revealed that this alteration was accompanied by the appearance of multilamellar body-like structures due to membrane stacking and accumulation, similar to that observed in cells from patients with Niemann–Pick disease.[63] Thus, zinc, which is mediated by ZNT5-6 and ZNT7, is required for sphingomyelin metabolism.

9.3.4 Contribution of ZNT5-6 and ZNT7 to GPI-anchor Biosynthesis

Glycosylphosphatidylinositol (GPI)-anchored proteins play crucial roles in cell signaling, adhesion, immune responses, and enzyme activity. GPI is attached to the C-terminus of proteins during posttranslational modifications, and over 150 GPI-anchored proteins have been identified in humans.[64] Biosynthesis of GPI-anchored proteins requires more than 20 phosphatidylinositol glycan (PIG) proteins and involves the sequential addition of sugar and other components to phosphatidylinositol in multiple reaction steps. These proteins are transported through a clathrin-mediated post-Golgi membrane trafficking pathway. Previous studies using zinc chelators showed that GPI-anchor biosynthesis requires zinc.[65,66] However, the underlying molecular mechanism remains unclear.

We recently demonstrated that GPI-anchor biosynthesis requires zinc through ZNT5-6 and ZNT7 using cultured cells and medaka fish.[67] In *Z5Z7*-DKO cells, the expression of GPI-anchored proteins was substantially reduced, similar to that in mutant cells lacking *PIG* genes such as *PIGA* and *PIGT*. However, chimeric mutants, in which the C-terminal GPI-anchor attachment portion of GPI-anchored proteins was substituted with a single membrane-spanning polypeptide-anchored portion, were expressed on the cell surface without degradation, similar to that in WT

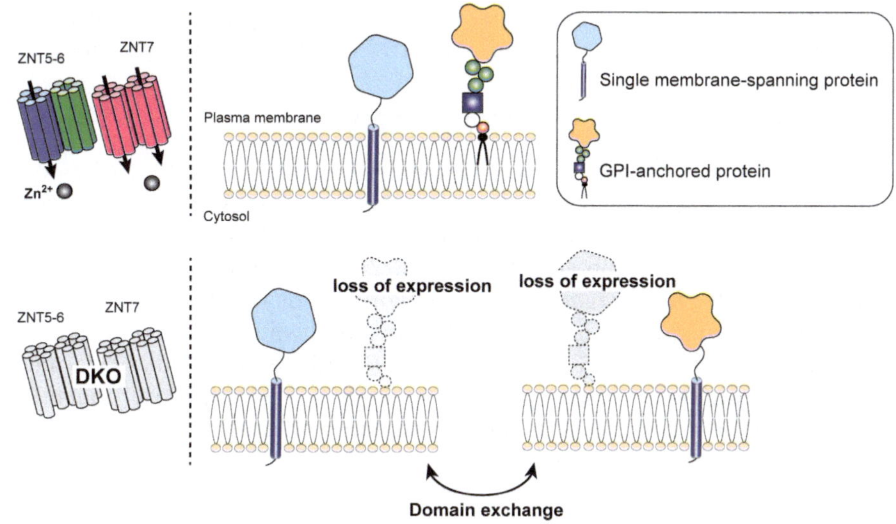

Figure 9.3 ZNT5-6 and ZNT7 functions are essential for the cell surface expression of GPI-anchored proteins. Both single membrane-spanning polypeptide-anchored proteins and GPI-anchored proteins are expressed on the cell surface in cells expressing ZNT5-6 or ZNT7 (upper row). In contrast, the cell surface expression of GPI-anchored proteins is lost in cells lacking ZNT5-6 and ZNT7 functions (lower row, left). If the GPI anchor attachment portion is exchanged with a single membrane-spanning polypeptide-anchored portion, the cell surface expression pattern is also exchanged, and *vice versa* (lower row, right).

cells. In contrast, membrane proteins anchored to the cell surface *via* a single membrane-spanning polypeptide portion, failed to be expressed in *Z5Z7*-DKO cells, if the single membrane-spanning polypeptide portion was substituted with a GPI anchor. Thus, zinc transported by ZNT5-6 and ZNT7 from the cytosol to the lumen of the early secretory pathway compartments plays an important role in the localization of GPI-anchored proteins to the plasma membrane (Figure 9.3).

A comparative analysis of *Z5Z7*-DKO cells and multiple KO cells lacking PIG proteins suggested that the defects in GPI-anchored protein expression were due to the loss of function of PIGN, PIGO, and PIGG (all GPI ethanolamine phosphate transferases), which are homologous to the zinc enzyme ALP, based on the conservation of the predicted metal-binding residues.[68] The involvement of ZNT5-6 and ZNT7 in GPI-anchor biosynthesis was further confirmed in medaka fish with disrupted *Znt5* and *Znt7*. Zebrafish *pig* mutants exhibit a touch-insensitive phenotype when poked with a needle,[69] and mutant medaka lacking Znt5-6 and Znt7 functions show a similar phenotype.[67] Mutations in multiple *PIG* genes are associated with defects in GPI-anchored proteins, resulting in human diseases (inherited GPI deficiency), including intellectual disability, facial dysmorphism, seizures, hypotonia, and dystonia.[64,70] Thus, the zinc-supplying functions of ZNT5-6 and ZNT7 are critical for the quality control of GPI-anchored proteins.

9.4 Conclusions and Perspectives

The significance of the biological functions of ZNT5-6 and ZNT7 has been elucidated as described above; however, there are many unanswered questions. The primary question remaining is: how are these zinc enzymes specifically regulated by zinc *via* ZNT5-6 and ZNT7? As reported in a previous study, there are zinc enzymes whose activity is not lost in *Z5Z7*-DKO cells; thus, they can acquire zinc in the secretory pathway *via* routes other than ZNT5-6 and ZNT7. The PP-motif of ZNT5-6 and ZNT7 is likely one mechanism that functions in the specific activation. However, other unclarified mechanisms are also likely to be significant and further investigation is required to clarify this point.

A second question is: why do vertebrate cells have two zinc transport complexes, ZNT5-6 and ZNT7, for zinc entry into the early secretory compartments? Finally, what are the critical differences in the biological functions of ZNT5-6 and ZNT7? The defects in enzyme activation and cellular phenotypes associated with loss of ZNT5-6 and ZNT7 functions are readily apparent in *Z5Z7*-DKO cells, and almost no defects (if any minor defects) are found in *Z5*- or *Z7*-single KO cells. However, recent reports indicate that ZNT5-6 transports zinc into the lumen of the medial Golgi, whereas ZNT7 transports zinc into the lumen of the *cis* Golgi[26,71] (see Chapter 6), and detailed analyses showed minor differences in biological functions (there are differences in the prevalence of specific glycan species between *Z5*- and *Z7*-single KO cells in *N*-glycomics[36]). The differences between their physiological functions are also found in the symptoms caused by mutations of *ZNT5/SLC30A5* or *ZNT7/SLC30A7* genes in human patients: the perinatal lethal cardiomyopathy is reported with mutations in ZNT5,[72] whereas stunted growth, testicular hypoplasia, and bone marrow failure is associated with mutations in ZNT7.[73] Mice and medaka deficient in *Znt5* or *Znt7* also show different phenotypes.[53,74,75] Elucidating the specific functions of ZNT5 and ZNT7 would lead to a deeper understanding of their differences and provide an answer to the question of why vertebrate cells have two zinc transport complexes, one of which is sufficient for yeast, nematode, fly, and plants.

Abbreviations

ER	Endoplasmic reticulum
DKO	Double knockout
ALP	Alkaline phosphatase
GPI	Glycosylphosphatidylinositol
GH	Glycoside hydrolase family
GMII	Golgi α-mannosidase II

LAMAN	Lysosomal α-mannosidase
TYR	Tyrosinase
TYRP	Tyrosinase-related protein
SMPD1	Sphingomyelin phosphodiesterase 1
PIG	Phosphatidylinositol glycan

Acknowledgements

This work was supported by a Grant-in-Aid for Scientific Research on Innovative Areas "Integrated Bio-metal Science" (MEXT KAKENHI Grant Number JP19H05768) from the Ministry of Education, Culture, Sports, Science, and Technology, Japan, a Grant-in-Aid for Scientific Research (B) (JSPS KAKENHI Grant Number JP22H02257) from the Japan Society for the Promotion of Science, and KOSÉ Cosmetology Research Foundation (to T. K.).

References

1. M. Brylinski and J. Skolnick, *Proteins*, 2011, **79**, 735–751.
2. B. L. Vallee and D. S. Auld, *Biochemistry*, 1990, **29**, 5647–5659.
3. W. N. Lipscomb and N. Strater, *Chem. Rev.*, 1996, **96**, 2375–2434.
4. K. J. Waldron, J. C. Rutherford, D. Ford and N. J. Robinson, *Nature*, 2009, **460**, 823–830.
5. J. Karges, R. W. Stokes and S. M. Cohen, *ACS Med. Chem. Lett.*, 2022, **13**, 428–435.
6. T. Tsuji, Y. Kurokawa, J. Chiche, J. Pouyssegur, H. Sato, H. Fukuzawa, M. Nagao and T. Kambe, *J. Biol. Chem.*, 2017, **292**, 2159–2173.
7. T. A. Takeda, S. Miyazaki, M. Kobayashi, K. Nishino, T. Goto, M. Matsunaga, M. Ooi, H. Shirakawa, F. Tani, T. Kawamura, M. Komai and T. Kambe, *Commun. Biol.*, 2018, **1**, 113.
8. T. Kambe, *Biosci., Biotechnol., Biochem.*, 2011, **75**, 1036–1043.
9. T. Kambe, T. A. Takeda and Y. Nishito, *Arch. Biochem. Biophys.*, 2016, **611**, 37–42.
10. T. Kambe, *Biochemistry*, 2020, **59**, 74–79.
11. T. Kambe, M. Matsunaga and T. A. Takeda, *Int. J. Mol. Sci.*, 2017, **18**, 2179.
12. T. Kambe, T. Tsuji, A. Hashimoto and N. Itsumura, *Physiol. Rev.*, 2015, **95**, 749–784.
13. T. Kambe, K. M. Taylor and D. Fu, *J. Biol. Chem.*, 2021, **296**, 100320.
14. A. Fukunaka, T. Suzuki, Y. Kurokawa, T. Yamazaki, N. Fujiwara, K. Ishihara, H. Migaki, K. Okumura, S. Masuda, Y. Yamaguchi-Iwai, M. Nagao and T. Kambe, *J. Biol. Chem.*, 2009, **284**, 30798–30806.
15. I. Lasry, Y. Golan, B. Berman, N. Amram, F. Glaser and Y. G. Assaraf, *J. Biol. Chem.*, 2014, **289**, 7275–7292.
16. H. B. Bui, S. Watanabe, N. Nomura, K. Liu, T. Uemura, M. Inoue, A. Tsutsumi, H. Fujita, K. Kinoshita, Y. Kato, S. Iwata, M. Kikkawa and K. Inaba, *Nat. Commun.*, 2023, **14**, 4770.

17. H. B. Bui and K. Inaba, *Int. J. Mol. Sci.*, 2024, **25**, 3045.
18. C. D. Ellis, C. W. Macdiarmid and D. J. Eide, *J. Biol. Chem.*, 2005, **280**, 28811–28818.
19. Y. Fang, R. Sugiura, Y. Ma, T. Yada-Matsushima, H. Umeno and T. Kuno, *Mol. Biol. Cell*, 2008, **19**, 1295–1303.
20. S. Choi, Y. M. Hu, M. E. Corkins, A. E. Palmer and A. J. Bird, *PLoS Genet.*, 2018, **14**, e1007262.
21. T. Fujiwara, M. Kawachi, Y. Sato, H. Mori, N. Kutsuna, S. Hasezawa and M. Maeshima, *FEBS J.*, 2015, **282**, 1965–1979.
22. S. Fujimoto, T. Tsuji, T. Fujiwara, T. A. Takeda, C. Merriman, A. Fukunaka, Y. Nishito, D. Fu, E. Hoch, I. Sekler, K. Fukue, Y. Miyamae, S. Masuda, M. Nagao and T. Kambe, *Biochem. J.*, 2016, **473**, 2611–2621.
23. T. Kambe, E. Suzuki and T. Komori, in *Zinc Signaling*, ed. T. Fukada and T. Kambe, Springer-Nature, 2nd edn, 2019, pp. 23–56.
24. K. Ishihara, T. Yamazaki, Y. Ishida, T. Suzuki, K. Oda, M. Nagao, Y. Yamaguchi-Iwai and T. Kambe, *J. Biol. Chem.*, 2006, **281**, 17743–17750.
25. S. Watanabe, Y. Amagai, S. Sannino, T. Tempio, T. Anelli, M. Harayama, S. Masui, I. Sorrentino, M. Yamada, R. Sitia and K. Inaba, *Nat. Commun.*, 2019, **10**, 603.
26. Y. Amagai, M. Yamada, T. Kowada, T. Watanabe, Y. Du, R. Liu, S. Naramoto, S. Watanabe, J. Kyozuka, T. Anelli, T. Tempio, R. Sitia, S. Mizukami and K. Inaba, *Nat. Commun.*, 2023, **14**, 2683.
27. T. Suzuki, K. Ishihara, H. Migaki, M. Nagao, Y. Yamaguchi-Iwai and T. Kambe, *J. Biol. Chem.*, 2005, **280**, 30956–30962.
28. T. Kambe and T. Wagatsuma, *Biophys. Rev.*, 2023, **4**, 041302.
29. E. Suzuki, N. Ogawa, T. A. Takeda, Y. Nishito, Y. K. Tanaka, T. Fujiwara, M. Matsunaga, S. Ueda, N. Kubo, T. Tsuji, A. Fukunaka, T. Yamazaki, K. M. Taylor, Y. Ogra and T. Kambe, *J. Biol. Chem.*, 2020, **295**, 5669–5684.
30. A. Fukunaka, Y. Kurokawa, F. Teranishi, I. Sekler, K. Oda, M. L. Ackland, V. Faundez, M. Hiromura, S. Masuda, M. Nagao, S. Enomoto and T. Kambe, *J. Biol. Chem.*, 2011, **286**, 16363–16373.
31. R. D. Cummings, *Glycoconjugate J.*, 2019, **36**, 241–257.
32. C. Reily, T. J. Stewart, M. B. Renfrow and J. Novak, *Nat. Rev. Nephrol.*, 2019, **15**, 346–366.
33. Y. Xiang, K. Karaveg and K. W. Moremen, *Proc. Natl. Acad. Sci. U. S. A.*, 2016, **113**, E7890–E7899.
34. J. Li and Y. Wang, *Cells*, 2022, **11**, 289.
35. K. W. Moremen and R. S. Haltiwanger, *Nat. Chem. Biol.*, 2019, **15**, 853–864.
36. H. Yuasa, N. Morino, T. Wagatsuma, M. Munekane, S. Ueda, M. Matsunaga, Y. Uchida, T. Katayama, T. Katoh and T. Kambe, *J. Biol. Chem.*, 2024, **300**, 107378.
37. N. Shah, D. A. Kuntz and D. R. Rose, *Proc. Natl. Acad. Sci. U. S. A.*, 2008, **105**, 9570–9575.
38. J. M. van den Elsen, D. A. Kuntz and D. R. Rose, *EMBO J.*, 2001, **20**, 3008–3017.

39. J. W. Dennis, K. Koch, S. Yousefi and I. VanderElst, *Cancer Res.*, 1990, **50**, 1867–1872.
40. P. E. Goss, J. Baptiste, B. Fernandes, M. Baker and J. W. Dennis, *Cancer Res.*, 1994, **54**, 1450–1457.
41. Z. Armstrong, C. L. Kuo, D. Lahav, B. Liu, R. Johnson, T. J. M. Beenakker, C. de Boer, C. S. Wong, E. R. van Rijssel, M. F. Debets, B. I. Florea, C. Hissink, R. G. Boot, P. P. Geurink, H. Ovaa, M. van der Stelt, G. M. van der Marel, J. D. C. Codee, J. Aerts, L. Wu, H. S. Overkleeft and G. J. Davies, *J. Am. Chem. Soc.*, 2020, **142**, 13021–13029.
42. I. Cenci di Bello, P. Dorling and B. Winchester, *Biochem. J.*, 1983, **215**, 693–696.
43. C. R. Huxtable and P. R. Dorling, *Am. J. Pathol.*, 1982, **107**, 124–126.
44. A. Slominski, D. J. Tobin, S. Shibahara and J. Wortsman, *Physiol. Rev.*, 2004, **84**, 1155–1228.
45. Y. Yamaguchi and V. J. Hearing, *Cold Spring Harbor Perspect. Med.*, 2014, **4**, a017046.
46. X. Lai, H. J. Wichers, M. Soler-Lopez and B. W. Dijkstra, *Chemistry*, 2018, **24**, 47–55.
47. F. Solano, *Int. J. Mol. Sci.*, 2018, **19**, 633.
48. M. J. Petris, D. Strausak and J. F. Mercer, *Hum. Mol. Genet.*, 2000, **9**, 2845–2851.
49. W. T. Ismaya, H. J. Rozeboom, A. Weijn, J. J. Mes, F. Fusetti, H. J. Wichers and B. W. Dijkstra, *Biochemistry*, 2011, **50**, 5477–5486.
50. N. Fujieda, K. Umakoshi, Y. Ochi, Y. Nishikawa, S. Yanagisawa, M. Kubo, G. Kurisu and S. Itoh, *Angew. Chem., Int. Ed.*, 2020, **59**, 13385–13390.
51. F. Aguilera, C. McDougall and B. M. Degnan, *BMC Evol. Biol.*, 2013, **13**, 96.
52. H. Decker and F. Tuczek, *Angew. Chem., Int. Ed.*, 2017, **56**, 14352–14354.
53. T. Wagatsuma, E. Suzuki, M. Shiotsu, A. Sogo, Y. Nishito, H. Ando, H. Hashimoto, M. J. Petris, M. Kinoshita and T. Kambe, *Commun. Biol.*, 2023, **6**, 403.
54. F. Murisier and F. Beermann, *Histol. Histopathol.*, 2006, **21**, 567–578.
55. E. E. Puckett, I. S. Davis, D. C. Harper, K. Wakamatsu, G. Battu, J. L. Belant, D. E. Beyer Jr., C. Carpenter, A. P. Crupi, M. Davidson, C. S. DePerno, N. Forman, N. L. Fowler, D. L. Garshelis, N. Gould, K. Gunther, M. Haroldson, S. Ito, D. Kocka, C. Lackey, R. Leahy, C. Lee-Roney, T. Lewis, A. Lutto, K. McGowan, C. Olfenbuttel, M. Orlando, A. Platt, M. D. Pollard, M. Ramaker, H. Reich, J. L. Sajecki, S. K. Sell, J. Strules, S. Thompson, F. van Manen, C. Whitman, R. Williamson, F. Winslow, C. B. Kaelin, M. S. Marks and G. S. Barsh, *Curr. Biol.*, 2023, **33**, 86–97 e10.
56. K. Bienias, A. Fiedorowicz, A. Sadowska, S. Prokopiuk and H. Car, *Pharmacol. Rep.*, 2016, **68**, 570–581.
57. E. H. Schuchman, *FEBS Lett.*, 2010, **584**, 1895–1900.
58. Y. F. Zhou, M. C. Metcalf, S. C. Garman, T. Edmunds, H. Qiu and R. R. Wei, *Nat. Commun.*, 2016, 7, 13082.
59. J. Kornhuber, C. Rhein, C. P. Muller and C. Muhle, *Biol. Chem.*, 2015, **396**, 707–736.

60. R. W. Jenkins, D. Canals, J. Idkowiak-Baldys, F. Simbari, P. Roddy, D. M. Perry, K. Kitatani, C. Luberto and Y. A. Hannun, *J. Biol. Chem.*, 2010, **285**, 35706–35718.
61. R. W. Jenkins, J. Idkowiak-Baldys, F. Simbari, D. Canals, P. Roddy, C. D. Riner, C. J. Clarke and Y. A. Hannun, *J. Biol. Chem.*, 2011, **286**, 3777–3788.
62. S. Ueda, Y. Manabe, N. Kubo, N. Morino, H. Yuasa, M. Shiotsu, T. Tsuji, T. Sugawara and T. Kambe, *Am. J. Physiol.: Cell Physiol.*, 2022, **322**, C948–C959.
63. E. Gabande-Rodriguez, P. Boya, V. Labrador, C. G. Dotti and M. D. Ledesma, *Cell Death Differ.*, 2014, **21**, 864–875.
64. T. Kinoshita, *Open Biol.*, 2020, **10**, 190290.
65. C. D. Ellis, F. Wang, C. W. MacDiarmid, S. Clark, T. Lyons and D. J. Eide, *J. Cell Biol.*, 2004, **166**, 325–335.
66. K. J. Mann and D. Sevlever, *Biochemistry*, 2001, **40**, 1205–1213.
67. T. Wagatsuma, K. Shimotsuma, A. Sogo, R. Sato, N. Kubo, S. Ueda, Y. Uchida, M. Kinoshita and T. Kambe, *J. Biol. Chem.*, 2022, **298**, 102011.
68. M. Y. Galperin and M. J. Jedrzejas, *Proteins*, 2001, **45**, 318–324.
69. Y. Nakano, M. Fujita, K. Ogino, L. Saint-Amant, T. Kinoshita, Y. Oda and H. Hirata, *Development*, 2010, **137**, 1689–1698.
70. K. Bellai-Dussault, T. T. M. Nguyen, N. V. Baratang, D. A. Jimenez-Cruz and P. M. Campeau, *Clin. Genet.*, 2019, **95**, 112–121.
71. T. Kowada, T. Watanabe, Y. Amagai, R. Liu, M. Yamada, H. Takahashi, T. Matsui, K. Inaba and S. Mizukami, *Cell Chem. Biol.*, 2020, **27**, 1521–1531 e1528.
72. J. K. Lieberwirth, P. Joset, A. Heinze, J. Hentschel, A. Stein, A. Iannaccone, K. Steindl, A. Kuechler and R. Abou Jamra, *Eur. J. Hum. Genet.*, 2021, **29**, 808–815.
73. L. Huang, Z. Yang, C. P. Kirschke, C. Prouteau, M. C. Copin, D. Bonneau, I. Pellier, R. Coutant, C. Miot and A. Ziegler, *Hum. Mol. Genet.*, 2023, 2016–2031.
74. K. Inoue, K. Matsuda, M. Itoh, H. Kawaguchi, H. Tomoike, T. Aoyagi, R. Nagai, M. Hori, Y. Nakamura and T. Tanaka, *Hum. Mol. Genet.*, 2002, **11**, 1775–1784.
75. L. Huang, Y. Y. Yu, C. P. Kirschke, E. R. Gertz and K. K. Lloyd, *J. Biol. Chem.*, 2007, **282**, 37053–37063.

Section III: Roles of Zinc in Health and Disease

CHAPTER 10

TRPC6-mediated Zn^{2+} Influx as a Cardiotonic Mediator

XINYA MI,[a] CHENLIN SU,[a] TOMOYA ITO,[a] YURI KATO,[a] AKIYUKI NISHIMURA[b,c,d] AND MOTOHIRO NISHIDA*[a,b,c,d]

[a] Department of Physiology, Graduate School of Pharmaceutical Sciences, Kyushu University, 3-1-1 Maidashi, Higashi-ku, Fukuoka 812-8582, Japan; [b] National Institute for Physiological Science (NIPS), National Institutes of Natural Sciences (NINS), Japan; [c] Exploratory Research Center on Life and Living Systems (ExCELLS), NINS, Japan; [d] SOKENDAI (The Graduate University for Advanced Studies), Japan
*Email: nishida@phar.kyushu-u.ac.jp

10.1 Introduction

Zinc ions (Zn^{2+}) are recognized as an indispensable mineral substance for human life.[1] The essential trace element Zn is a structural constituent in numerous proteins, including growth factors, cytokines, receptors, enzymes, and transcription factors belonging to cellular signaling pathways, and is essential for their biological activity.[2,3] Emphasizing Zn's physiological relevance to life, a human genome bioinformatics study revealed that approximately 10% of all proteins may bind with Zn^{2+}.[4] The most abundant class of Zn^{2+}-binding proteins in humans is that of Zn^{2+}-fingers, with 4 cysteine (Cys) or 2 Cys and 2 histidine (His) being the most common types of coordination environment.[4] Zn^{2+} is primarily stored in muscles and bones, comprising approximately 85% of the body's total Zn^{2+} content. The skin and liver contain around 11% of the total Zn^{2+}.[5] This vital micronutrient is integral to a multitude of physiological processes, including cellular proliferation, programmed cell death (apoptosis), DNA synthesis

Coordination Chemistry Perspectives No. 2
Zinc in Biology: Molecular Structures, Cellular Processes and Living Systems
Edited by Hiroyuki Yasui and Taiho Kambe

Published by the Royal Society of Chemistry, www.rsc.org

and repair, immune system function, and neural activity.[5,6] These diverse roles are executed through Zn^{2+} interaction with a range of Zn^{2+}-dependent proteins, such as transcription factors and enzymes that are pivotal in essential cellular signaling pathways.

Around 90% of Zn^{2+} are believed to be tightly bound to intracellular molecules, while the remaining 10% of intracellular Zn^{2+} exists in a labile form.[3] Intracellular Zn^{2+} levels are tightly regulated to maintain precise physiological concentrations, due to the extensive range of Zn^{2+}-dependent functions. Zn^{2+} uptake, subcellular distribution, storage, and excretion are stringently regulated. In contrast, some kinds of metals such as iron, copper, and mercury can accumulate excessively in the organism and reach toxic levels.

Homeostasis of intracellular Zn^{2+} dynamics generally includes the following processes: (1) Zn^{2+} incorporation into intracellular organelles and molecules such as mitochondria and metallothioneins (MTs). Additionally, the storage of this essential element is predominantly mediated by MTs;[7,8] (2) Zn^{2+} is transported across the plasma membrane under the control of two major transporter families:[9] SLC30 (zinc transports (ZnT)) for the excretion of zinc and SLC39 (Zrt/Irt-like proteins (ZIP)) for the zinc intake;[5,9] (3) Zn^{2+}-permeable ion channels, such as voltage-dependent Ca^{2+} channels (VDCCs)[10,11] and some ion channels of the TRP family;[12] (4) Zn^{2+} extrusion into extracellular space or secretion by vesicles.[13,14]

MT is regulated by the redox state and intracellular pH.[12] When a cell is under oxidative stress, reactive oxygen species (ROS) can oxidize the thiol groups on MT, leading to the dissociation of Zn^{2+} from MTs and an increase in the concentration of free Zn^{2+} in the cytoplasm.[15,16] This release of Zn^{2+} can activate metal-responsive transcription factor 1 (MTF-1), which in turn induces the expression of genes encoding Zn^{2+} transporters and MT, thereby regulating intracellular Zn^{2+} homeostasis.[16] The Zn^{2+}-buffering capacity of MT is crucial for maintaining intracellular Zn^{2+} homeostasis.[5,7] MT can bind excess Zn^{2+} to prevent their toxic effects on the cell and can release Zn^{2+} when there is a deficiency to meet the physiological needs of the cell.[5,17]

Intracellular Zn^{2+} homeostasis is primarily maintained by two major families of zinc ion transport proteins: the ZnT family and the ZIP family.[5,7,17] The ZnT family is responsible for transporting zinc ions from the cytoplasm to the extracellular space or into intracellular organelles, whereas the ZIP family transports Zn^{2+} from the extracellular space or organelles into the cytoplasm.[5,17,18] These two families of transport proteins work synergistically to precisely regulate Zn^{2+} concentrations in different cellular compartments, thereby maintaining Zn^{2+} homeostasis.[19–21]

Zn^{2+}-permeable ion channels, mainly expressed in the cytoplasmic membrane, regulate the maintaining Zn^{2+} homeostasis. Notably, the cytoplasm expressed ion channels are not only involved in the modulation of electrochemical signals but are also a critical method to increase the metal ion influx. However, the trait of ion channels is low metal ion selectivity, which allows passive diffusion of metal ions. For example, Zn^{2+} cross the cell membrane through VDCCs to inverse the intracellular Zn^{2+} concentration and activate the related signaling pathway. VDCCs are primarily expressed in excitable cells, such as mammalian neurons, cardiomyocytes, pancreatic β-cells, and chromaffin cells,[22,23] which are also permeable to Zn^{2+}.[24] Inhibitory effects on Zn^{2+} influx indicate that the Zn^{2+} permeability is relayed by different types of VDCCs.[25]

The Transient Receptor Potential (TRP) channel family consists of seven subfamilies: TRPC (canonical), TRPV (vanilloid), TRPM (melastatin), TRPA (ankyrin repeat), TRPP (polycystin), TRPML (mucolipin), and TRPN (no mechanoreceptor potential C).[6,26] The TRP channel family comprises a group of non-selective cation channels widely expressed in various cell types [6,27] and which are involved in multiple physiological processes, including sensory transduction, cell proliferation, cell death, and ion homeostasis.[28] The properties of TRP channels make a prompt and dynamic response to the regulation of Zn^{2+} homeostasis. VDCCs are mainly located on conductive cell membranes and rely on the voltage change to regulate Zn^{2+} influx in the specific window. On the other hand, TRP channels are associated with the mechanism of pain, inflammation, cancer and so on in disease, but VDCCs contribute to cardiovascular disease. Thus, TRP channels could diversify the different signaling responses and provide a potential target in pathological disease.

10.2 The Contribution of Zn^{2+} Permeable Mammalian TRP Channels in Physiological and Pathological Conditions

10.2.1 TRPM Subfamily

Increasing research indicates that TRP channels, particularly those in the TRPM subfamily, play a critical role in regulating cellular Zn^{2+} homeostasis.[6,27] TRPM7 is considered the TRP channel most closely associated with Zn^{2+} homeostasis.[27] It is the only known TRP channel with a higher permeability to Zn^{2+} than to Ca^{2+} (permeability: $Zn^{2+} > Mg^{2+} > Ca^{2+}$).[29,30] Zn^{2+} influx mediated by TRPM7 is a key factor in neuronal death associated with neurological disorders such as ischemic stroke, traumatic brain injury, and epilepsy.[25,27] TRPM7 can sense oxidative stress and facilitate the release of Zn^{2+} from intracellular glutathione-rich vesicles (M7Vs), which are organelles distinct from endosomes, lysosomes, and other common vesicles or organelles. When cytoplasmic Zn^{2+} concentrations increase, M7Vs accumulate Zn^{2+} within their reducing lumens.[31,32] Cellular stress hormones, such as corticotropin-releasing factor (CRF)-related peptide urocortin, upregulate the expression of TRPM7 channels, leading to increased Zn^{2+} influx and neurotoxicity.[33] In certain cancer cell lines and patient tumor tissues (*e.g.*, human pancreatic cancer, breast cancer, and glioblastoma), TRPM7 expression is abnormally upregulated, indicating a role for TRPM7 in promoting cell proliferation.[32] Overexpression of TRPM7 in HEK293 cells increases intracellular Zn^{2+} accumulation, whereas knockout of TRPM7 leads to a significant decrease in intracellular Zn^{2+} concentration.[12] TRPM7 is expressed in mouse oocytes, eggs, and embryos, playing an important role in fertilization and early embryonic development. Pharmacological inhibition of TRPM7 channels delays preimplantation embryonic development and reduces progression to the blastocyst stage.[27,34] TRPM6 is a homolog of TRPM7 and is involved in Mg^{2+} homeostasis in the intestines and kidney.[6,35] TRPM6 can also mediate Zn^{2+} influx but has a lower permeability to Zn^{2+} compared to TRPM7.[30] However, the sources do not provide further details on how TRPM6 channels specifically interact with zinc or how zinc regulates the function of TRPM6 channels. While TRPM3 channels can be activated by extracellular Zn^{2+},[36] there is ongoing debate about whether TRPM3 directly mediates Zn^{2+} influx.[37] TRPM3 channels are expressed in several tissues and cell types, including pancreatic β-cells and somatosensory neurons.[38,39] Under physiologically relevant ionic conditions, TRPM3 channels mediate zinc uptake in pancreatic β-cells, and this process can be controlled by TRPM3 agonists. Moreover, even when spontaneously actived without agonists, TRPM3 channels can conduct Zn^{2+} without the need for stimulation by high concentrations of steroidal agonists. In addition, the depolarization caused by the activation of TRPM3 channels leads to the activation of voltage-gated calcium channels (VGCCs) and increases the Zn^{2+} influx through these channels.[36] Unlike the closely related channels TRPM6 and TRPM7, which primarily participate in cellular and systemic magnesium homeostasis, TRPM3 channels have high permeability to Zn^{2+}. The divalent cation conductance profile of TRPM3 channels is like that of TRPM6 and TRPM7, but there are quantitative and qualitative differences among these channels. The conductance sequence for TRPM3 channels is $Ca^{2+} > Zn^{2+} > Mg^{2+} > Ni^{2+}$.[36]

10.2.2 TRPC6 Channel

The TRPC6 channel is a non-selective cation channel capable of transporting zinc ions. Zinc uptake can be regulated by activating TRPC6 channels, and the expression levels of TRPC6 affect intracellular zinc content and storage capacity.[36,40,41] When HEK-TRPC6 cells (HEK cells overexpressing TRPC6 channels) are placed in saline solutions containing 2 to 100 μM zinc acetate, activation of TRPC6 channels induces an inward zinc current[42] and have higher intracellular zinc content compared to their parental HEK cells, which suggests that TRPC6 channels may play a role in controlling Zn^{2+} homeostasis. The expression of actin-associated proteins ezrin and cofilin-1 was up-regulated by the TRPC6 channel which also changed the actin cytoskeleton without the actin content, suggesting the TRPC6 channel participates in Zn^{2+} transport and storage and buffering capacities of cells.[43] The TRPC6 channel is considered a mediator of pathological cardiac remodeling.

Zn^{2+} influx through TRPC6 channels stimulated by α_1-adrenergic receptors (α_1AR) can enhance β-adrenergic receptor (βAR)-dependent positive inotropic effects in mouse hearts and improve chronic heart failure (Figure 10.1).[40,44] TRPC6-deficient mice showed low level accumulation of zinc and iron contents in the placenta, liver, kidneys and brain during embryonic development.

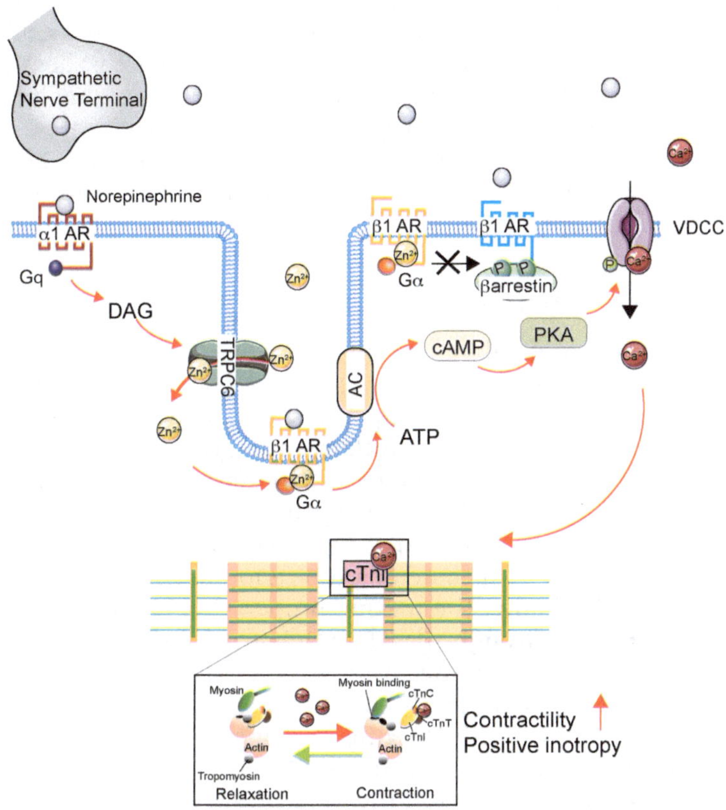

Figure 10.1 Schema of the role of TRPC6-mediated Zn^{2+} influx in adrenergic receptor-stimulated cardiac contractility and inotropy. Cardiac inotropy through the sympathetic nervous system has been predominantly mediated by βAR activation, although the chemical binding affinity of NE to β1AR is much lower than that to α_1AR. α_1AR-mediated Zn^{2+} influx *via* TRPC6 promotes NE-induced cardiac contractility and inotropy by local Zn^{2+}-dependent enhancement of β1ARGs signaling. Activated β1AR elevated intracellular cAMP levels which activates PKA to voltage dependent calcium channels. The increased Ca^{2+} through voltage dependent calcium channels bind to the Ca^{2+}-sensor of the sarcomere, cardiac troponin C (cTnC), which forms together with cTnl and cTnT the troponin complex, releasing cTnl from actin and occurring force production. TRPC6-mediated Zn^{2+} influx negatively regulates β-arrestin-dependent β1AR internalization. Downregulation of the α_1AR-TRPC6-Zn^{2+} axis may reduce sympathetic nerve-mediated cardiac inotropy.

Additionally, these metal imbalances (excess or deficiency) develop in an age- and tissue-dependent manner, and elevated levels of TRPC6 protein are found in the placentas of patients with preeclampsia.[45] TRPC6-mediated Zn^{2+} influx plays a crucial role in resisting intestinal inflammation by maintaining redox homeostasis and the gut microbiota.[46,47] TRPC6 deficiency exacerbates the progression of dextran sulfate sodium (DSS)-induced colitis, while activation of TRPC6 can alleviate the severity of colitis.[46] The increase of Zn^{2+} influx activated by the TRPC6 channel in renal tubular epithelial cells could inhibit activity of the NLRP3 inflammasome and attenuate renal ischemia-reperfusion injury.[48] The TRPC6 channel significantly influences zinc homeostasis and is involved in various physiological and pathological processes. Understanding the mechanisms by which TRPC6 channels regulate Zn^{2+} uptake and storage can provide insights into therapeutic strategies for diseases associated with zinc imbalance.

10.2.3 TRPA1 Channel

TRPA1 is widely expressed in neuronal and non-neuronal cells, and it acts as a biological sensor that is activated by environmental irritants and oxidative- and thiol-reactive compounds.[49,50] The activation of TRPA1 induced under HIF-1[51] mediated inflammation can synergistically change intracellular zinc concentration in inflammatory FLSs,[52] even though Zn^{2+} is known to be highly potent against TRPA1.[49,50] When the concentration is higher than 300 μM, zinc also exhibited an inhibitory effect on TRPA1, suggesting that there might be a low affinity inhibitory site for zinc in TRPA1.[50] Of note, the unique property of TRPA1 is extracellular zinc can enter cells through the TRPA1 channel and interact with intracellular cysteine and histidine residues, thereby activating TRPA1.[50] In the present study, it was identified that two sets of four amino acids in TRPA1 constitute intracellular zinc binding domains (IZDs) in both hTRPA1 and chicken TRPA1 (gTRPA1). Interestingly, gTRPA1 does not respond to external zinc even in the presence of external calcium ions.[53] Research indicates that intracellular zinc can enhance pain signals by activating TRPA1. For example, in animal studies, intraplantar injection of sulfide (a known TRPA1 agonist that increases intracellular zinc levels) induces pain behaviors in wild-type mice but not in TRPA1 knockout mice. These evidences suggest that Zn^{2+} is considered a modulator of pain signals, and the TRPA1 channel plays a crucial role in mediating pain,[54] especially inflammatory pain[52,55,56] and pain-sensing neurons.[56] The TRPA1 channel is involved in regulating inflammatory responses. During inflammation, the release of cytokines such as TNF-α and IL-1α upregulates TRPA1 expression. Thus, intracellular zinc can activate TRPA1 channels, thereby affecting the release of inflammatory mediators like IL-6 and IL-8.[52,53] Zn^{2+} has been shown to induce itch[57] in mice, an effect that requires the involvement of TRPA1 channels. The use of TRPA1 blockers can inhibit zinc-induced itch behaviors. The use of TRPA1 blockers (such as ruthenium red, A967079, and HC030031) can inhibit zinc-induced itch behaviors.[58] Zn^{2+}, as a novel stimulant for TRPA1, could react on cholecystokinin secretion, which involves intracellular Zn^{2+} mobilization.[59]

10.2.4 TRPML1 Channel

TRPML1 is a non-selective cation channel located at the membranes of late endosomes and lysosomes.[60] Mutations in TRPML1 are the genetic cause of the disease Mucolipidosis type IV (MLIV).[61] It participates in regulating their pH balance, transport, membrane fusion, autophagy,[62–64] and other functions.[65] Studies have shown that loss of TRPML1 function is associated with disruption of intracellular Zn^{2+} homeostasis.[63,66] Deletion of TRPML1 leads to the accumulation of zinc ions in both the cytoplasm and lysosomes.[62,67,68] TRPML1 may regulate cellular Zn^{2+} dynamics through its interaction with Transmembrane Protein 163 (TMEM163), a putative Zn^{2+} transporter that is highly expressed in the brain, lungs, and testes. The interaction between TMEM163[69,70] and TRPML1 is disrupted in fibroblasts from MLIV patients, which is associated with increased lysosomal Zn^{2+} level.[67,69] Activation of TRPML1 enhances the initiation of autophagy in cancer cells.[64] Overexpression of TRPML1 makes normal cells more susceptible to attack by ML-SA (a TRPML1 agonist), whereas inhibition or silencing of TRPML1 in metastatic melanoma cells can prevent such cell death.[68]

10.2.5 TRPV6 Channel

TRPV6 is a TRP channel predominantly expressed in epithelial tissues such as the intestine and placenta, playing a crucial role in calcium absorption.[71] Except for calcium, TRPV6 could transport Zn^{2+} and is vital for maintaining systemic zinc homeostasis and regulating various physiological and pathological processes.[71,72] TRPV6 is mainly localized on the apical membrane of intestinal epithelial cells and participates in the body's absorption of Zn^{2+}.[73] Particularly under conditions of low dietary calcium intake, the expression of TRPV6 is upregulated, possibly serving as a primary pathway for zinc absorption.[71,72] Zinc absorption mediated by TRPV6 may exacerbate the negative effects associated with zinc overload.[71,72] Excessive zinc may also promote tumor development and progression.[74] Low concentration of Zn^{2+} could enhance TRPV6-mediated calcium currents and calcium absorption.[71] However, high concentration of Zn^{2+} could inhibit TRPV6 activity, blocking the influx of calcium and other cations, including Zn^{2+} itself.[71] Some studies suggest that TRPV6 may play a role in certain cancers, such as prostate cancer, although the exact mechanisms remain unclear.[71] The relationship between TRPV6 and Zn^{2+} in tumorigenesis requires further investigation.

10.3 TRPC6-mediated Zn^{2+} Influx as a New Therapeutic Target of Heart Failure

10.3.1 TRPC6 Channel Activation for the Treatment of Heart Failure

Chronic heart failure is treatment-resistant and carries a high risk of acute exacerbations. There have been no reports of cardiac function recovery except through transplantation. We previously demonstrated that TRPC6-dependent Zn^{2+} influx enhances the myocardial baroreflex-dependent positive inotropic effect of normal mouse hearts (Figure 10.1).[40] Therefore, we investigated whether the activation of TRPC6-dependent Zn^{2+} influx could contribute to the recovery of cardiac function and survival prognosis in chronic heart failure. We developed a pressure overload-induced chronic heart failure model in mice by performing transverse aortic constriction

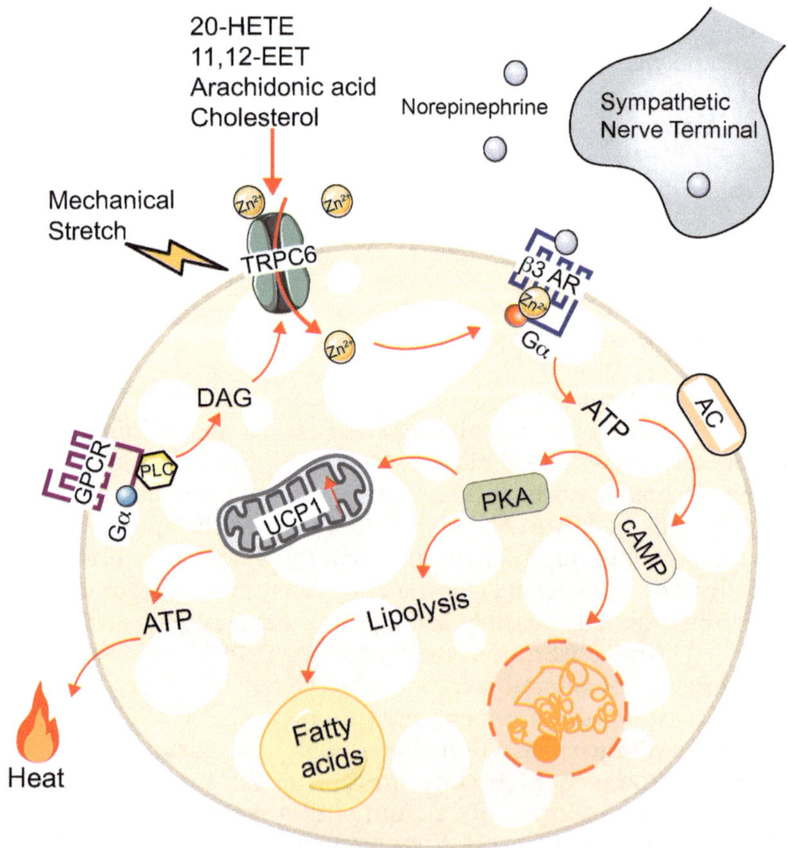

Figure 10.2 An increase of TRPC6-activated Zn^{2+} influx enhances β2AR-mediated vasodilation in peripheral blood vessels. Zn^{2+} through the TRPC6 channel contributes to β2AR activation. Stimulated β2AR enhances the function of vasodilation in peripheral blood vessels.

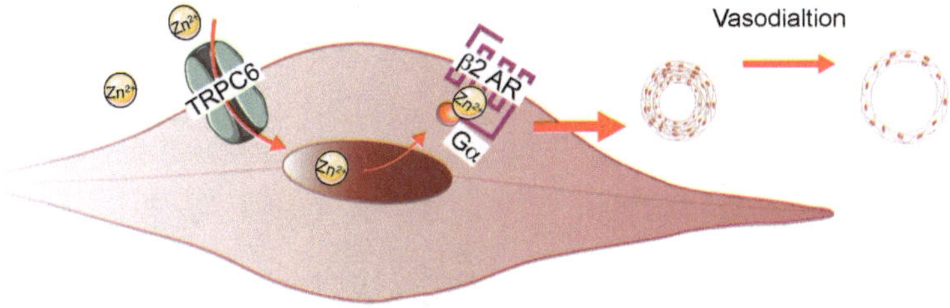

Figure 10.3 An increase in Zn^{2+} influx by stimulation of TRPC6 channels contributes to β3AR-mediated thermogenesis in brown adipocytes. Intracellular diacylglycerol (DAG) and extracellular lipid metabolites (such as 20-HETE, 11,12-EET, arachidonic acid and cholesterol), NE released from the sympathetic nerve, as well as mechanical stretch of the cell membrane activate the TRPC6 channel to increase Zn^{2+}-influx which contributes to the activation of β3AR. Stimulated β3AR elevates intracellular cAMP levels which activates PKA to lipolysis *via* phosphorylation, with subsequent release of free fatty acids which promote mitochondrial β-oxidation and direct activation of UCP1.

(TAC). One week after this procedure, an osmotic pump including TRPC3/6/7 channel activator, PPZ2 (2-[4-(2,3-dimethylphenyl)-piperazin-1-yl]-*N*-(2-ethoxyphenyl) acetamide), was implanted intraperitoneally and chronically administered for four weeks. A significant recovery of left ventricular functions was observed in the PPZ2-treated mice compared to vehicle-treated mice. However, this recovery was not observed in TRPC6-deficient mice treated with PPZ2 after TAC. Additionally, PPZ2 treatment suppressed TAC-induced myocardial hypertrophy and interstitial fibrosis, which are associated with heart failure. Furthermore, a marked increase in Zn^{2+} levels was observed in TAC-operated hearts treated with PPZ2, compared with those with vehicle or sham-operated hearts with PPZ2. These results indicate that TRPC6 activation improves chronic heart failure by promoting Zn^{2+} mobilization in *in vivo* mice. Globally, TRPC3/6 channels are considered mediators of cardiovascular remodeling, and research is advancing on the development of heart failure therapies that primarily inhibit TRPC3/6 channel activities. The increased expression of *TRPC6* gene in the hearts of heart failure patients further supports the validity of TRPC6 as a drug target. However, we failed to obtain any protective effect by TRPC6 deficiency from pressure overload-induced heart failure.[75,76] These observations clearly suggest that TRPC6 activation enhances the positive inotropic effect on the heart *via* Zn^{2+} influx, showing a stronger effect in improving chronic heart failure than TRPC6 inhibition.

10.3.2 The Potential of Drug Development Strategies Targeting Zn^{2+}

As both TRPC6 and β-adrenoceptors are also expressed in peripheral smooth muscle cells and adipocytes, the activation of TRPC6-mediated Zn^{2+} influx by physico-chemical stimuli has potential to promote smooth muscle dilatation (Figure 10.2), or thermogenesis and lipid metabolism (Figure 10.3).

Table 10.1 Zn^{2+}-permeable TRP channels and pathological roles.

Ion channel	Cells/organs	Physiological roles of Zn^{2+}	Pathophysiological roles of Zn^{2+}	Ref.
TRPM7	Mouse oocytes, eggs, and embryos; cancer cell lines (*e.g.*, human pancreatic cancer, breast cancer, glioblastoma); neurons	**Fertilization & early embryonic development:** essential for fertilization and early embryonic development. **Zinc homeostasis:** regulates intracellular Zn^{2+} levels through uptake and release. **Maintenance of redox balance:** facilitates release of Zn^{2+} from glutathione-rich vesicles (M7Vs) under oxidative stress. **Cell proliferation:** promotes cell proliferation in cancer cells. **Zinc accumulation:** overexpression increases intracellular Zn^{2+} accumulation.	**Neuronal death:** contributes to neuronal death in neurological disorders such as ischemic stroke, traumatic brain injury, and epilepsy. **Cancer progression:** Aberrant upregulation promotes uncontrolled cell proliferation in various cancers. **Neurotoxicity:** increased Zn^{2+} influx leads to toxic effects on neurons. **Embryonic development delay:** pharmacological inhibition delays preimplantation and blastocyst formation. **Zinc deficiency:** knockout leads to significant decrease in intracellular Zn^{2+} concentration.	6, 12, 23, 25, 27–32 and 34
TRPM6	Intestines; kidneys	**Magnesium homeostasis:** primarily involved in Mg^{2+} regulation in the intestines and kidneys. **Zinc influx:** can mediate Zn^{2+} influx, albeit to a lesser extent compared to TRPM7.	**Limited pathological information:** current studies do not provide detailed insights into specific pathological roles related to Zn^{2+}.	6, 28, 33 and 35

TRPM3	Pancreatic β-cells; somatosensory neurons	**Zinc uptake in pancreatic β-cells:** facilitates Zn^{2+} uptake under physiologically relevant ionic conditions, which can be modulated by TRPM3 agonists. **Independent Zn^{2+} conduction:** can conduct Zn^{2+} ions without agonist stimulation when spontaneously active. **Regulation of calcium homeostasis:** activation leads to VGCC activation and increased Zn^{2+} influx.	**Neurotoxicity:** activation induces depolarization and VGCC-mediated Zn^{2+} influx, contributing to neurotoxic effects. **Debated role in Zn^{2+} influx:** ongoing discussions about whether TRPM3 directly mediates Zn^{2+} influx. **Enhanced Zn^{2+} influx *via* VGCCs:** activation induces VGCC-mediated Zn^{2+} entry.	34–39
TRPC6	Heart; placenta; liver, kidneys, brain; intestines	**Zinc homeostasis:** regulates intracellular Zn^{2+} content and storage capacity. **Cardiac function:** enhances β-adrenergic receptor (βAR)-dependent positive inotropic effects in mouse hearts and improves chronic heart failure. **Resistance to intestinal inflammation:** maintains redox homeostasis and gut microbiota by mediating Zn^{2+} influx. **Actin cytoskeleton regulation:** upregulates actin-associated proteins ezrin and cofilin-1, influencing Zn^{2+} transport and storage.	**Pathological cardiac remodeling:** mediates remodeling processes in the heart. **Metal imbalances:** TRPC6 deficiency alters zinc and iron levels in placenta, liver, kidneys, and brain during embryonic development, with no changes in the lungs. Imbalances develop in an age- and tissue-dependent manner. **Preeclampsia:** elevated TRPC6 levels in placentas of preeclampsia patients. **Intestinal inflammation:** TRPC6 deficiency exacerbates dextran sulfate sodium (DSS)-induced colitis; activation alleviates colitis severity. **Renal ischemia-reperfusion injury:** increased Zn^{2+} influx *via* TRPC6 inhibits NLRP3 inflammasome and attenuates injury. **Actin cytoskeleton changes:** participation in Zn^{2+} transport and buffering capacities.	34, 38–46

Table 10.1 (*Continued*)

Ion channel	Cells/organs	Physiological roles of Zn^{2+}	Pathophysiological roles of Zn^{2+}	Ref.
TRPA1	Neuronal cells; non-neuronal cells; pancreatic β-cells; somatosensory neurons; respiratory tissues	**Pain signal modulation:** acts as a biological sensor activated by environmental irritants and oxidative- and thiol-reactive compounds, facilitating pain signal transduction. **Inflammatory response regulation:** synergistically changes intracellular Zn^{2+} concentration with Zn^{2+} carrier proteins (ZIP8, ZIP14) under HIF-1 mediated inflammation. **Itch induction:** mediates zinc-induced itch behaviors in mice. **Cholecystokinin secretion:** zinc mobilization involving intracellular Zn^{2+}.	**Pain and inflammation:** intracellular Zn^{2+} enhances pain signals by activating TRPA1. **Inflammatory mediator release:** activated TRPA1 affects the release of cytokines like IL-6 and IL-8. **Respiratory toxicity:** involved in sensing toxic chemicals, including zinc chloride fumes, contributing to toxic lung injury. **Itch behaviors:** Zn^{2+}-induced itch *via* TRPA1 can be inhibited by TRPA1 blockers. **Inflammatory diseases:** regulation of inflammatory responses *via* cytokine release. **Pain-sensing neurons:** crucial role in mediating pain, especially inflammatory pain.	47–57
TRPML1	Late endosomes; lysosomes; brain; lungs; testes	**Intracellular zinc homeostasis:** regulates cellular zinc dynamics through interaction with TMEM163, a putative zinc transporter. **Autophagy initiation:** activation enhances autophagy in cancer cells.	**Mucolipidosis type IV (MLIV):** mutations cause MLIV, associated with disrupted zinc homeostasis. **Zinc accumulation:** loss of TRPML1 function leads to Zn^{2+} accumulation in cytoplasm and lysosomes.	58–68

		pH balance & membrane transport: maintains pH balance and transport functions in endosomes and lysosomes. **Autophagy & membrane fusion:** participates in membrane fusion and autophagy processes.	**Cancer resistance:** overexpression increases susceptibility to TRPML1 agonists (*e.g.*, ML-SA), whereas inhibition prevents cell death in metastatic melanoma. **Disrupted TMEM163 interaction:** leads to increased lysosomal Zn^{2+} levels in MLIV patients. **Genetic disorders:** associated with MLIV and disrupted intracellular zinc dynamics.	
TRPV6	Intestinal epithelial cells (apical membrane); placenta; prostate (in certain cancers); other epithelial tissues	**Calcium and zinc absorption:** predominantly involved in Ca^{2+} absorption, also transports Zn^{2+}, maintaining systemic zinc homeostasis. **Adaptation to dietary calcium levels:** upregulated under low dietary calcium to facilitate Zn^{2+} absorption. **Regulation of systemic zinc levels:** critical for maintaining zinc levels necessary for various physiological functions. **Zinc transport:** vital for maintaining systemic zinc homeostasis and regulating various physiological and pathological processes.	**Heavy metal toxicity:** can transport toxic heavy metals like cadmium, leading to accumulation and bone lesions under low-calcium conditions. **Cancer progression:** may play a role in certain cancers (*e.g.*, prostate cancer) by promoting tumor development and progression. **Zinc overload:** TRPV6-mediated Zn^{2+} absorption can exacerbate zinc overload effects. **Regulation by zinc concentrations:** low Zn^{2+} enhances TRPV6-mediated Ca^{2+} and Zn^{2+} absorption; high Zn^{2+} inhibits TRPV6 activity, blocking influx of Ca^{2+} and Zn^{2+}. **Tumorigenesis:** excessive zinc may promote tumor development and progression.	69–74

We also found that the administration of PPZ2 in mice suppressed the excessive increase in the expression of inflammation-related genes induced by TAC, and prevents colitis progression in mice,[46] suggesting that TRPC6-dependent Zn^{2+} influx also contributes to anti-inflammatory effects. Although it is unclear whether the β-adrenoceptors participate in the anti-inflammatory effects, demonstrating the beneficial effects of PPZ2 on inflammation occurred in smooth muscle and adipose tissue will help the establishment of new drug discovery strategies targeting TRPC6-Zn^{2+} influx.

It is unlikely that the Zn^{2+} influx dependent on TRPC6 is only targeting β-adrenoceptors. Given the improvements in not only cardiac function but also heart remodeling and the extended lifespan of TAC-treated mice, it is highly possible that the Zn^{2+} entering the cell has multiple targets, contributing to the improvement of heart failure through various mechanisms.[5] It has been reported that zinc supplementation may prevent diarrhea and pneumonia,[77,78] and there is potential for zinc to improve systemic immunity against viral infections.[79] Some of these findings have already been applied in clinical settings.[80] Aside from TRPC6, there are other TRP channels that are permeable to Zn^{2+} that play physiological and pathophysiological roles as described in Section 1 (Table 10.1). For example, in nematodes, TRPM channels have been reported to function in maintaining mitochondrial homeostasis and promoting growth during larval development *via* Zn^{2+} influx.[81] However, the expression and physiological roles of Zn^{2+}-permeable TRP channels, other than TRPC6, in the heart are not well understood. Future research focusing on these channels may lead to the elucidation of the physiological roles of localized Zn^{2+} dynamics in the heart.

10.4 Conclusion

In the field of bioinorganic chemistry, the most critical challenge for expanding research horizons is finding simple and effective ways to detect trace metals in the body and apply this to evaluation systems. By expanding the evaluation system from the cellular level to the level of myocardial tissue, we have successfully proposed a new medical application centered around Zn^{2+}, specifically targeting TRPC channel drug development. Moving forward, by applying this approach to the study of Zn^{2+}-permeable TRP channels and Zn^{2+} transporters that are highly expressed in cells and organs with strong functional connections to the myocardium, such as the kidneys and adipose tissue, we expect to uncover new insights into inter-organ and intercellular interactions mediated by Zn^{2+}. This could ultimately lead to the proposal of comprehensive therapeutic strategies for disease treatment.

References

1. A. S. Prasad, J. A. Halsted and M. Nadimi, *Am. J. Med.*, 1961, **31**, 532–546.
2. A. S. Prasad, *Nutrition*, 1995, **11**, 93–99.
3. B. L. Vallee and K. H. Falchuk, *Physiol. Rev.*, 1993, 73, 79–118.

4. C. Andreini, L. Banci, I. Bertini and A. Rosato, *J. Proteome Res.*, 2006, **5**, 196–201.
5. B. Chen, P. Yu, W. N. Chan, F. Xie, Y. Zhang, L. Liang, K. T. Leung, K. W. Lo, J. Yu, G. M. K. Tse, W. Kang and K. F. To, *Signal Transduction Targeted Ther.*, 2024, **9**, 6.
6. V. Chubanov, L. Mittermeier and T. Gudermann, *Pharmacol. Ther.*, 2018, **184**, 159–176.
7. N. Levaot and M. Hershfinkel, *Cell Calcium*, 2018, **75**, 53–63.
8. F. Pierrel, P. A. Cobine and D. R. Winge, *Biometals*, 2007, **20**, 675–682.
9. C. Pang, J. Chai, P. Zhu, J. Shanklin and Q. Liu, *Nat. Commun.*, 2023, **14**, 3404.
10. A. V. Gyulkhandanyan, S. C. Lee, G. Bikopoulos, F. Dai and M. B. Wheeler, *J. Biol. Chem.*, 2006, **281**, 9361–9372.
11. T. Priel and M. Hershfinkel, *Biochem. Biophys. Res. Commun.*, 2006, **346**, 205–212.
12. K. Inoue, Z. O'Bryant and Z. G. Xiong, *Curr. Med. Chem.*, 2015, **22**, 1248–1257.
13. D. Beyersmann and H. Haase, *Biometals*, 2001, **14**, 331–341.
14. M. Capasso, J. M. Jeng, M. Malavolta, E. Mocchegiani and S. L. Sensi, *J. Alzheimers Dis.*, 2005, **8**, 93–108; discussion 209-115.
15. T. Fukada, S. Yamasaki, K. Nishida, M. Murakami and T. Hirano, *J. Biol. Inorg. Chem.*, 2011, **16**, 1123–1134.
16. W. Maret, *Int. J. Mol. Sci.*, 2017, **18**, 2285.
17. M. I. Costa, A. B. Sarmento-Ribeiro and A. C. Gonçalves, *Int. J. Mol. Sci.*, 2023, **24**, 4822.
18. T. Kambe, *Biosci., Biotechnol., Biochem.*, 2011, **75**, 1036–1043.
19. D. J. Eide, *Biochim. Biophys. Acta*, 2006, **1763**, 711–722.
20. L. A. Lichten and R. J. Cousins, *Annu. Rev. Nutr.*, 2009, **29**, 153–176.
21. J. P. Liuzzi and R. J. Cousins, *Annu. Rev. Nutr.*, 2004, **24**, 151–172.
22. W. A. Catterall, *Cold Spring Harbor Perspect. Biol.*, 2011, **3**, a003947.
23. R. W. Tsien, *Annu. Rev. Physiol.*, 1983, **45**, 341–358.
24. K. Kawa, *J. Membr. Biol.*, 1979, **49**, 325–344.
25. K. Inoue, D. Branigan and Z. G. Xiong, *J. Biol. Chem.*, 2010, **285**, 7430–7439.
26. B. Colsoul, R. Vennekens and B. Nilius, *Rev. Physiol., Biochem. Pharmacol.*, 2011, **161**, 87–110.
27. J. Jin, B. N. Desai, B. Navarro, A. Donovan, N. C. Andrews and D. E. Clapham, *Science*, 2008, **322**, 756–760.
28. M. Zhang, Y. Ma, X. Ye, N. Zhang, L. Pan and B. Wang, *Signal Transduction Targeted Ther.*, 2023, **8**, 261.
29. Y. Komiya, Z. Bai, N. Cai, L. Lou, N. Al-Saadi, C. Mezzacappa, R. Habas and L. W. Runnels, *Sci. Rep.*, 2017, **7**, 15623.
30. M. Li, J. Jiang and L. Yue, *J. Gen. Physiol.*, 2006, **127**, 525–537.
31. S. A. Abiria, G. Krapivinsky, R. Sah, A. G. Santa-Cruz, D. Chaudhuri, J. Zhang, P. Adstamongkonkul, P. G. DeCaen and D. E. Clapham, *Proc. Natl. Acad. Sci. U. S. A.*, 2017, **114**, E6079–E6088.

32. Y. Xing, X. Wei, M. M. Wang, Y. Liu, Z. Sui, X. Wang, Y. Zhang, Y. H. Fei, Y. Jiang, C. Lu, P. Zhang, R. Chen, N. Liu, M. Wu, L. Ding, Y. Wang, F. Guo, J. L. Cao, J. Qi and W. Wang, *Cancer Lett.*, 2022, **525**, 179–197.
33. T. D. Leng, J. Lin, H. W. Sun, Z. Zeng, Z. O'Bryant, K. Inoue and Z. G. Xiong, *CNS Neurosci. Ther.*, 2015, **21**, 32–39.
34. I. Carvacho, G. Ardestani, H. C. Lee, K. McGarvey, R. A. Fissore and K. Lykke-Hartmann, *Sci. Rep.*, 2016, **6**, 34236.
35. K. P. Schlingmann, S. Weber, M. Peters, L. Niemann Nejsum, H. Vitzthum, K. Klingel, M. Kratz, E. Haddad, E. Ristoff, D. Dinour, M. Syrrou, S. Nielsen, M. Sassen, S. Waldegger, H. W. Seyberth and M. Konrad, *Nat. Genet.*, 2002, **31**, 166–170.
36. T. F. Wagner, A. Drews, S. Loch, F. Mohr, S. E. Philipp, S. Lambert and J. Oberwinkler, *Pflugers Arch.*, 2010, **460**, 755–765.
37. L. Loviscach, T. M. Backes, D. S. Langfermann, M. Ulrich and G. Thiel, *Metallomics*, 2020, **12**, 1735–1747.
38. K. Held, T. Kichko, K. De Clercq, H. Klaassen, R. Van Bree, J. C. Vanherck, A. Marchand, P. W. Reeh, P. Chaltin, T. Voets and J. Vriens, *Proc. Natl. Acad. Sci. U. S. A.*, 2015, **112**, E1363–E1372.
39. I. Vandewauw, K. De Clercq, M. Mulier, K. Held, S. Pinto, N. Van Ranst, A. Segal, T. Voet, R. Vennekens, K. Zimmermann, J. Vriens and T. Voets, *Nature*, 2018, **555**, 662–666.
40. S. Oda, K. Nishiyama, Y. Furumoto, Y. Yamaguchi, A. Nishimura, X. Tang, Y. Kato, T. Numaga-Tomita, T. Kaneko, S. Mangmool, T. Kuroda, R. Okubo, M. Sanbo, M. Hirabayashi, Y. Sato, Y. Nakagawa, K. Kuwahara, R. Nagata, G. Iribe, Y. Mori and M. Nishida, *Nat. Commun.*, 2022, **13**, 6374.
41. M. Trebak, G. Vazquez, G. S. Bird and J. W. Putney, Jr., *Cell Calcium*, 2003, **33**, 451–461.
42. J. Gibon, P. Tu, S. Bohic, P. Richaud, J. Arnaud, M. Zhu, G. Boulay and A. Bouron, *Biochim. Biophys. Acta*, 2011, **1808**, 2807–2818.
43. M. Chevallet, L. Jarvis, A. Harel, S. Luche, S. Degot, V. Chapuis, G. Boulay, T. Rabilloud and A. Bouron, *Metallomics*, 2014, **6**, 1269–1276.
44. A. Dietrich, *Cell Calcium*, 2023, **109**, 102674.
45. J. Hasna, R. Abi Nahed, F. Sergent, N. Alfaidy and A. Bouron, *Cell. Physiol. Biochem.*, 2019, **52**, 455–467.
46. K. Nishiyama, Y. Kato, A. Nishimura, X. Mi, R. Nagata, Y. Mori, Y. T. Azuma and M. Nishida, *Int. J. Mol. Sci.*, 2024, **25**, 2401.
47. S. Sawamura, M. Hatano, Y. Takada, K. Hino, T. Kawamura, J. Tanikawa, H. Nakagawa, H. Hase, A. Nakao, M. Hirano, R. Rotrattanadumrong, S. Kiyonaka, M. X. Mori, M. Nishida, Y. Hu, R. Inoue, R. Nagata and Y. Mori, *Mol. Pharmacol.*, 2016, **89**, 348–363.
48. B. Shen, M. Mei, S. Ai, X. Liao, N. Li, S. Xiang, C. Wen, Y. Tao and H. Dai, *FASEB J.*, 2022, **36**, e22527.
49. D. A. Andersson, C. Gentry, S. Moss and S. Bevan, *Proc. Natl. Acad. Sci. U. S. A.*, 2009, **106**, 8374–8379.
50. H. Hu, M. Bandell, M. J. Petrus, M. X. Zhu and A. Patapoutian, *Nat. Chem. Biol.*, 2009, **5**, 183–190.

51. N. Hatano, M. Matsubara, H. Suzuki, Y. Muraki and K. Muraki, *Int. J. Mol. Sci.*, 2021, **22**, 6349.
52. N. Hatano, Y. Itoh, H. Suzuki, Y. Muraki, H. Hayashi, K. Onozaki, I. C. Wood, D. J. Beech and K. Muraki, *J. Biol. Chem.*, 2012, **287**, 31962–31972.
53. M. Matsubara, Y. Muraki, H. Suzuki, N. Hatano and K. Muraki, *J. Biol. Chem.*, 2024, **300**, 107302.
54. F. Sekiguchi, M. Tsubota and A. Kawabata, *J. Pharmacol. Sci.*, 2024, **155**, 113–120.
55. T. R. Büch, E. A. Schäfer, M. T. Demmel, I. Boekhoff, H. Thiermann, T. Gudermann, D. Steinritz and A. Schmidt, *Chem. - Biol. Interact.*, 2013, **206**, 462–471.
56. T. G. Banke and A. D. Wickenden, *Nat. Chem. Biol.*, 2009, **5**, 141–142.
57. Y. Hu, Q. Y. Fu, D. N. Fu, X. L. Wang, Z. H. Wang, J. T. Zhang, W. J. Xu, G. K. Zhou, L. H. Chen and T. Liu, *Front. Mol. Neurosci.*, 2021, **14**, 768731.
58. D. Steinritz, F. Zehfuß, B. Stenger, A. Schmidt, T. Popp, K. Kehe, H. Mückter, H. Thiermann and T. Gudermann, *Toxicol. Lett.*, 2018, **293**, 133–139.
59. S. Nakajima, T. Hira, H. Iwaya and H. Hara, *Mol. Cell. Endocrinol.*, 2016, **430**, 108–114.
60. A. Scotto Rosato, S. Montefusco, C. Soldati, S. Di Paola, A. Capuozzo, J. Monfregola, E. Polishchuk, A. Amabile, C. Grimm, A. Lombardo, M. A. De Matteis, A. Ballabio and D. L. Medina, *Nat. Commun.*, 2019, **10**, 5630.
61. B. Nilius, G. Owsianik, T. Voets and J. A. Peters, *Physiol. Rev.*, 2007, **87**, 165–217.
62. B. Kim, G. Kim, S. Jeon, W. S. Cho, H. P. Jeon and J. Jung, *Toxicol. Rep.*, 2023, **10**, 529–536.
63. T. F. Minckley, C. Zhang, D. H. Fudge, A. M. Dischler, K. D. LeJeune, H. Xu and Y. Qin, *Nat. Commun.*, 2019, **10**, 4806.
64. J. Qi, Y. Xing, Y. Liu, M. M. Wang, X. Wei, Z. Sui, L. Ding, Y. Zhang, C. Lu, Y. H. Fei, N. Liu, R. Chen, M. Wu, L. Wang, Z. Zhong, T. Wang, Y. Liu, Y. Wang, J. Liu, H. Xu, F. Guo and W. Wang, *Autophagy*, 2021, **17**, 4401–4422.
65. X. Cheng, D. Shen, M. Samie and H. Xu, *FEBS Lett.*, 2010, **584**, 2013–2021.
66. J. L. Eichelsdoerfer, J. A. Evans, S. A. Slaugenhaupt and M. P. Cuajungco, *J. Biol. Chem.*, 2010, **285**, 34304–34308.
67. I. Kukic, J. K. Lee, J. Coblentz, S. L. Kelleher and K. Kiselyov, *Biochem. J.*, 2013, **451**, 155–163.
68. W. Du, M. Gu, M. Hu, P. Pinchi, W. Chen, M. Ryan, T. Nold, A. Bannaga and H. Xu, *Cell Rep.*, 2021, **37**, 109848.
69. M. P. Cuajungco, L. C. Basilio, J. Silva, T. Hart, J. Tringali, C. C. Chen, M. Biel and C. Grimm, *Traffic*, 2014, **15**, 1247–1265.
70. M. P. Cuajungco and K. Kiselyov, *Front. Biosci.*, 2017, **22**, 1330–1343.
71. G. Kovacs, T. Danko, M. J. Bergeron, B. Balazs, Y. Suzuki, A. Zsembery and M. A. Hediger, *Cell Calcium*, 2011, **49**, 43–55.
72. G. Kovacs, N. Montalbetti, M. C. Franz, S. Graeter, A. Simonin and M. A. Hediger, *Cell Calcium*, 2013, **54**, 276–286.

73. P. Holzer, *Pharmacol. Ther.*, 2011, **131**, 142–170.
74. M. Ishizawa, A. Hirayu and M. Makishima, *Anticancer Res.*, 2021, **41**, 5453–5459.
75. T. Numaga-Tomita, N. Kitajima, T. Kuroda, A. Nishimura, K. Miyano, S. Yasuda, K. Kuwahara, Y. Sato, T. Ide, L. Birnbaumer, H. Sumimoto, Y. Mori and M. Nishida, *Sci. Rep.*, 2016, **6**, 39383.
76. S. Oda, T. Numaga-Tomita, N. Kitajima, T. Toyama, E. Harada, T. Shimauchi, A. Nishimura, T. Ishikawa, Y. Kumagai, L. Birnbaumer and M. Nishida, *Sci. Rep.*, 2017, 7, 7511.
77. R. Black, *Bull. W. H. O.*, 2003, **81**, 79.
78. C. F. ischer Walker and R. E. Black, *Annu. Rev. Nutr.*, 2004, **24**, 255–275.
79. S. A. Read, S. Obeid, C. Ahlenstiel and G. Ahlenstiel, *Adv. Nutr.*, 2019, **10**, 696–710.
80. I. Wessels, B. Rolles and L. Rink, *Front. Immunol.*, 2020, **11**, 1712.
81. P. Georgiev, H. Okkenhaug, A. Drews, D. Wright, S. Lambert, M. Flick, V. Carta, C. Martel, J. Oberwinkler and P. Raghu, *Cell Metab.*, 2010, **12**, 386–397.

CHAPTER 11

Role of Zinc in Brain Functions, and Brain Zinc Dyshomeostasis

Y. HIGASHI*[a] AND T. ARATAKE[b]

[a] Department of Pharmacology, Kochi Medical School, Kochi University, Kohasu, Okoh-cho, Nankoku 783-8505, Japan; [b] Institut für Biochemie, Friedrich-Alexander-Universität Erlangen-Nürnberg, Erlangen, Fahrstrasse 17, 91054, Germany
*Email: higasi@kochi-u.ac.jp

11.1 Role of Zinc in Neurons

The concentration of zinc in brain cells, including neurons, is tightly regulated by zinc importers from the Zrt-Irt-like protein (ZIP) family and exporters from the zinc transporter (ZnT) family.[1] Most ZIP family members are localized on the plasma membrane, where they function in the uptake of zinc from the extracellular space into the cytoplasm. In the central nervous system (CNS), ZIP family members ZIP1 and ZIP3 are considered the primary regulators of zinc uptake among the ZIP family.[2] However, the role of the ZIP family in the function of neurons remains largely unclear. The ZnT family, in contrast, facilitates zinc export and intracellular storage, with ZnT3 being indispensable for storing labile zinc in synaptic vesicles of glutamatergic neurons throughout the forebrain, particularly in the hippocampus and cerebral cortex. The stored zinc is released into the synaptic space together with the neurotransmitter glutamate during calcium-dependent neuronal activity, where it acts as a neuromodulator of membrane receptors, ion channels, and transporters on postsynaptic neurons.[3]

Coordination Chemistry Perspectives No. 2
Zinc in Biology: Molecular Structures, Cellular Processes and Living Systems
Edited by Hiroyuki Yasui and Taiho Kambe

Published by the Royal Society of Chemistry, www.rsc.org

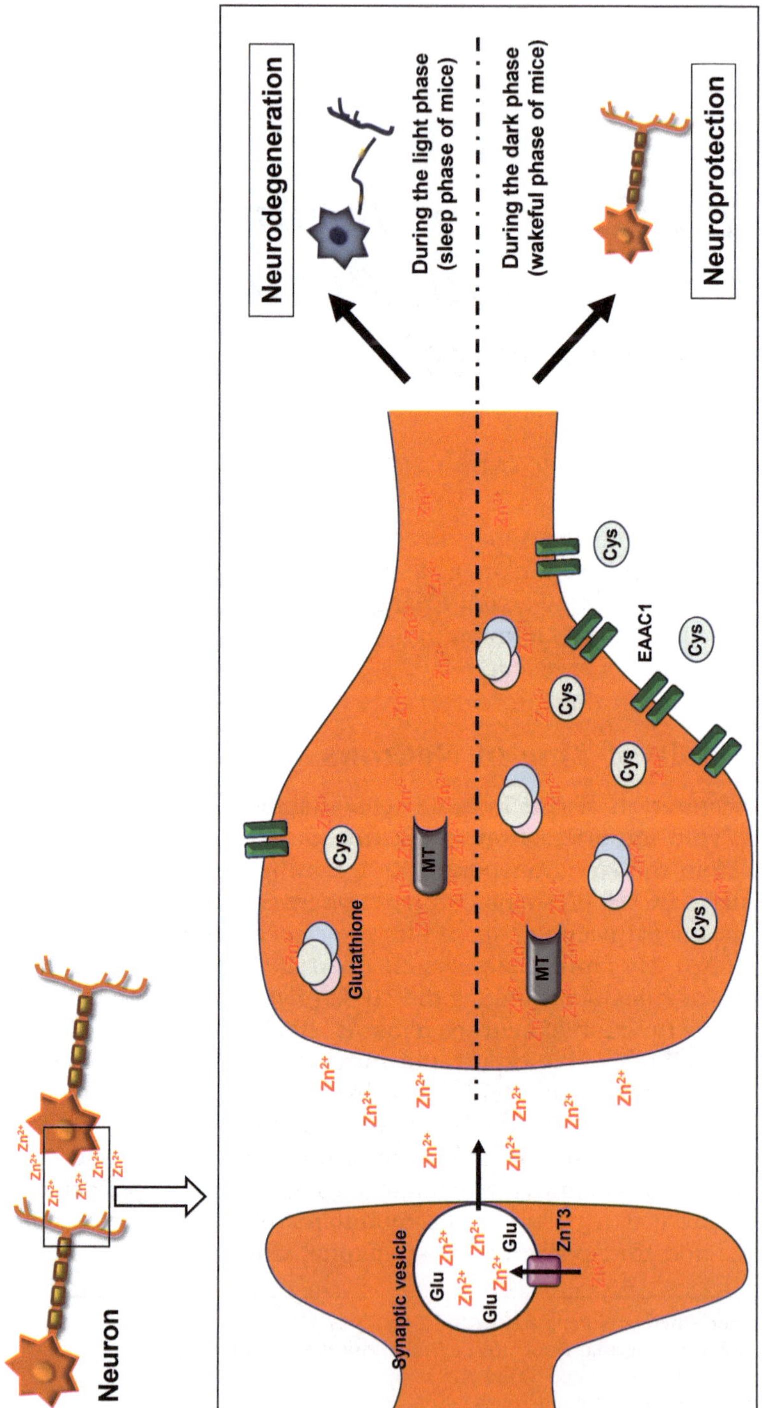
Neurodegeneration
During the light phase
(sleep phase of mice)
During the dark phase
(wakeful phase of mice)
Neuroprotection
EAAC1
Cys
MT
Glutathione
Zn2+
Synaptic vesicle
Glu
ZnT3
Neuron

The *N*-methyl-D-aspartate (NMDA) receptor is one of the most well-studied receptors regarding the effects of zinc. Early studies revealed that exposing primary cultured neurons to zinc inhibited NMDA receptor currents. Low zinc concentrations exhibited a voltage-independent inhibition mode, while high concentrations showed a voltage-dependent inhibition mode.[4,5] However, NMDA receptors are heterogeneous with multiple subclasses and the subtypes have distinct sensitivities to zinc.[6,7] Takeda *et al.* demonstrated that NMDA receptor-dependent long-term potentiation at hippocampal CA1 is positively modulated by low micromolar concentrations of zinc.[8] In addition to NMDA receptors, zinc inhibits γ-aminobutyric acid type A ($GABA_A$) receptors, and promotes responses of α-amino-3-hydroxy-5-methyl-4-isoxazolepropionic acid (AMPA) receptors and glycine amino acid receptors.[9–13] Furthermore, zinc has been reported to act on kainate receptors, serotonin receptors, dopamine receptors, acetylcholine receptors, and voltage-dependent ion channels (Na^+, K^+, Ca^{2+}, Cl^-).[11,14] Thus, zinc is known to modulate synaptic transmission through various receptors, including excitatory receptors and inhibitory receptors, suggesting that zinc released into the synaptic space alongside neurotransmitters may act as the neuromodulator.

In cases of severe brain injury induced by cerebral ischemia, status epilepticus, hypoglycemia, or traumatic brain injury, an excessive amount of zinc is released from presynaptic neurons in the hippocampus. Under these conditions, neurodegeneration is triggered in postsynaptic neurons, accompanied by zinc accumulation in the affected cells (Figure 11.1). The administration of zinc chelators suppresses both zinc accumulation and neurodegeneration, suggesting that excessive zinc influx into neurons is a primary cause of neurodegeneration. During this neurodegenerative process, zinc uptake predominantly occurs through Ca^{2+}-permeable AMPA and kainate receptors.[15–17] Intracellular zinc is known to activate various signaling pathways, among which

Figure 11.1 Mechanisms of synaptic zinc homeostasis, neuroprotection, and circadian modulation. In glutamatergic neurons of the forebrain, including the hippocampus, zinc is stored in synaptic vesicles *via* the zinc transporter ZnT3. Upon stimulation such as cerebral ischemia, excessive zinc is released into the synaptic cleft, taken up by postsynaptic neurons, and accumulates intracellularly, leading to an increase in the intracellular concentration of zinc ions (Zn^{2+}), which ultimately triggers neurodegeneration. Meanwhile, in postsynaptic neurons, intracellular zinc homeostasis is maintained by the action of metallothionein (MT) and glutathione, which are present within the cytoplasm and bind to the imported Zn^{2+}, thereby preventing excessive increases in its intracellular levels and contributing to neuroprotection. The synthesis of glutathione is supported by excitatory amino acid transporter 1 (EAAC1), which is expressed in neurons and facilitates cysteine uptake necessary for glutathione production. It has been reported that EAAC1 expression increases during the active phase (dark phase) in mice. This upregulation contributes to a significant reduction in hippocampal neuronal death in ischemic model mice subjected to ischemia during the active phase compared to those subjected to ischemia during the sleep phase (light phase).

oxidative stress plays a central role in zinc-induced neuronal cell death. Specifically, zinc activates 12-lipoxygenase and nicotinamide adenine dinucleotide phosphate (NADPH) oxidase, causing the production of reactive oxygen species (ROS).[18–20] Subsequently, ROS produced by zinc activates the MAP kinase cascade, which then triggers the activation of poly(ADP-ribose) polymerase (PARP)-1 and mitochondrial dysfunction, ultimately leading to cell death.[21]

Against zinc toxicity, neurons possess several protective mechanisms. One of these is metallothionein (MT), a small-molecule protein composed of 61–68 amino acids. MT is a protein family with four isoforms, containing 20 cysteine residues that allow it to bind divalent metals, including up to seven zinc ions.[22–26] Numerous studies have revealed that metallothionein plays a key physiological role in the homeostasis and transport of essential elements like zinc and copper. Specifically, when intracellular zinc concentration is at the basal level (~100 pM), most metallothioneins form complexes binding five zinc ions. If the intracellular zinc concentration reaches 10 nM, equivalent to the extracellular zinc concentration, MT binds up to seven zinc ions, quickly reducing the intracellular zinc concentration.[27–29] However, this zinc-capturing activity of metallothionein can become saturated by excessive zinc uptake, as induced by extracellular zinc increase from brain ischemia, oxidative stress, or Aβ1–42, ultimately leading to neuronal cell death.[28] Another protective mechanism involves intracellular glutathione and its precursor, cysteine. These molecules utilize their thiol groups to non-enzymatically bind (chelate) zinc ions, thereby mitigating the increase in intracellular zinc concentration caused by extracellular zinc uptake. Glutathione is synthesized using cysteine as a substrate, and its synthesis rate is largely determined by intracellular cysteine availability (Figure 11.1).[30,31] In neurons, cysteine uptake is predominantly mediated by excitatory amino acid carrier 1 (EAAC1).[32] Interestingly, Aratake *et al.* found that EAAC1 expression in the mouse hippocampus increases during the dark phase, when mice are most active.[33] Additionally, mice subjected to transient brain ischemia during the dark phase showed lower zinc ion accumulation and less neuronal cell death in the hippocampal hilus compared to those treated during the light phase. When an EAAC1 inhibitor was administered, zinc accumulation and neuronal cell death in the hilus worsened following ischemia. These results suggest that circadian variations in EAAC1 expression regulate zinc accumulation in the hilus following ischemia, which contributes to time-dependent differences in ischemic neuronal injury (Figure 11.1).[33] It has long been known that the incidence of stroke is influenced by circadian rhythms, with ischemic stroke during sleep associated with higher mortality and poorer outcomes than during wakefulness.[34] The findings demonstrated by Aratake *et al.* may help elucidate the mechanisms underlying the diurnal variations in ischemic neuronal injury.

11.2 Role of Zinc in Microglia

Microglia are the resident immune cells of the CNS, and are also called brain macrophages. They are broadly distributed throughout the brain and the spinal cord.[35] Under physiological conditions, microglia exist as resting microglia. They form a "ramified" shape characterized by multiple branched processes and maintain the homeostasis of the CNS by surveillance of the extracellular microenvironment.[36] When activated by pathophysiological stimuli (*e.g.* ischemia, tissue injury, or other signals), their characters are altered pro-inflammatory M1 type or anti-inflammatory M2 type.[37,38] M1 microglia secrete pro-inflammatory cytokines such as interleukin (IL)-1β, IL-6, and tumor necrosis factor (TNF)α.[38] On the other hand, M2 microglia express CD206 (also known as mannose receptor type C), produce anti-inflammatory factors IL-10 and arginase-1 (Arg-1), and contribute to neuroprotection *via* production of neuroprotective factors such as insulin-like growth factor-1 (IGF-1) and brain-derived neurotrophic factor.[39,40] In morphological changes, M1 microglia are characterized by amoeboid form with large somata, and M2 microglia exhibit long-processes and spindle-like shapes.[41,42] We found that resting primary microglia change into an "amoeboid" form when exposed to extracellular zinc (15–60 μM).[43–45] Moreover, the morphological changes in microglia observed in the hippocampus of mice after transient brain ischemia were suppressed by pre-treatment with zinc chelators,[43,45] suggesting that the excessive release of zinc from neurons following brain ischemia contributes to these changes in microglia. Zinc transporters such as ZIP and ZnT play an important role in regulating the functions of peripheral immune cells. In microglia, ZIP1 expression has been reported to be higher compared to other members of the ZIP family.[44] Furthermore, suppression of ZIP1 expression using siRNA has been shown to reduce zinc uptake into the cells.[46] Higashi *et al.* demonstrated that exposure to zinc increases intracellular zinc concentration in primary microglia, which is followed by ATP release from the cells. This triggers the activation of P2X7R in an autocrine or paracrine manner, leading to the activation of PARP-1 and NF-κB *via* NADPH oxidase.[44] These processes ultimately cause the morphological changes in microglia, a mechanism known as the zinc-induced microglial activation pathway (Figure 11.2). However, these zinc-induced morphological changes did not trigger the strong pro-inflammatory response characteristic of M1 microglia.[43] Based on these findings, we established an *in vitro* brain ischemia model by pre-exposing primary microglia to zinc, followed by treatment with lipopolysaccharide (LPS), an M1 inducer.[45] Compared to M1 microglia treated with LPS alone, zinc pretreatment exacerbated secretion of M1 microglial pro-inflammatory cytokines such as IL-1β, IL-6, and TNF-α. Additionally, when mice were pre-treated with a zinc chelator before transient brain ischemia, the production of pro-inflammatory cytokines was reduced, and the memory impairment caused by ischemia was prevented.[45] These findings suggest that excess zinc released from neurons following transient brain ischemia not only induces morphological changes in microglia but also acts as a priming factor, enhancing the pro-inflammatory response of M1 microglia and potentially contributing to memory impairments (Figure 11.2). Park *et al.* demonstrated that cytosolic zinc levels increased in

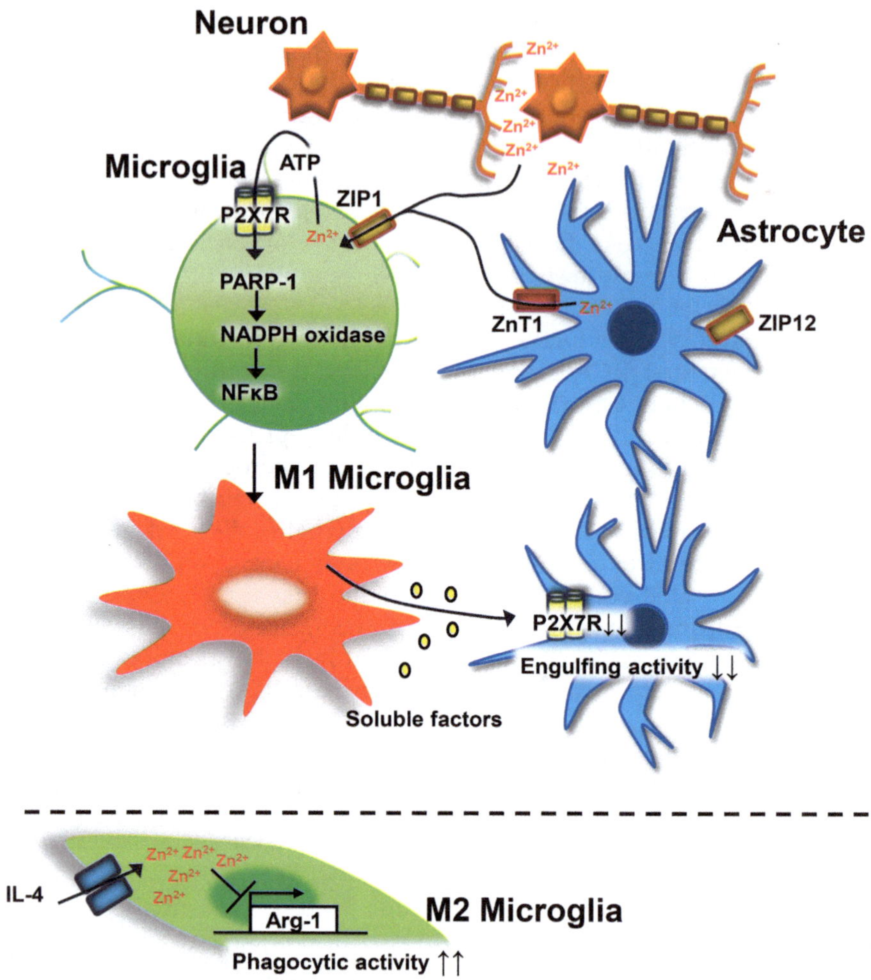

Figure 11.2 Zinc signaling in microglia and crosstalk with astrocytes. Microglia take up extracellular zinc *via* Zrt-Irt-like protein (ZIP1), and then zinc signaling (Following ATP release into the extracellular space, activation of the P2X7 receptor (P2X7R) initiates downstream activation of NADPH oxidase, PARP-1, and NF-κB) is activated. Zinc signaling activated microglia secretes soluble factors which downregulate astrocytic P2X7R expression, resulting in suppressing engulfing activity in astrocytes. IL-4 increases M2 microglial cytosolic zinc concentration from intracellular sources. Increased cellular zinc concentration suppresses arginase-1 (Arg-1) overexpression, contributing to maintenance of phagocytic activity in M2 microglia.

the mouse microglial cell line BV-2 following 2 hours of oxygen-glucose deprivation (OGD), which are *in vitro* models of brain ischemia, followed by 6 hours of reoxygenation.[47] Under OGD conditions, no significant changes were observed in the levels of inflammasome components such as NOD-like receptor family pyrin domain containing 3 (NLRP3), IL-1β, and IL-18. However, under reoxygenation conditions, these protein levels were significantly elevated. Moreover, the membrane-permeable zinc chelator TPEN effectively inhibited these increases.[47] These results support the hypothesis that zinc acts as a priming factor in microglia. Additionally, the zinc-enhanced pro-inflammatory response of M1 microglia was blocked by TPEN, the P2X7R antagonist A438079, and the ROS scavenger Trolox. These findings suggest that zinc primes microglia through the zinc-induced microglial activation pathway (Figure 11.2).[43–48]

We next considered glial crosstalk between M1 microglia and astrocytes. To examine this, conditioned media were collected from M1 microglia and treated with primary astrocytes. We found that M1 microglia induced by LPS secrete soluble factors that increase P2X7R expression in astroglia, thereby enhancing astrocytic engulfing activity. However, LPS-induced M1 microglia aggravated by zinc suppress the increase in P2X7R expression in astrocytes, which leads to a reduction in astrocytic engulfing activity.[49] Nevertheless, the precise mechanism underlying the regulation of astroglial P2X7R expression by soluble factors from M1 microglia remains unclear, and further experiments are needed to fully elucidate this process (Figure 11.2).

On the other hand, we have also reported the role of zinc in M2 microglia. Treatment of primary microglia with IL-4, an M2 inducer, resulted in the activation of the M2 phenotype. In IL-4-treated microglia, an increase in cytosolic Zn^{2+} concentration was observed, which was inhibited by the intracellular zinc chelator TPEN, but not by the extracellular zinc chelator CaEDTA.[50] These results indicate that the increase in cytosolic Zn^{2+} induced by IL-4 originates from the release of intracellular zinc. Furthermore, IL-4-induced M2 microglia showed upregulation of M2 markers such as Arg-1, CD206, and IGF-1. However, pre-treatment with TPEN in IL-4-induced M2 microglia led to excessive expression of Arg-1 and a reduction in phagocytic activity. This decrease in phagocytosis was reversed by the addition of L-arginine, the substrate for Arg-1.[50] These findings suggest that the release of intracellular zinc in response to IL-4 acts as a negative regulator of Arg-1 expression and plays an essential role in maintaining the normal phagocytic activity of M2 microglia (Figure 11.2). To our knowledge, this is the first report on the role of zinc in M2 microglia, and there have been very few studies on this topic to date. Therefore, further investigation is still needed to fully elucidate the role of zinc in M2 microglia.

11.3 Role of Zinc in Astrocytes

Astrocytes are a predominant glial cell type in the CNS and are mainly characterized by glial fibrillary acid protein (GFAP) expression. Neuroanatomists in the late 19th century recognized two morphologically distinct populations of astroglia with fibrous (nonstellate) and protoplasmic (stellate) features in the white and grey matter, respectively.[51,52] Astrocytes play a critical role in brain function by supplying neurons with energy, nutrients, trophic factors, and substrates necessary for neurotransmission. Additionally, they modulate synaptic activity through the formation of the tripartite synapse.

Extracellular zinc (100–400 μM) has been reported to affect the morphological changes and functional activity of astrocytes. Similar to activated microglia, reactive astrocytes are categorized into two phenotypes: the neurotoxic A1 astrocytes and the neuroprotective A2 astrocytes. Huiliang and colleagues demonstrated that extracellular zinc at a concentration of 100 μM activates ERK and JAK2, which subsequently leads to the activation of STAT3. This activation triggers morphological changes in astrocytes and promotes the expression of A1-specific genes, such as H2-D1, Amiogo2, and Ugt1a, as well as pro-inflammatory cytokine genes, such as IL-1β, IL-6, and TNFα.[53] Furthermore, exposure to zinc at concentrations between 100 and 400 μM has been shown to inhibit glutamate uptake in astrocytes *via* PARP-1 activation.[54] Glutamate uptake by astrocytes is not only involved in the regulation of neurotransmission, but also plays a role in alleviating glutamate-induced excitotoxicity. Thus, Zn^{2+} released from neurons immediately after cerebral ischemia/reperfusion may contribute to the exacerbation of neuronal injury by promoting A1 induction in astrocytes and inhibiting glutamate uptake. Under hypoosmotic stress-loaded conditions, intra- and extracellular zinc levels in astrocytes increased in a time-dependent manner, accompanied by elevated expression levels of ZnT1, which is responsible for zinc transport to the extracellular space.[55] When microglia were treated with the conditioned medium from astrocytes subjected to hypoosmotic stress, microglial activation was observed.[55] These findings suggest that under hypoosmotic stress-loaded conditions, astrocytes release zinc as a gliotransmitter through ZnT1, which subsequently induces microglial activation.[55] In addition, astrocytes are the main cells which express ZIP12 in the CNS.[56] Growing single cell RNA-seq data in mouse and human are supporting this notice.[57–60] These evidences suggest that zinc is involved in various astrocytic activity.

On the other hand, extracellular zinc at concentrations of 50–400 μM has been reported to induce astroglial cell death,[61–63] and this toxic effect is shown to be suppressed by preconditioning with low concentrations of zinc (priming) or overexpression of ZnT1.[62] The signaling cascades involved in zinc-induced astroglial cell death have been relatively well studied, with evidence suggesting the involvement of ROS/PKC-α/Ras/Raf/ERK and ROS/Src/Ras/Raf/ERK pathways, as well as cPLA2.[62,63] These findings indicate that the zinc concentrations affecting astrocyte function and those inducing cell death may overlap. Further research is needed to elucidate the physiological roles of zinc in astrocytes, as well as the impact of zinc homeostasis disruptions on astrocytes under pathological conditions, including stroke and other brain injuries.

11.4 Role of Zinc in Oligodendrocytes

Oligodendrocytes (OL) constitute the second prevalent macroglial cell type of the CNS and are mainly located in the white matter. The major roles of OL

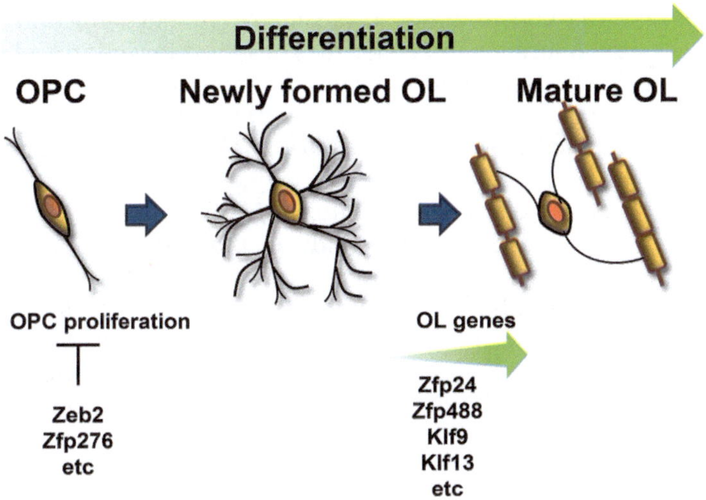

Figure 11.3 Zinc binding transcription factors in oligodendrocyte (OL) differentiation. Zeb2, Zfp276, *etc.* inhibit proliferating genes in OL precursor cells (OPC), leading to differentiation. OL genes are promoted by Zfp24, Zfp488, Klf9, Klf13, *etc.*

are to produce myelin sheaths to speed up nerve conduction and provide metabolic support to neuronal axons. OL are generated from OL precursor cells (OPC). They require a serial transcriptional regulation according to the differentiation process. Many studies reported that zinc finger transcription factors are involved in the regulation of OL differentiation (Figure 11.3). These include, for instance, Zeb2/Sip1, Yy1, Zfp24 (formerly known as Zfp191) and Zfp488.[64] In addition, Wegner and colleagues have reported the promoting effect of Zfp276, KLF9 and KLF13 on OL differentiation.[65,66] They further demonstrated that Zfp488 regulates myelination-associated G protein coupled receptor 37 expression with myelin-associated transcription factor Nkx2.2 and oligodendroglial transcription factor SOX10 during OL differentiation.[67] Zinc-dependent histone deacetylases also promote differentiation in OL.[68] In addition, zinc is a critical structural component for myelin-related proteins such as MBP and MAG.[69,70] Zinc binds both MBP and MAG for their stabilization effects on membrane interactions and increased surface hydrophobicity, respectively.[69,71] Zinc homeostasis in oligodendroglia, like that in other cells, is thought to be regulated by zinc transporters. Many zinc transporters, both ZIPs and ZnTs, are expressed in oligodendroglia, and several ZIPs have dramatically decreased expression levels during OL differentiation according to RNAseq data.[60] A recent study indicated that mutations in TMEM163, a new member of the ZnT transporter family (ZnT11), cause zinc dyshomeostasis, resulting in hypomyelination/leukodystrophy.[72,73] In addition, maternal marginal zinc deficiency caused less oligodendrogenesis and myelination.[74] Growing evidence indicates that regulation of zinc homeostasis is critical in OL differentiation, while there is little known about intracellular Zn^{2+} dynamics in oligodendroglia. Recently, Bourassa and colleagues observed lower intracellular Zn^{2+} levels in mature OL, compared to premyelinating OL.[75,76] Their group also reported transient increase in Zn^{2+} concentration during differentiation.[75] However, the detailed mechanisms of how Zn^{2+} affects oligodendroglia remain unknown.

Excess Zn^{2+} causes oligodendroglial cell death as well as death of other cells in the CNS. Several publications indicate that AMPA receptor activation induces cytosolic Zn^{2+} accumulation, resulting in oligodendroglial cell death through ERK phosphorylation.[77–79] Furthermore, although glial crosstalk with zinc between microglia and astrocytes has been studied, there is no evidence in the case of oligodendroglia. As described above, there are much less detailed mechanisms about zinc signaling compared to in microglia and astrocytes.

11.5 Role of Zinc in Pericytes

Pericytes are critical components of the neurovascular unit and the blood-brain barrier (BBB), encircling endothelial cells in capillaries, precapillary arterioles, and postcapillary venules.[80] They play essential roles in maintaining the BBB, regulating cerebral blood flow, and controlling immune cell infiltration into the CNS.[81]

However, research on the role of zinc in pericytes remains limited. Low concentrations of zinc (30 μM) induce the expression of tissue plasminogen activator (tPA), a serine protease, and its substrate plasminogen (PLG), whereas high concentrations of zinc (120 μM) induce cell death in pericytes, similar to other glial cells.[82,83] Interestingly, while low concentrations of zinc promote tPA induction, treatment with tPA elevates intracellular zinc levels in pericytes to cytotoxic levels, subsequently triggering the generation of reactive oxygen species (ROS).[83] This tPA-induced zinc toxicity has been shown to be inhibited by pyruvate,[84] which exerts protective effects against zinc-induced neurotoxicity, as well as by cilostazol (a phosphodiesterase inhibitor) and dibutyryl cAMP (a cAMP analog).[83] These findings suggest that the interaction between tPA and zinc is complex and involves bi-directional effects, yet the underlying mechanisms remain unclear. Further research is necessary to elucidate the relationship between tPA and zinc in pericytes. Moreover, compared to other glial cells, basic research on the role of zinc in pericytes has significantly lagged behind that of other glial cells, underscoring the need to deepen our understanding of this topic. Comprehensive studies are crucial to uncover the physiological and pathological significance of zinc in pericytes.

11.6 Conclusion

Zinc plays a dual role in the brain: it acts both as a crucial neuromodulator that supports various brain functions and, when accumulated in excess, contributing to the pathogenesis of neurological disorders such as neurodegeneration, ischemia, and inflammation. Its involvement in glial cell signaling highlights the complexity and versatility of zinc's regulatory functions in maintaining CNS homeostasis. This duality also opens avenues for therapeutic strategies aimed at modulating zinc levels to mitigate neurodegenerative processes. However, a deeper and more comprehensive understanding of the molecular mechanisms underlying zinc signaling, particularly in the context of glial cell interactions, is essential to develop targeted treatments for neurodegenerative diseases.

References

1. A. Takeda, *Brain Res. Rev.*, 2000, **34**, 137–148.
2. S. D. Portbury and P. A. Adlard, *Int. J. Mol. Sci.*, 2017, **18**, 2506.
3. L. J. Blakemore and P. Q. Trombley, *Front. Cell. Neurosci.*, 2017, **11**, 297.
4. C. W. Christine and D. W. Choi, *J. Neurosci.*, 1990, **10**, 108–116.
5. P. Legendre and G. L. Westbrook, *J. Physiol.*, 1990, **429**, 429–449.
6. N. Chen, A. Moshaver and L. A. Raymond, *Mol. Pharmacol.*, 1997, **51**, 1015–1023.
7. K. Williams, *Neurosci. Lett.*, 1996, **215**, 9–12.
8. A. Takeda, S. Fuke, M. Ando and N. Oku, *Neuroscience*, 2009, **158**, 585–591.
9. A. S. Nakashima and R. H. Dyck, *Brain Res. Rev.*, 2009, **59**, 347–373.
10. B. K. Y. Bitanihirwe and M. G. Cunningham, *Synapse*, 2009, **63**, 1029–1049.
11. T. G. Smart, X. Xie and B. J. Krishek, *Prog. Neurobiol.*, 1994, **42**, 393–441.
12. G. L. Westbrook and M. L. Mayer, *Nature*, 1987, **328**, 640–643.
13. M. L. Mayer, L. Vyklicky Jr and G. L. Westbrook, *J. Physiol.*, 1989, **415**, 329–350.
14. C. J. Frederickson, J. Y. Koh and A. I. Bush, *Nat. Rev. Neurosci.*, 2005, **6**, 449–462.
15. S. W. Suh, E. T. Gum, A. M. Hamby, P. H. Chan and R. A. Swanson, *J. Clin. Invest.*, 2007, **117**, 910–918.
16. S. W. Suh, J. W. Chen, M. Motamedi, B. Bell, K. Listiak, N. F. Pons, G. Danscher and C. J. Frederickson, *Brain Res.*, 2000, **852**, 268–273.
17. J. H. Weiss and S. L. Sensi, *Trends Neurosci.*, 2000, **23**, 365–371.
18. Y. Zhang, *J. Neurosci.*, 2004, **24**, 10616–10627.
19. R. F. Krall, T. Tzounopoulos and E. Aizenman, *Neuroscience*, 2021, **457**, 235–258.
20. K. M. Noh and J. Y. Koh, *J. Neurosci.*, 2000, **20**, RC111.
21. Y. H. Kim and J. Y. Koh, *Exp. Neurol.*, 2002, **177**, 407–418.
22. D. Juárez-Rebollar, C. Rios, C. Nava-Ruíz and M. Méndez-Armenta, *Oxid. Med. Cell. Longevity*, 2017, **2017**, 5828056.
23. A. K. West, J. Hidalgo, D. Eddins, E. D. Levin and M. Aschner, *Neurotoxicology*, 2008, **29**, 489–503.
24. A. T. Miles, G. M. Hawksworth, J. H. Beattie and V. Rodilla, *Crit. Rev. Biochem. Mol. Biol.*, 2000, **35**, 35–70.

25. M. Vašák, *J. Trace Elem. Med. Biol.*, 2005, **19**, 13–17.
26. M. Nordberg and G. F. Nordberg, *Cell. Mol. Biol.*, 2000, **46**, 451–463.
27. C. J. Frederickson, L. J. Giblin, A. Krezel, D. J. McAdoo, R. N. Muelle, Y. Zeng, R. V. Balaji, R. Masalha, R. B. Thompson, C. A. Fierke, J. M. Sarvey, M. de Valdenebro, D. S. Prough and M. H. Zornow, *Exp. Neurol.*, 2006, **198**, 285–293.
28. Y. Sato, M. Takiguchi, H. Tamano and A. Takeda, *Biol. Trace Elem. Res.*, 2021, **199**, 53–61.
29. A. Krężel and W. Maret, *Int. J. Mol. Sci.*, 2017, **18**, 1237.
30. A. Bendahan, A. Armon, N. Madani, M. P. Kavanaugh and B. I. Kanner, *J. Biol. Chem.*, 2000, **275**, 37436–37442.
31. T. Himi, M. Ikeda, T. Yasuhara, M. Nishida and I. Morita, *J. Neural Transm.*, 2003, **110**, 1337–1348.
32. S. J. Won, B. H. Yoo, A. M. Brennan, B. S. Shin, T. M. Kauppinen, A. E. Berman, R. A. Swanson and S. W. Suh, *J. Neurosci.*, 2010, **30**, 15409–15418.
33. T. Aratake, Y. Higashi, T. Hamada, Y. Ueba, T. Shimizu, S. Shimizu, T. Yawata, T. Ueba and M. Saito, *Exp. Neurol.*, 2021, **336**, 113538.
34. J. Jiménez-Conde, A. Ois, A. Rodríguez-Campello, M. Gomis and J. Roquer, *J. Neurol.*, 2007, **254**, 782–788.
35. L. J. Lawson, V. H. Perry, P. Dri and S. Gordon, *Neuroscience*, 1990, **39**, 151–170.
36. U. K. Hanisch and H. Kettenmann, *Nat. Neurosci.*, 2007, **10**(11), 1387–1394.
37. M. Antonietta Ajmone-Cat, M. Mancini, R. De Simone, P. Cilli and L. Minghetti, *Glia*, 2013, **61**, 1698–1711.
38. S. David and A. Kroner, *Nat. Rev. Neurosci.*, 2011, **12**, 388–399.
39. Z. Chen and B. D. Trapp, *J. Neurochem.*, 2016, **136**, 10–17.
40. Y. Nakagawa and K. Chiba, *Pharmacol. Ther.*, 2015, **154**, 21–35.
41. J. Yuan, H. Ge, W. Liu, H. Zhu, Y. Chen, X. Zhang, Y. Yang, Y. Yin, W. Chen, W. Wu, Y. Yang, J. Lin, J. Yuan, H. Ge, W. Liu, H. Zhu, Y. Chen, X. Zhang, Y. Yang, Y. Yin, W. Chen, W. Wu, Y. Yang and J. Lin, *Oncotarget*, 2017, **8**, 19855–19865.
42. T. Tanaka, K. Murakami, Y. Bando and S. Yoshida, *Glia*, 2015, **63**, 595–610.
43. T. M. Kauppinen, Y. Higashi, S. W. Suh, C. Escartin, K. Nagasawa and R. A. Swanson, *J. Neurosci.*, 2008, **28**, 5827–5835.
44. Y. Higashi, S. Segawa, T. Matsuo, S. Nakamura, Y. Kikkawa, K. Nishida and K. Nagasawa, *Glia*, 2011, **59**, 1933–1945.
45. Y. Higashi, T. Aratake, S. Shimizu, T. Shimizu, K. Nakamura, M. Tsuda, T. Yawata, T. Ueba and M. Saito, *Sci. Rep.*, 2017, 7, 43778.
46. S. Segawa, N. Tatsumi, A. Ohishi, K. Nishida and K. Nagasawa, *Metallomics*, 2015, 7, 1067–1077.
47. H. S. Park, M. H. Yoo and J. Y. Koh, *Neurobiol. Dis.*, 2020, **137**, 104771.
48. T. M. Kauppinen, S. W. Suh, Y. Higashi, A. E. Berman, C. Escartin, S. J. Won, C. Wang, S.-H. Cho, L. Gan and R. A. Swanson, *J. Neuroinflammation*, 2011, **8**, 152.

49. T. Hamada, T. Aratake, Y. Higashi, Y. Ueba, T. Shimizu, S. Shimizu, T. Yawata, T. Ueba, R. Nakamura, T. Akizawa, M. Fujieda and M. Saito, *J. Trace Elem. Med. Biol.*, 2020, **61**, 126518.
50. T. Aratake, Y. Higashi, Y. Ueba, T. Hamada, T. Shimizu, S. Shimizu, T. Yawata, T. Ueba and M. Saito, *Metallomics*, 2018, **10**, 1501–1509.
51. J. A. Hewett, *J. Neurochem.*, 2009, **110**, 1717–1736.
52. W. Lloyd Andriezen, *Br. Med. J.*, 1893, **2**, 227–230.
53. Z. Huiliang, Y. Mengzhe, W. Xiaochuan, W. Hui, D. Min, W. Mengqi, W. Jianzhi, C. Zhongshan, P. Caixia and L. Rong, *J. Neurochem.*, 2021, **159**, 1016–1027.
54. W. S. Sang, K. Aoyama, C. C. Alano, C. M. Anderson, A. M. Hamby and R. A. Swanson, *Mol. Med.*, 2007, **13**, 344–349.
55. S. Segawa, T. Nishiura, T. Furuta, Y. Ohsato, M. Tani, K. Nishida and K. Nagasawa, *Life Sci.*, 2014, **94**, 137–144.
56. M. Nishikawa, H. Mori and M. Hara, *Environ. Toxicol. Pharmacol.*, 2017, **53**, 81–88.
57. T. E. Bakken, N. L. Jorstad, Q. Hu, B. B. Lake, W. Tian, B. E. Kalmbach, M. Crow, R. D. Hodge, F. M. Krienen, S. A. Sorensen, J. Eggermont, Z. Yao, B. D. Aevermann, A. I. Aldridge, A. Bartlett, D. Bertagnolli, T. Casper, R. G. Castanon, K. Crichton, T. L. Daigle, R. Dalley, N. Dee, N. Dembrow, D. Diep, S. L. Ding, W. Dong, R. Fang, S. Fischer, M. Goldman, J. Goldy, L. T. Graybuck, B. R. Herb, X. Hou, J. Kancherla, M. Kroll, K. Lathia, B. van Lew, Y. E. Li, C. S. Liu, H. Liu, J. D. Lucero, A. Mahurkar, D. McMillen, J. A. Miller, M. Moussa, J. R. Nery, P. R. Nicovich, S. Y. Niu, J. Orvis, J. K. Osteen, S. Owen, C. R. Palmer, T. Pham, N. Plongthongkum, O. Poirion, N. M. Reed, C. Rimorin, A. Rivkin, W. J. Romanow, A. E. Sedeño-Cortés, K. Siletti, S. Somasundaram, J. Sulc, M. Tieu, A. Torkelson, H. Tung, X. Wang, F. Xie, A. M. Yanny, R. Zhang, S. A. Ament, M. M. Behrens, H. C. Bravo, J. Chun, A. Dobin, J. Gillis, R. Hertzano, P. R. Hof, T. Höllt, G. D. Horwitz, C. D. Keene, P. V. Kharchenko, A. L. Ko, B. P. Lelieveldt, C. Luo, E. A. Mukamel, A. Pinto-Duarte, S. Preissl, A. Regev, B. Ren, R. H. Scheuermann, K. Smith, W. J. Spain, O. R. White, C. Koch, M. Hawrylycz, B. Tasic, E. Z. Macosko, S. A. McCarroll, J. T. Ting, H. Zeng, K. Zhang, G. Feng, J. R. Ecker, S. Linnarsson and E. S. Lein, *Nature*, 2021, **598**, 111–119.
58. Z. Yao, C. T. J. van Velthoven, T. N. Nguyen, J. Goldy, A. E. Sedeno-Cortes, F. Baftizadeh, D. Bertagnolli, T. Casper, M. Chiang, K. Crichton, S. L. Ding, O. Fong, E. Garren, A. Glandon, N. W. Gouwens, J. Gray, L. T. Graybuck, M. J. Hawrylycz, D. Hirschstein, M. Kroll, K. Lathia, C. Lee, B. Levi, D. McMillen, S. Mok, T. Pham, Q. Ren, C. Rimorin, N. Shapovalova, J. Sulc, S. M. Sunkin, M. Tieu, A. Torkelson, H. Tung, K. Ward, N. Dee, K. A. Smith, B. Tasic and H. Zeng, *Cell*, 2021, **184**, 3222–3241.e26.
59. Y. Zhang, S. A. Sloan, L. E. Clarke, C. Caneda, C. A. Plaza, P. D. Blumenthal, H. Vogel, G. K. Steinberg, M. S. B. Edwards, G. Li, J. A. Duncan, S. H. Cheshier, L. M. Shuer, E. F. Chang, G. A. Grant, M. G. H. Gephart and B. A. Barres, *Neuron*, 2016, **89**, 37–53.

60. Y. Zhang, K. Chen, S. A. Sloan, M. L. Bennett, A. R. Scholze, S. O'Keeffe, H. P. Phatnani, P. Guarnieri, C. Caneda, N. Ruderisch, S. Deng, S. A. Liddelow, C. Zhang, R. Daneman, T. Maniatis, B. A. Barres and J. Q. Wu, *J. Neurosci.*, 2014, **34**, 11929–11947.
61. C. Nolte, A. Gore, I. Sekler, W. Kresse, M. Hershfinkel, A. Hoffmann, H. Kettenmann and A. Moran, *Glia*, 2004, **48**, 145–155.
62. C.-Y. Chang, Y.-C. Ou, T.-K. Kao, H.-C. Pan, S.-Y. Lin, S.-L. Liao, W.-Y. Wang, H.-C. Lu and C.-J. Chen, *Toxicol. Lett.*, 2010, **199**, 102–109.
63. S.-L. Liao, Y.-C. Ou, S.-Y. Lin, T.-K. Kao, H.-C. Pan, C.-Y. Chang, C.-Y. Lai, H.-C. Lu, W.-Y. Wang and C.-J. Chen, *Toxicol. Lett.*, 2011, **204**, 108–117.
64. E. Sock and M. Wegner, *Dev. Neurobiol.*, 2021, **81**, 892–901.
65. T. Aberle, S. Piefke, S. Hillgärtner, E. R. Tamm, M. Wegner and M. Küspert, *Nucleic Acids Res.*, 2022, **50**, 1951–1968.
66. C. Bernhardt, E. Sock, F. Fröb, S. Hillgärtner, M. Nemer and M. Wegner, *Nucleic Acids Res.*, 2022, **50**, 11509–11528.
67. A. L. Schmidt, M. Kremp, T. Aratake, S. Cui, Y. Lin, X. Zhong, Q. R. Lu, C. Zhang, M. Qiu, T. Aberle and M. Wegner, *Glia*, 2024, **72**, 1304–1318.
68. G. D. Conway, M. A. O'Bara, B. H. Vedia, S. U. Pol and F. J. Sim, *Glia*, 2012, **60**, 1944–1953.
69. P. Kursula, G. Meriläinen, V. P. Lehto and A. M. Heape, *J. Neurochem.*, 1999, **73**, 2110–2118.
70. P. Riccio, S. Giovannelli, A. Bobba, E. Romito, A. Fasano, T. Bleve-Zacheo, R. Favilla, E. Quagliariello and P. Cavatorta, *Neurochem. Res.*, 1995, **20**, 1107–1113.
71. D. Tsang, Y. S. Tsang, W. K. K. Ho and R. N. S. Wong, *Neurochem. Res.*, 1997, **22**, 811–819.
72. H. Yan, S. Yang, Y. Hou, S. Ali, A. Escobar, K. Gao, R. Duan, T. Kubisiak, J. Wang, Y. Zhang, J. Xiao, Y. Jiang, T. Zhang, Y. Wu, M. Burmeister, Q. Wang, M. P. Cuajungco and J. Wang, *Cells*, 2022, **11**, 1285.
73. M. C. Do Rosario, G. R. Bey, B. Nmezi, F. Liu, T. Oranburg, A. S. A. Cohen, K. A. Coffman, M. R. Brown, K. Kiselyov, Q. Waisfisz, M. T. Flohil, S. Siddiqui, J. A. Rosenfeld, A. Iglesias, K. M. Girisha, N. I. Wolf, Q. S. Padiath and A. Shukla, *Brain*, 2022, **145**, 4202–4209.
74. X. Liu, A. M. Adamo and P. I. Oteiza, *J. Nutr.*, 2023, **153**, 2778–2796.
75. D. Bourassa, C. M. Elitt, A. M. McCallum, S. Sumalekshmy, R. L. McRae, M. T. Morgan, N. Siegel, J. W. Perry, P. A. Rosenberg and C. J. Fahrni, *ACS Sens.*, 2018, **3**, 458–467.
76. C. M. Elitt, M. M. Ross, J. Wang, C. J. Fahrni and P. A. Rosenberg, *Neurosci. Lett.*, 2024, 137727.
77. S. Mato, M. V. Sánchez-Gómez, A. Bernal-Chico and C. Matute, *Glia*, 2013, **61**, 750–764.
78. M. Domercq, S. Mato, F. N. Soria, M. V. Sánchez-gómez, E. Alberdi and C. Matute, *Glia*, 2013, **61**, 383–393.
79. Y. Zhang, H. Wang, J. Li, L. Dong, P. Xu, W. Chen, R. L. Neve, J. J. Volpe and P. A. Rosenberg, *J. Biol. Chem.*, 2006, **281**, 9460–9470.

80. A. Armulik, G. Genové and C. Betsholtz, *Dev. Cell*, 2011, **21**, 193–215.
81. D. Attwell, A. Mishra, C. N. Hall, F. M. O'Farrell and T. Dalkara, *J. Cereb. Blood Flow Metab.*, 2016, **36**, 451–455.
82. M.-K. Cho, E.-S. Sun and Y.-H. Kim, *Exp. Neurobiol.*, 2013, **22**, 315–321.
83. H. N. Kim, T. Y. Kim, Y. H. Yoon and J. Y. Koh, *Brain Res.*, 2015, **1628**, 317–326.
84. C. T. Sheline, M. M. Behrens and D. W. Choi, *J. Neurosci.*, 2000, **20**, 3139–3146.

CHAPTER 12

Regulation of Skin Wound Healing by Zinc and Zinc Transporters

KEIGO NISHIDA*[a,b] AND NAOYA NAKAGAWA[b]

[a] Laboratory of Immune Regulation, Graduate School of Pharmaceutical Sciences, Suzuka University of Medical Science, 3500-3 Minamitamagaki, Suzuka, Mie 513-8607, Japan; [b] Laboratory of Immune Regulation, Faculty of Pharmaceutical Sciences, Suzuka University of Medical Science, 3500-3 Minamitamagaki-cho, Suzuka, Mie 513-8670, Japan

*Email: knishida@suzuka-u.ac.jp

12.1 Introduction

Zinc is an essential trace metal element that is crucial for human survival, and the total zinc concentration in the body is maintained at approximately 2–3 g in adults. A daily consumption (through food or drink) of approximately 10–15 mg of zinc is required to maintain these zinc levels. Ingested zinc is first taken up by small intestinal cells, followed by binding to albumin and macroglobulin and circulation in the blood, where it is transported to tissues and cells. Zinc homeostasis is strictly regulated by zinc transport and storage in and out of cells, and its disruption due to zinc deficiency or excess intake causes disorders in the body. Zinc deficiency causes mild (such as taste disorder, night blindness, and decreased reproductive function) and severe (including growth disorder, mental disorders, and infections and dermatitis due to decreased immune function) disorders.[1] Similarly, excessive zinc intake causes metal poisoning symptoms such as vomiting and diarrhea, as well as hemolytic anemia.[2]

Coordination Chemistry Perspectives No. 2
Zinc in Biology: Molecular Structures, Cellular Processes and Living Systems
Edited by Hiroyuki Yasui and Taiho Kambe

Published by the Royal Society of Chemistry, www.rsc.org

The widespread functional disorders owing to disrupted zinc homeostasis are believed to reflect the significance of zinc in the formation of the active centers of over 300 enzymes and in maintaining the three-dimensional structure of transcription factors and signal transduction molecules that have zinc ion-binding motifs, such as zinc fingers. Bioinformatics suggests that approximately 4–10% of all human genes may encode zinc-binding proteins.[3] Therefore, zinc is believed to play a role in early development, immune function, cancer cell metastasis, wound healing, and other aspects of wound healing by regulating various molecules that mediate biological responses such as cell proliferation, differentiation, functional expression, survival, and motility.[4–7] However, the mechanisms by which zinc regulates the above-mentioned biological phenomena are partially understood. In this article, we focus on the relationship between zinc and skin wound healing and discuss our research findings on a novel regulatory mechanism by which zinc controls the inflammatory phase of skin wound healing.

12.2 Zinc and Wound Healing

When a tissue is injured, specific highly complex biological reactions occur, starting with hemostasis and inflammation, and the process that leads to healing is called wound healing.[8,9] The wound healing process comprises three major phases, namely inflammatory, cell proliferation, and maturation and remodeling (Table 12.1). The inflammatory phase induces the removal of microorganisms, foreign bodies, and necrotic cells, as well as the activation of factors that are crucial to the subsequent proliferation process, leading to the repair or replacement of the damaged tissue. Inflammation is accompanied by vasodilation and plasma exudation, as well as the

Table 12.1 Wound healing process and associated parameters.

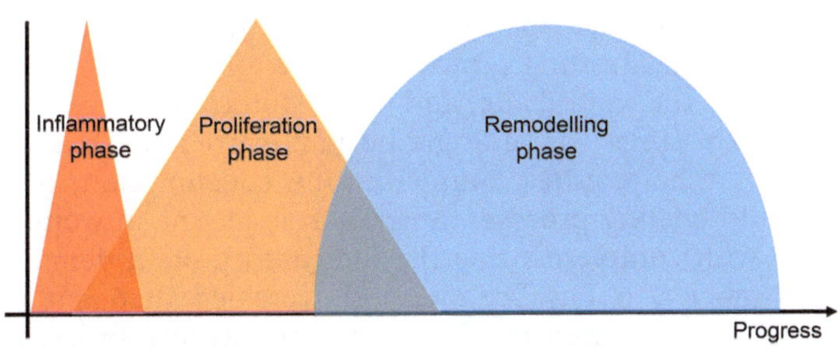

Wound Healing Phase	Functions	Cell Types	Cytokines
Inflammatory	Inflammatory cell infiltration	Platelet, Vascular endothelial cells, Neutrophils, Macrophages	TGF-α, EGF, PDGF, IL-1, IL-6 TNF-α, TGF-β, and FGF
Proliferation	Granulation	Vascular endothelial cells, Fibroblasts, Epidermal cells	EGF, PDGF, IL-1, TNF-α, TGF-β, FGF, IL-2, IL-6, and VEGF
Remodelling	Wound contraction, Epithelialization	Fibroblasts, Epidermal cells	EGF, TGF-α, FGF, PDGF, TGF-β, and IL-1

proliferation of macrophages and mononuclear cells with phagocytic capacity, which, together with neutrophil granulocytes, cleanse the wound. In the cell proliferation phase, angiogenesis and the formation of the extracellular matrix (ECM) occur, leading to granulation. Rapidly after an injury, significant fibroblast proliferation promotes the secretion of hyaluronic acid, an effective component in the formation of collagen fibers. Fibroblasts constitute almost all the cells in the wound, and they function during the time required for the collagen produced to fill the wound. At this point, the fibroblasts cease to function and differentiate into myofibroblasts, acquiring smooth muscle actin expression. These myofibroblasts synthesize and deposit ECM components that replace the provisional matrix. In the maturation and remodeling phases, ECM maturation and skin cell regeneration and migration are the primary processes. Each phase overlaps, and in humans, the healing process takes many months. Many cytokines participate in these processes in a complex and multifaceted manner, forming a so-called cytokine network that is believed to be crucial.[10,11]

In recent years, bedsores have become a challenge in the long-term care of older people and those in home medical care settings. Zinc deficiency may cause decreased Langerhans cells in pressure ulcers, possibly contributing to prolonged treatment by promoting vascular disorders and oxidative stress.[12] Thus, zinc is significant in the treatment of pressure ulcers; however, its role in the phenomenon is partially understood. Considering the effective treatment of pressure ulcers, a comprehensive understanding of the detailed mechanisms of the wound healing process described above is essential. Zinc acts on skin metabolism and promotes wound repair.[13,14] In addition, zinc is used in the private sector as an eczema and burn ointment, and its effectiveness in healing skin wounds and keeping the skin normal has been demonstrated. New wound dressings containing zinc ions are being investigated, which reportedly accelerate wound healing by promoting wound re-epithelialization and angiogenesis.[15] In addition, it has been noted recently that the inclusion of zinc nanoparticles in wound dressings can achieve the wound healing promotion effect of zinc and the infection prevention effect at the wound site, and many studies have been conducted on new wound dressings containing zinc nanoparticles.[16–18] Protein synthesis, a process that requires abundant zinc consumption, is activated during the wound healing process; thus, the role of zinc in wound healing was associated with complementing the increased protein demand during the process. However, it has been recently revealed that zinc is closely involved in the cytokine network in the wound-healing process through enzyme activity. Matrix metalloproteinase 8 (MMP8) is a zinc-containing enzyme, and when fibroblast-derived MMP8 is activated by zinc, it promotes wound healing by suppressing transforming growth factor-β (TGF-β) signaling.[19] In addition, as described below, zinc participates in the production of cytokines involved in wound healing. Thus, zinc is believed to be intricately involved at each stage of the wound-healing process. The role of zinc in wound healing remains partially understood; thus, further research is required.

12.3 Wound Healing and Mast Cells

Mast cells are immune cells and have long been recognized as the primary immune cells involved in the development of allergic diseases by releasing abundant chemical mediators such as histamine (degranulation).[20,21] Similarly, various physiologically active substances released from mast cells, such as proteases, cytokines, chemokines, and lipid mediators, have been implicated in inflammatory diseases other than allergic diseases, and the relationship between mast cells and inflammatory diseases has garnered attention.[22] Weller *et al.* (2006) studied experimentally induced skin wounds in mast cell-deficient *KitW/KitW-v*, normal *Kit+/+*, and mast cell-reconstituted *KitW/KitW-v* mice.[23] The absence of mast cells during the first 6 days of wound healing significantly inhibited wound closure, and histological analysis of mast cell degranulation at the wound edge revealed distance-dependent mast cell activation. Furthermore, neutrophil extravasation and recruitment to the wound site were inhibited in the *KitW/KitW-v* mice. Notably, wound closure, extravasation, and neutrophil recruitment were normal in the MC-reconstituted *KitW/KitW-v* mice. Similarly, wound closure was reduced in the mice treated with a histamine H_1 receptor antagonist but not in those treated with a histamine H_2 receptor antagonist or in the absence of tumor necrosis factor (TNF-α).

Other studies have shown that the mast cell products histamine and serotonin exert mitogenic effects on mouse epidermal keratinocytes *in situ*.[24] *In vitro* studies have demonstrated that mast cells promote the transformation of fibroblasts into a myofibroblast phenotype,[25] stimulate fibroblast proliferation and migration,[26] and induce neovascularization when they accumulate at neovascularization sites.[27] Moreover, Egozi *et al.* demonstrated that mast cells control crucial events in wound healing, such as the inflammatory response.[28] Similarly, other studies have demonstrated the role of mast cells in the revascularization, re-epithelialization, and deposition of connective tissues and subsequent remodeling of damaged tissues.[29,30] Mast cells are activated by tissue injury release mediators that induce vasodilation and increase vascular permeability.[23] Endothelial cells affect the functional state of mast cells by releasing the stem cell factor, interleukin-3 (IL-3), and thrombin, which promote mast cell migration, proliferation, and local differentiation. At the wound edge, keratinocytes secrete several cytokines and LL37, which affect mast cell recruitment and function. These research trends, including our observations, indicate that mast cells participate in skin wound healing, and various physiologically active substances they release are suggested to be associated with skin wound healing.

Cytokines are significant in wound healing.[10,31] Inflammatory cytokines such as interleukin-6 (IL-6) play a central role in the early inflammatory response (Table 12.1). An analysis involving *il6*-deficient mice has demonstrated the significance of IL-6 in wound healing, as it regulates leukocyte infiltration, angiogenesis, and collagen accumulation.[32] TGF-β also plays a role in regulating myofibroblast activation during the wound-healing process. Myofibroblasts are significant cells in wound healing, and when activated, they restore tissue integrity and promote the replacement of parenchymal cells.[33] The basic fibroblast growth factor reportedly participates in all stages of the wound healing process. During the inflammatory phase, it enhances the migration of inflammatory cells, such as neutrophils and macrophages, to the wound site and promotes their proliferation. Similarly, during the proliferation phase, it facilitates angiogenesis (by promoting the steps involved) and the formation of benign granulation tissues rich in neovascularization by directly stimulating and proliferating fibroblasts.[34] Furthermore, during the remodeling phase, it promotes epithelialization by promoting keratinocyte proliferation.

12.4 Intracellular Granule Zinc and Zinc Transporters

Cellular zinc levels are regulated by zinc uptake, release, and storage, and zinc exchange across the cell membrane is mediated by zinc transporters (Figure 12.1). There are 14 types of ZIP (Zrt-, Irt-like protein: Slc39) family zinc transporters that have a directional tendency to increase cytoplasmic zinc, and nine types of ZnT (zinc transporter: Slc30) family zinc transporters that reduce cytoplasmic zinc.[35–37] Histamine-containing granules in mast cells contain zinc, an essential trace metal element,[38] which is released extracellularly upon stimulation. In addition to mast cells, this secretory granule zinc has been observed in secretory vesicles in mammary epithelial cells, secretory granules in intestinal Paneth cells, synaptic vesicles in nerve cells, and insulin-containing granules in pancreatic β cells (Figure 12.2).

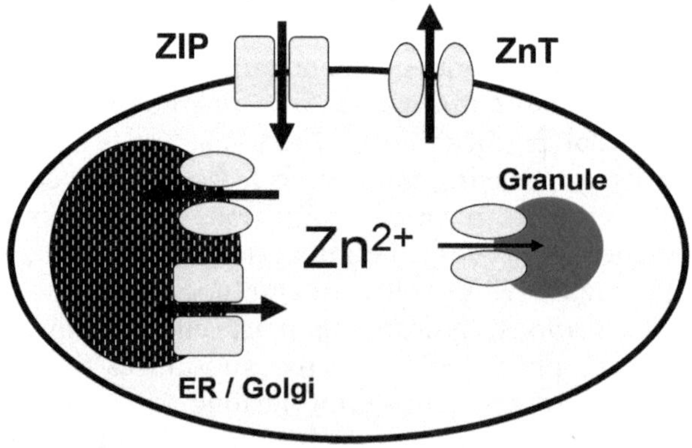

Figure 12.1 Intracellular zinc is regulated by zinc transporters. Zinc concentration in the cytoplasm is regulated by ZIP family proteins, which function as transporters that increase zinc concentration, and ZnT family proteins, which function as transporters that decrease intracellular zinc concentration. Fourteen types of ZIP family proteins, from ZIP1 to ZIP14, and nine types of ZnT family proteins, from ZnT1 to ZnT10, excluding ZnT9, have been reported.

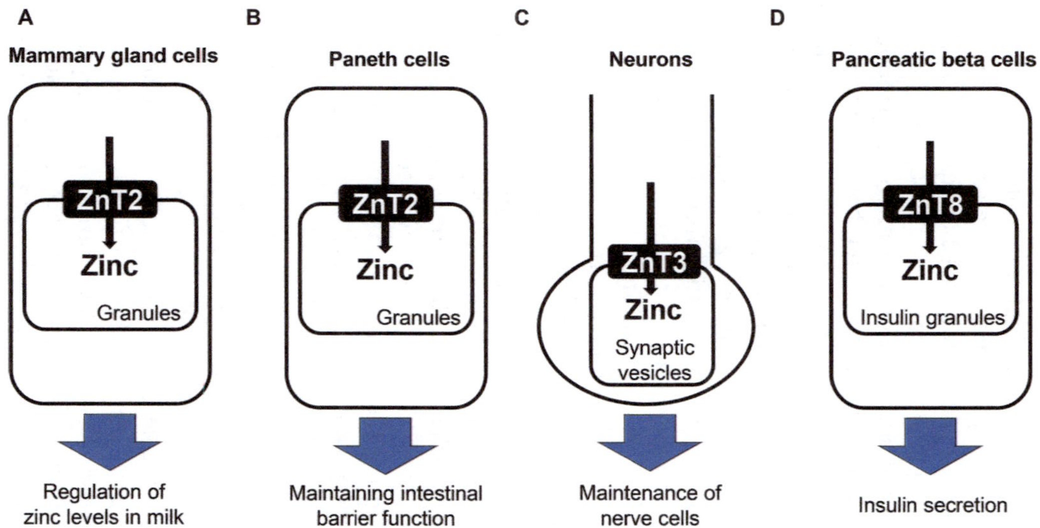

Figure 12.2 Physiological functions of granule/vesicle zinc. (A) ZnT2 expressed in the granules uptakes zinc from the cytosol in the mammary gland, modulating normal zinc content in breast milk. (B) ZnT2 expressed on the granules in the Paneth cells uptakes zinc into the granules, maintaining intestinal barrier function. (C) ZnT3 expressed in the synaptic vesicles uptakes zinc into the vesicles. Zn is released from vesicles into the surrounding milieu upon exocytotic presynaptic stimulation and acts on postsynaptic neurons to modulate neurotransmission. (D) ZnT8 expressed in the insulin granules in the pancreatic beta cells uptakes zinc into the granules, modulating proper insulin secretion from the beta cells.

ZnT2 is crucial in the mammary gland. ZnT2 transports zinc to secretory vesicles in mammary epithelial cells, which secrete zinc into breast milk.[39] Zinc is an essential trace element for infant growth and development, and its concentration in breast milk directly affects infant health. Moreover, genetic loss of ZnT2 function reduces zinc levels in breast milk and causes zinc deficiency and related symptoms in infants.[40] The intestine surface has a pleated structure, and the part at the bottom of the valley is called the intestinal crypt. The bottom of the intestinal crypt contains a cell group called Paneth cells. Paneth cells contain numerous secretory granules and produce antimicrobial peptides such as defensins. ZnT2 is expressed in the secretory granules of these Paneth cells and mediates zinc supply.[41] Zinc in Paneth cells is essential for normal secretion of antimicrobial peptides, and zinc deficiency may impair Paneth cell and intestinal barrier functions.[42] ZnT3 is expressed in the synaptic vesicles of nerve cells and participates in zinc uptake into the vesicles. In addition, zinc in synaptic vesicles is released with neurotransmitters such as glutamate when nerve cells are excited and acts on receptors in the postsynaptic membrane. Zinc transport *via* ZnT3 is essential for maintaining memory and cognitive function.[43] ZnT8 is crucial in pancreatic β cells. ZnT8 transports zinc into insulin secretory granules, facilitating insulin crystallization and stability.[44] This process is essential for normal insulin secretion and function. The gene encoding *ZNT8* (*SLC30A8*) is also known as a risk gene for Type 2 diabetes.[45] Malfunction or deficiency of ZnT8 causes abnormal insulin secretion and glucose metabolism disorders. For example, in *ZnT8*-deficient mice, insulin secretion is increased; however, the inadequate zinc secretion causes excessive insulin degradation in the liver, resulting in insufficient insulin supply to peripheral tissues.[46]

Therefore, the zinc released from mast cells was suggested to participate in wound healing, prompting the identification of the zinc transporter that controls mast cell secretory granule zinc and the investigation of the relationship between released zinc and skin wound healing. To investigate whether the zinc in mast cell secretory granules participates in skin wound healing, we examined the expression of zinc transporters in mast cells. Among the ZnT family members, ZnT1, 2, 5, 6, and 7 were expressed, with ZnT2 exhibiting high expression levels. Thus, we suggested that ZnT2 may participate in zinc regulation in mast cell granules and conducted further analyses. The zinc transporter ZnT2 was identified to be localized in intracellular vesicles such as endosomes and lysosomes and encapsulates excess zinc from the cytoplasm into the vesicles.[47] Based on its characteristics, such as its intracellular localization, ZnT2 was believed to play a role in zinc detoxification in the cytoplasm by encapsulating excess cytoplasmic zinc in endosomes. Hence, we performed immunoelectron microscopy using the ZnT2 antibody and found that ZnT2 was localized in the secretory granule membrane of mast cells. Subsequently, to investigate whether ZnT2 participates in zinc accumulation in mast cell secretory granules and zinc release, we established a *ZnT2*-knockout (*ZnT2*-KO) mouse model, prepared mast cells from the *ZnT2*-KO mice, and investigated the presence or absence of released zinc. In the mast cells derived from the *ZnT2*-KO mice, no stimulation-dependent release of zinc was observed. These results indicate that the zinc transporter ZnT2 is responsible for zinc accumulation in mast cell secretory granules.[48]

12.5 Role of Zinc/Zinc Transporter ZnT2 in Skin Wound Healing

As ZnT2 has been shown to participate in the regulation of zinc release from mast cells, we used *ZnT2*-KO mice to investigate the role of zinc in skin wound healing. Skin wound healing was found to be significantly delayed in *ZnT2*-KO mice compared with that in the wild-type mice. This result implies that for the first time, a zinc transporter involved in skin wound healing has been identified. We further investigated whether the zinc released from mast cells participates in skin wound healing. We transplanted mast cells derived from *ZnT2*-KO mice (which exhibited no zinc release) and control mast cells into *KitW-sh/KitW-sh* mice and observed skin wound healing. Normal skin wound healing was observed in *KitW-sh/KitW-sh* mice transplanted with control mast cells; however, skin wound healing was significantly delayed in *KitW-sh/KitW-sh* mice transplanted with mast cells derived from *ZnT2*-KO mice that exhibited no zinc release. These transplantation experiments suggest that the zinc released from mast cells participates in skin wound healing. Furthermore, we evaluated cytokine activation in skin tissues to determine whether the delayed skin wound healing observed in *ZnT2*-KO mice is related to the inflammatory phase. IL-6 is a vital cytokine in the skin wound healing process, which is activated early after injury and promotes increased vascular permeability and migration of inflammatory cells. We quantified IL-6 transcriptional activation around the wound on the first day after injury in mice and found that it was significantly reduced in *ZnT2*-KO mice.

These results suggest that the zinc released from mast cells participates in IL-6 production during the inflammatory phase of skin wound healing.[48] Thus, we have, for the first time, demonstrated that the zinc released from mast cells positively regulates skin wound healing.

12.6 Zinc Receptor GPR39 and Cytokine Production

To examine whether extracellular free zinc induces cytokine expression in fibroblasts, mouse fibroblasts were treated with zinc, and the cytokine transcription rates were measured. Zinc treatment increased the transcription rates of IL-6 and TNF-α. Consistent with this result, zinc treatment also increased the protein expression of IL-6 and TNF-α in the culture supernatant.

Extracellular zinc is a ligand for the GPR39 receptor, a type of G protein-coupled receptor.[49–51] Therefore, we investigated the involvement of GPR39 in zinc-induced cytokine transcription activation by knockdown experiments using siRNA. Zinc-induced IL-6 transcription activation was suppressed by *Gpr39* knockdown in fibroblasts. To clarify the mechanism of IL-6 transcription activation *via* the zinc receptor GPR39, we performed a detailed analysis using fibroblasts. When zinc binds to the zinc receptor GPR39, GPR39 activates the PKC-ERK signaling pathway. ERK phosphorylates C/EBP, which translocates into the nucleus and finally binds to the IL-6 promoter region. A model of the mechanism of IL-6 transcriptional activation by zinc is shown in Figure 12.3. These results indicate that the zinc receptor GPR39 is required for zinc-induced cytokine production.[48]

The mechanism of zinc-induced inflammatory layer control in skin wound healing has been demonstrated in a model (Figure 12.4).

12.7 Perspective

Since the 1960s, reports suggesting the existence of zinc deficiency in humans[52] have been published, and various symptoms of zinc deficiency have been identified. These case reports of zinc deficiency have suggested that zinc, an essential trace metal element in the body, at approximately 2–3 g, plays a crucial role in regulating various biological functions. The skin is a zinc-rich organ, with only muscles and bones having higher zinc levels, and dermatitis observed in zinc deficiency appears relatively early. In our study, we explained the mechanism of action of mast cell secretory granule zinc in skin wound healing as an inflammatory response. In addition, the significance of zinc stored in the secretory granules of mast cells has long been

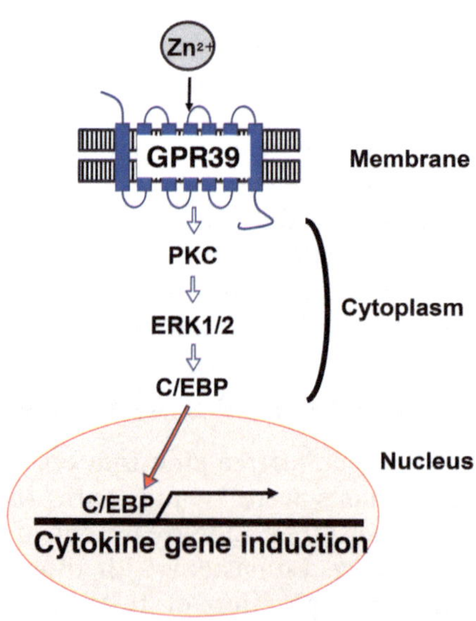

Figure 12.3 Model for zinc-mediated cytokine gene induction. Zinc binds to the zinc receptor GPR39, activating PKC, which phosphorylates and activates ERK1/2. Phosphorylated ERK1/2 activates C/EBP, which controls the cytokine gene expression.

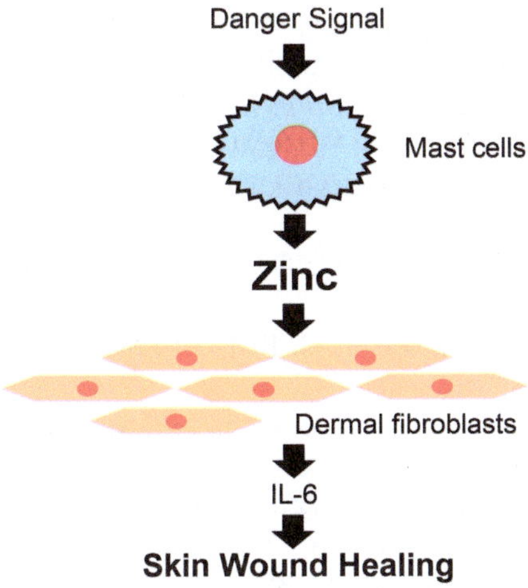

Figure 12.4 Mechanism by which zinc released from mast cells functions in skin wound healing. Danger signals are generated from damaged skin tissue. These danger signals activate skin mast cells, prompting zinc release from the activated mast cells. The released zinc acts on skin fibroblasts, inducing the production of inflammatory cytokines. Inflammatory cytokines such as IL-6 regulate the inflammatory phase of wound healing. Zinc released from mast cells plays a role in modulating the amplification of inflammatory cytokines in local skin tissue.

unknown; however, the fact that zinc released upon activation significantly contributes to skin wound healing is consistent with the observational results of empirical treatment where zinc is added to wounds.

In this chapter, we elucidated a mechanism by which the zinc released upon mast cell activation influences inflammatory layer control during skin wound healing. Granule/vesicular zinc has also been found in other cells (Figure 12.2), suggesting that the released zinc may participate in various physiological phenomena. Notably, zinc is essential for maintaining protein structure, as in zinc fingers. In addition to this, zinc significantly participates in intercellular communication.

References

1. A. S. Prasad, *Nutrition*, 1995, **11**(1 Suppl.), 93–99.
2. G. J. Fosmire, *Am. J. Clin. Nutr.*, 1990, **51**(2), 225–227.
3. G. K. Andrews, *Biometals*, 2001, **14**(3–4), 223–237.
4. N. Z. Gammoh and L. Rink, *Nutrients*, 2017, **9**(6), 624.
5. T. Kambe, T. Tsuji, A. Hashimoto and N. Itsumura, *Physiol. Rev.*, 2015, **95**(3), 749–784.
6. M. Murakami and T. Hirano, *Cancer Sci.*, 2008, **99**(8), 1515–1522.
7. K. Nishida and R. Uchida, *J. Immunol. Res.*, 2018, **2018**, 5749120.
8. R. Grose and S. Werner, *Mol. Biotechnol.*, 2004, **28**(2), 147–166.
9. O. A. Pena and P. Martin, *Nat. Rev. Mol. Cell Biol.*, 2024, **25**(8), 599–616.

10. S. Barrientos, O. Stojadinovic, M. S. Golinko, H. Brem and M. Tomic-Canic, *Wound Repair Regen.*, 2008, **16**(5), 585–601.
11. M. Schafer and S. Werner, *Annu. Rev. Cell Dev. Biol.*, 2007, **23**, 69–92.
12. H. Nakamura, A. Sekiguchi, Y. Ogawa, T. Kawamura, R. Akai, T. Iwawaki, T. Makiguchi, S. Yokoo, O. Ishikawa and S. I. Motegi, *J. Dermatol. Sci.*, 2019, **95**(2), 62–69.
13. W. J. Pories, J. H. Henzel, C. G. Rob and W. H. Strain, *Lancet*, 1967, **1**(7482), 121–124.
14. W. J. Pories, J. H. Henzel, C. G. Rob and W. H. Strain, *Ann. Surg.*, 1967, **165**(3), 432–436.
15. F. Yang, Y. Xue, F. Wang, D. Guo, Y. He, X. Zhao, F. Yan, Y. Xu, D. Xia and Y. Liu, *Bioact. Mater.*, 2023, **26**, 88–101.
16. R. Haghniaz, A. Rabbani, F. Vajhadin, T. Khan, R. Kousar, A. R. Khan, H. Montazerian, J. Iqbal, A. Libanori, H. J. Kim and F. Wahid, *J. Nanobiotechnol.*, 2021, **19**(1), 38.
17. C. Malakar, B. Kashyap, M. C. Kalita and S. Deka, *Exp. Dermatol.*, 2023, **32**(2), 154–164.
18. M. S. Saddik, M. M. A. Elsayed, M. A. El-Mokhtar, H. Sedky, J. A. Abdel-Aleem, A. M. Abu-Dief, M. F. Al-Hakkani, H. L. Hussein, S. A. Al-Shelkamy, F. Y. Meligy, A. Khames and H. A. Abou-Taleb, *Pharmaceutics*, 2022, **14**(1), 111.
19. R. C. Stone, O. Stojadinovic, A. P. Sawaya, G. D. Glinos, L. E. Lindley, I. Pastar, E. Badiavas and M. Tomic-Canic, *Wound Repair Regen.*, 2020, **28**(2), 164–176.
20. S. J. Galli, S. Nakae and M. Tsai, *Nat. Immunol.*, 2005, **6**(2), 135–142.
21. H. S. Kim, Y. Kawakami, K. Kasakura and T. Kawakami, *F1000Research*, 2020, **9**, Faculty Rev-196.
22. S. J. Galli, N. Gaudenzio and M. Tsai, *Annu. Rev. Immunol.*, 2020, **38**, 49–77.
23. K. Weller, K. Foitzik, R. Paus, W. Syska and M. Maurer, *FASEB J.*, 2006, **20**(13), 2366–2368.
24. M. Maurer, M. Opitz, B. M. Henz and R. Paus, *J. Dermatol. Sci.*, 1997, **16**(1), 79–84.
25. J. Gailit, M. J. Marchese, R. R. Kew and B. L. Gruber, *J. Invest. Dermatol.*, 2001, **117**(5), 1113–1119.
26. F. Levi-Schaffer and A. Kupietzky, *Exp. Cell Res.*, 1990, **188**(1), 42–49.
27. K. Norrby, A. Jakobsson and J. Sorbo, *Virchows Arch. B*, 1986, **52**(3), 195–206.
28. E. I. Egozi, A. M. Ferreira, A. L. Burns, R. L. Gamelli and L. A. Dipietro, *Wound Repair Regen.*, 2003, **11**(1), 46–54.
29. M. Artuc, B. Hermes, U. M. Steckelings, A. Grutzkau and B. M. Henz, *Exp. Dermatol.*, 1999, **8**(1), 1–16.
30. C. L. Baum and C. J. Arpey, *Dermatol. Surg.*, 2005, **31**(6), 674–686; discussion 686.
31. S. Werner and R. Grose, *Physiol. Rev.*, 2003, **83**(3), 835–870.
32. B. Z. Johnson, A. W. Stevenson, C. M. Prele, M. W. Fear and F. M. Wood, *Biomedicines*, 2020, **8**(5), 101.

33. R. W. D. Gilbert, M. K. Vickaryous and A. M. Viloria-Petit, *J. Dev. Biol.*, 2016, **4**(2), 21.
34. V. Y. Syromiatnikova, A. I. Kvon, I. G. Starostina and M. O. Gomzikova, *Arch. Dermatol. Res.*, 2024, **316**(7), 405.
35. T. Fukada, S. Yamasaki, K. Nishida, M. Murakami and T. Hirano, *J. Biol. Inorg. Chem.*, 2011, **16**(7), 1123–1134.
36. T. Hirano, M. Murakami, T. Fukada, K. Nishida, S. Yamasaki and T. Suzuki, *Adv. Immunol.*, 2008, **97**, 149–176.
37. T. Kambe, K. M. Taylor and D. Fu, *J. Biol. Chem.*, 2021, **296**, 100320.
38. G. T. Gustafson, *Lab. Invest.*, 1967, **17**(6), 588–598.
39. S. Lee, S. R. Hennigar, S. Alam, K. Nishida and S. L. Kelleher, *J. Biol. Chem.*, 2015, **290**(21), 13064–13078.
40. N. Itsumura, Y. Kibihara, K. Fukue, A. Miyata, K. Fukushima, R. Tamagawa-Mineoka, N. Katoh, Y. Nishito, R. Ishida, H. Narita, H. Kodama and T. Kambe, *Pediatr. Res.*, 2016, **80**(4), 586–594.
41. M. E. Elmes and J. G. Jones, *Histochem. J.*, 1981, **13**(2), 335–337.
42. A. B. Podany, J. Wright, R. Lamendella, D. I. Soybel and S. L. Kelleher, *Cell. Mol. Gastroenterol. Hepatol.*, 2016, **2**(3), 369–383.
43. M. S. Hildebrand, A. M. Phillips, S. A. Mullen, P. A. Adlard, K. Hardies, J. A. Damiano, V. Wimmer, S. T. Bellows, J. M. McMahon, R. Burgess, R. Hendrickx, S. Weckhuysen, A. Suls, P. De Jonghe, I. E. Scheffer, S. Petrou, S. F. Berkovic and C. A. Reid, *Sci. Rep.*, 2015, **5**, 17816.
44. D. A. Scott, *Biochem. J.*, 1934, **28**(4), 1592–1602.
45. F. Takeuchi, M. Serizawa, K. Yamamoto, T. Fujisawa, E. Nakashima, K. Ohnaka, H. Ikegami, T. Sugiyama, T. Katsuya, M. Miyagishi, N. Nakashima, H. Nawata, J. Nakamura, S. Kono, R. Takayanagi and N. Kato, *Diabetes*, 2009, **58**(7), 1690–1699.
46. M. Tamaki, Y. Fujitani, A. Hara, T. Uchida, Y. Tamura, K. Takeno, M. Kawaguchi, T. Watanabe, T. Ogihara, A. Fukunaka, T. Shimizu, T. Mita, A. Kanazawa, M. O. Imaizumi, T. Abe, H. Kiyonari, S. Hojyo, T. Fukada, T. Kawauchi, S. Nagamatsu, T. Hirano, R. Kawamori and H. Watada, *J. Clin. Invest.*, 2013, **123**(10), 4513–4524.
47. R. D. Palmiter, T. B. Cole and S. D. Findley, *EMBO J.*, 1996, **15**(8), 1784–1791.
48. K. Nishida, A. Hasegawa, S. Yamasaki, R. Uchida, W. Ohashi, Y. Kurashima, J. Kunisawa, S. Kimura, T. Iwanaga, H. Watarai, K. Hase, H. Ogura, M. Nakayama, J. I. Kashiwakura, Y. Okayama, M. Kubo, O. Ohara, H. Kiyono, H. Koseki, M. Murakami and T. Hirano, *Sci. Rep.*, 2019, **9**(1), 10842.
49. M. Hershfinkel, *Int. J. Mol. Sci.*, 2018, **19**(2), 439.
50. A. Laitakari, L. Liu, T. M. Frimurer and B. Holst, *Int. J. Mol. Sci.*, 2021, **22**(8), 3872.
51. S. Yasuda, T. Miyazaki, K. Munechika, M. Yamashita, Y. Ikeda and A. Kamizono, *J. Recept. Signal Transduction Res.*, 2007, **27**(4), 235–246.
52. J. A. Halsted and A. S. Prasad, *Isr. Med. J.*, 1963, **22**, 307–315.

CHAPTER 13

Role of Zinc Homeostasis in the Pathophysiology of Diabetes

AYAKO FUKUNAKA* AND YOSHIO FUJITANI*

Institute for Molecular & Cellular Regulation, Gunma University, Gunma, Japan
*Emails: fukunaka@gunma-u.ac.jp; fujitani@gunma-u.ac.jp

13.1 Role of Zinc in Pancreatic β-cells

Zinc was first identified as an essential factor for insulin crystallization in the 1930s.[1] A few years later, it was also found that zinc content in the pancreas of diabetic patients was reduced by 50%.[2] These early findings suggested a close association between zinc and diabetes. Subsequent studies demonstrated that insulin exists as a hexamer composed of six insulin molecules and two zinc molecules[3,4] (Figure 13.1). Under an electron-microscope, crystallized insulin in insulin secretory granules is observed as "dense core granules".[5,6]

Proinsulin is the precursor molecule of insulin, synthesized in the pancreas's β cells. Its structure consists of three domains: A-chain, B-chain, and C-peptide (Figure 13.1A). The proinsulin protein forms dimers upon folding, through electrostatic interactions. The residues B24 to 26, which are located on the B-chain of the proinsulin molecule, are thought to play a crucial role in this dimer formation.[7] Amino acid substitutions of these residues have led to the development of ultrarapid-acting insulin analogues, which are now widely used in clinical practice in diabetes patients.[8,9] The electrostatically coupled proinsulin dimers and zinc bind to the B10

Coordination Chemistry Perspectives No. 2
Zinc in Biology: Molecular Structures, Cellular Processes and Living Systems
Edited by Hiroyuki Yasui and Taiho Kambe

Published by the Royal Society of Chemistry, www.rsc.org

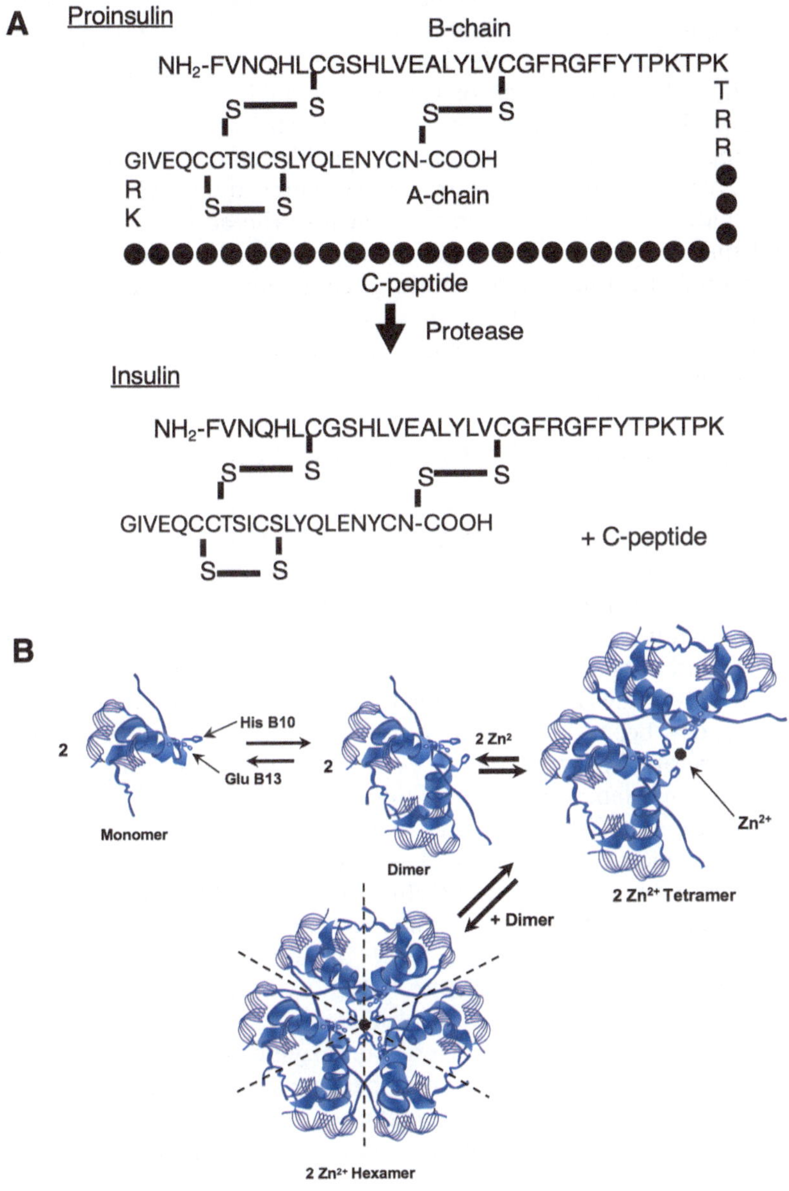

Figure 13.1 Biosynthesis of insulin. (A) Amino acid sequences of proinsulin and insulin peptides. The insulin precursor preproinsulin contains a signal sequence that is proteolytically cleaved to yield proinsulin, in which the C-peptide links the future A- and B-chains of mature insulin. Cleavage of the C-peptide converts proinsulin to insulin. (B) Association of insulin monomers in the presence of zinc. Insulin readily associates into dimers, and then aggregates into hexameric forms in the presence of zinc. His B10 is implicated in zinc-mediated insulin hexamer formation.

histidine residue of insulin, resulting in formation of the proinsulin hexamer[10] (Figure 13.1B). This hexamer formation is facilitated by prohormone convertase-mediated C-peptide dissociation.[11] As the C-peptide portion of proinsulin is located outside of the proinsulin hexamer, it has been suggested that the formation of the hexamer contributes to C-peptide dissociation.[4] Insulin crystals form under specific conditions within insulin granules, in which both insulin and zinc are present at high concentrations, and an acidic pH is maintained.[12,13] Because the addition of zinc to commercial insulin products improves their stability, hexamer formation and crystallization are thought to similarly contribute to insulin stabilization *in vivo*. Once insulin crystals are secreted from pancreatic β-cells, they are believed to rapidly dissociate into monomers, and flow into the bloodstream.

13.2 Role of Zinc in Peripheral Tissues

A chronic low intake of zinc is associated with an increased risk of diabetes.[14] Therefore, zinc supplementation is thought to be an effective method for preventing metabolic syndrome and diabetes. A previous study analyzed the effect of zinc supplementation on insulin resistance and metabolic syndrome in prepubertal obese children, suggesting that zinc supplementation is a useful and safe additional intervention.[15] However, to date, the effectiveness of zinc supplementation in treating obesity and diabetes has not been demonstrated in large-scale studies, particularly in adults. Therefore, it is important to establish the safety, efficacy, and optimal dose of zinc supplementation for adults.

Zinc is often referred to as an "insulin-mimetic," as it stimulates lipogenesis and glucose uptake in isolated adipocytes. Zinc ions act as an insulin-mimetic through their direct effect on the insulin-signaling pathway.[16] The insulin-sensitizing effect of zinc is thought to be due to inhibition of the tyrosine phosphatase activity of protein tyrosine phosphatase 1B (PTP1B).[17,18] Zinc ions inactivate PTP1B by noncovalently binding to its cysteine residues, which are crucial for the enzyme's activity. Reactive oxygen species are also known to inactivate the enzyme by a similar mechanism. Oxidative stress is upregulated in most patients with diabetes and in diabetic animals. Therefore, further mechanistic analyses are needed to clarify the risks and benefits of zinc supplementation as a potential treatment for obesity and T2D.

13.3 Structure and Function of ZnT8

Zinc homeostasis is controlled by zinc transporters, which are classified into two categories according to the direction of their transport of zinc. The Zn transporter (ZnT) (SLC30A) family, which comprises transporters that transport zinc from the cytoplasm to the extracellular space or to intracellular organelles, and the Zrt-, Irt-like protein (ZIP) (SLC39A) family, which comprises transporters that transport zinc from the extracellular space or intracellular organelles to the cytoplasm. There are 10 types of ZnT proteins and 14 types of ZIP proteins in mammals, and it has been clarified that each works in cooperation with other transporters or in a tissue-specific manner. The phenotypes of knockout mice lacking these zinc transporters, and humans expressing variants of these transporters have been characterized previously.[19,20]

ZnT8 is a member of the ZnT family, and has a predicted membrane topology of six membrane-spanning domains with four conserved amino acid residues within transmembrane domains (TMDs) II and V, and a cytoplasmic His-rich loop between TMDs IV and V.[21] A previous study demonstrated that the expression level of ZnT8 is 2 to 3 orders of magnitude higher in the pancreatic islets than in all other tissues analyzed.[22] Thus, ZnT8 is responsible for transporting zinc into insulin-containing granules in islet β-cells, where insulin is stored as a hexamer bound with two zinc ions before secretion (Figure 13.1B). ZnT8 is also expressed in other tissues, including islet α-cells, pancreatic polypeptide cells, and a subset of endocrine cells in the thyroid and adrenal glands.[23] The islet-specific expression of ZnT8 was shown to be regulated by the β-cell enriched transcriptional factor Pdx1 (pancreatic and duodenal homeobox 1) through an intronic enhancer.[24]

The expression of ZnT transporters is transcriptionally regulated by various stimuli. In β-cells, it has been shown that glucose and cytokines play roles in the regulation of zinc transporter expression. Cytokines (interleukin 1 β (IL-1β) and tumor necrosis factor α (TNF-α)) decrease the expression of zinc transporters, particularly ZnT8,[25] suggesting that the expression of ZnT8 might be decreased during the cytokine simulation. Moreover, ZnT8 expression was found to be substantially downregulated in the early stages of diabetes in diabetic model mice, such as *db/db* and *Akita* mice.[23] Regulation of the expression and activity of human ZnT8 appears to be similar to that of mouse ZnT8. High levels of glucose and free fatty acids did not alter cellular human ZnT8 levels, but proinflammatory cytokines acutely, reversibly, and gradually downregulated hZnT8 levels. The cytokine-induced downregulation of the endoplasmic reticulum (ER) ZnT8 level promotes the adaptive unfolded protein response (UPR), acting as a protective mechanism that decongests the ER burden of ZnT8 to protect β-cells from proapoptotic UPR during chronic low-grade inflammation.[26] We and other groups also observed that the zinc contents in islets were reduced at an earlier time point than the onset of T2D.[27,28] These findings suggested that the expression of ZnT8 and zinc levels in islets might be reduced before the onset of T2D.

Several studies have shown that ZnT8 is a *bona fide* zinc transporter, using ZnT8-overexpressed cells,[29] ZnT8-knockdown cells,[30] or ZnT8-KO mice.[22] The mechanism of zinc transport is still incompletely understood. However, experiments using fluorescence-based functional measurements have shown that some ZnT transporters function as Zn^{2+}/H^{+} exchangers.[31] ZnT8 may also use the same mechanism, because ZnT8 is located within insulin granules where the pH is lower than in the cytoplasm. Homology modelling studies based on the bacterial zinc transporter and ZnT homolog YiiP have shown that ZnT8 forms homodimers held together by four zinc ions located at the interface of the cytoplasmic domains, with two transmembrane domains swinging out to yield a Y-shaped structure.[22] Genome-wide association studies demonstrated that the nonsynonymous single nucleotide polymorphism rs1326634 in the *SLC30A8* gene, which results in substitution of the tryptophan with arginine at position 325 of ZnT8, increases the risk of T2D.[32–34] More recently, re-examination of zinc transport activity by liposome analysis using the purified human ZnT8 protein as well as analysis using human islets led to the conclusion that the risk allele ZnT8 type has higher zinc transport activity.[35]

13.4 Role of ZnT8 in Glucose Homeostasis

As mentioned above, recent genome-wide association studies have indicated *ZnT8* as a gene associated with an increased risk of T2D. To investigate the physiological role of ZnT8, several research groups have generated whole-body *ZnT8*-deficient mice or β-cell specific *ZnT8*-deficient mice.[6,22,36–41] However, there is a lack of consensus regarding the effects of *ZnT8* deletion on glucose tolerance, insulin secretion, and circulating insulin levels in mice, as summarized in Table 13.1.[42] Some studies reported that deletion of *SLC30A8*/*ZnT8* results in impaired glucose tolerance,[22,39] whereas others reported that loss of *ZnT8* results in essentially no change in glucose tolerance.[36,38] These differences might have been caused by differences in study designs or the genetic backgrounds of the mice generated.[6,37]

We also decided to create and analyze mice that specifically lack ZnT8 in pancreatic β-cells (βZnT8-KO mice).[6] Consistent with previous reports, insulin crystallization was incomplete in pancreatic β-cells of βZnT8-KO mice. Thus, ZnT8-mediated zinc transport into insulin granules was shown to be important for insulin crystallization. The glucose tolerance test demonstrated that βZnT8-KO mice have mild glucose intolerance, and hence it was initially predicted that insulin secretion was impaired together with the disruption of insulin crystallization. In fact, insulin concentrations in peripheral blood after glucose loading was reduced in βZnT8-KO mice. However, when investigated using the isolated islets of mice, unexpectedly, insulin secretion was rather enhanced in βZnT8-KO mice. To explain this discrepancy, a pancreas perfusion experiment was performed. Enhanced insulin secretion was still noted in βZnT8-KO islets upon pancreas perfusion, further supporting that insulin secretion is actually increased *in vivo* in βZnT8-KO islets. Finally, a comparison of the results obtained by pancreas perfusion *versus* pancreas–liver perfusion demonstrated that a large proportion of the secreted insulin was degraded during the first liver passage. Mechanistically, zinc secreted together with insulin suppressed hepatic insulin clearance by inhibiting clathrin-dependent insulin endocytosis. In addition, in βZnT8-KO mice and in humans with rs13266634, which is a major risk allele of *SLC30A8*, the insulin degradation rate calculated from

Table 13.1 Summary of ZnT8-KO mouse studies.

Study	Ref.	Deletion location	Deletion strategy	Genetic background	Insulin crystallization	Proinsulin concentration	Glucose intolerance	Glucose-stimulated insulin secretion	Insulin level *in vivo*
Nicolson *et al.* (2009)	22	Whole body	Exon 1	Mixed	Loss of dense core granules/rod-shaped cores	Unchanged	Impaired (only in male mice)	Increased	Impaired
Pound *et al.* (2009)	36	Whole body	Exon 3	Mixed	Not analyzed	Not analyzed	Normal	Decreased	Impaired
Lemaire *et al.* (2009)	38	Whole body	Exon 1	Mixed	Loss of dense core granules	Unchanged	Normal	Unchanged	Unchanged
Wijesekara *et al.* (2010)	39	Pancreatic β-cell-specific (Ins2Cre)	Exon 1	Mixed	Loss of dense core granules/rod-shaped cores	Increased	Impaired	Decreased	Unchanged
Pound *et al.* (2012)	37	Whole body	Exon 3	C57BL/6J	Unchanged	Decreased	Impaired	Decreased	Unchanged
Tamaki *et al.* (2013)	23	Pancreatic β-cell- specific (Ins2Cre)	Exon 5	C57BL/6J	Loss of dense core granules	Increased	Impaired	Increased	Impaired
Mitchell *et al.* (2016)	40	Pancreatic β-cell- specific (Ins1Cre)	Exon 1	C57BL/6J	Loss of dense core granules/rod-shaped cores	Not analyzed	Impaired	Unchanged	Impaired
Kleiner *et al.* (2018)	45	Knock in loss of function (R183X)			Not analyzed	Not analyzed	Normal	Not analyzed	Increased (under S961 treatment)

the C peptide/insulin ratio was increased.[6] Our results suggest that *SLC30A8*, which is expressed in pancreatic β-cells, regulates insulin degradation in the liver, and that the dysregulation of this system may be involved in the increased risk of developing T2D.

13.5 Reconsideration of *SLC30A8* Activity and Diabetes Risk

Recently, the analysis of several rare variants of *SLC30A8* have demonstrated that haploinsufficiency of ZnT8 reduces the risk of developing diabetes by 65%.[43,44] Functional analysis of each variant and the corresponding ZnT8 proteins showed a functional decline, such as decreased expression level and protein stability. Given the findings that the risk allele of ZnT8 is associated with higher zinc transport activity, as shown in studies using liposomes expressing purified ZnT8,[35] it is reasonable to conclude that, at least in humans, ZnT8 with reduced function is linked to a lower risk of diabetes (Figure 13.2). Consistent with this idea, in 2018, Kleiner *et al.* developed a knock-in mouse model of ZnT8 with the R138X mutation, which expresses loss-of-function (LOF) *SLC30A8* alleles (p138* variant (R138X): which means LOF *SLC30A8* alleles). This model showed increased insulin secretion compared with control mice. Interestingly, increased insulin secretion was not observed under basal conditions or during the oral glucose tolerance test, but rather under conditions of severe acute insulin resistance during the administration of S961, a biosynthetic insulin receptor antagonist,

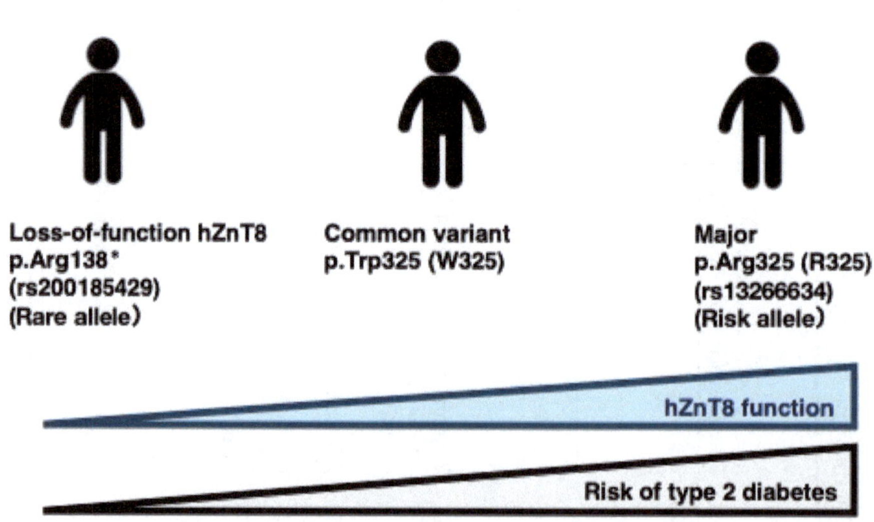

Figure 13.2 Association between hZnT8 activity and type 2 diabetes risk. Type 2 diabetes risk is increased in individuals who have risk alleles with high ZnT8 activity (encoding ZnT8 with high activity), whereas type 2 diabetes risk is reduced in individuals who have rare alleles causing ZnT8 haploinsufficiency.

Table 13.2 Recent advances in ZnT8 research.

Study	Ref.	Mouse information	Mouse phenotype
Syring *et al.* (2020)	48	Same mice as in Pound *et al.* (2009) in the high-fat diet (HFD) group	Protection against diet-induced obesity, resulting in improved glucose tolerance
Xu *et al.* (2021)	41	Same mice as in Wijesekara *et al.* (2010) for crossing with hIAPP-Tg mice	βZnT8-KO mice with *hIAPP*-expressing mice on a high-fat diet (HFD) had increased amyloid deposition, and further impaired glucose tolerance and insulin secretion compared with Ctrl, βZnT8-KO, and *hIAPP* mice
Su *et al.* (2021)	54	Whole body of ZnT8-KO mice with administration of acetaminophen	ZnT8 deficiency protected the liver from acetoaminophen toxicity
Piro *et al.* (2024)	49	Same mice as in Pound *et al.* (2009) in high-fat diet (HFD) group with administration of S961	Glucose-stimulated insulin secretion from isolated islets from βZnT8-KO mice demonstrated enhanced insulin secretion capacity

Study	Ref.	Human embryonic stem cell-derived β-cells
Ma *et al.* (2022)	46	Insulin secretion using human embryonic stem cell-derived β-cells
Jaffredo *et al.* (2024)	47	Multi-electrode arrays (MEAs) using human embryonic stem cell-derived β-cells

suggesting that hZnT8 deficiency may enhance insulin secretion, particularly under metabolic stress, and this may contribute to a reduced risk of T2D.[45] Another study using a human embryonic stem-cell derived β-cell model expressing ZnT8 LOF alleles showed enhanced glucose-stimulated insulin secretion and stronger electrical coupling activity by using multi-electrode arrays (MEA).[46,47]

In mouse studies, as mentioned above, there have been multiple reports of impaired glucose tolerance in mice expressing ZnT8 with loss-of-function mutations, under normal conditions. However, few studies have reported phenotypes that show improvements in diabetes-associated phenotypes (Table 13.2). Syring *et al.* suggested that *SLC30A8* haploinsufficiency in mice leads to normal glucose tolerance, and protects against diet-induced obesity, resulting in improved glucose tolerance.[48] More recently, Piro *et al.* reported that βZnT8-KO mice demonstrate significantly increased plasma insulin levels and improved glucose tolerance following the induction of acute insulin resistance with S961.[49]

Another explanation for the discrepancy between human and mouse ZnT8 function involves specific factors, such as human islet amyloid polypeptide (hIAPP). Pancreatic islet amyloid consists of human IAPP, a 37-amino acid polypeptide synthesized in pancreatic β-cells, which is co-secreted with insulin to regulate blood glucose levels. The physiological functions of IAPP have been proposed to include inhibiting insulin secretion, delaying gastric emptying, reducing appetite, suppressing glucagon release, and inhibiting tumorigenesis. Islet amyloid deposits are present in more than 90% of T2D patients, and several studies have linked IAPP aggregation with β-cell loss and the progression of T2D.[50] The IAPP amino acid sequence, particularly at the amino and carboxyl termini, is highly conserved, although there are crucial interspecies differences, particularly in residues 20 to 29, which are considered to comprise the domain responsible for amyloidogenicity. It has been reported that rodent insulin does not form amyloid, due to a substitution of three proline residues (a well-known β-sheet-breaking amino acid) in the 20 to 29 amyloidogenic region. This proline substitution reduces the propensity of rodent IAPP to misfold and aggregate compared with hIAPP.[50] A recent computational analysis showed that zinc concentration determines insulin oligomer equilibrium, and that the hIAPP monomer preferentially binds to both the insulin monomer and dimer, as opposed to forming hIAPP homodimers. Thus, in the context of ZnT8 loss of function, zinc ion deficiency shifts the equilibrium of the insulin oligomers toward monomers and dimers. This isolates the hIAPP monomers (in their nontoxic form), and prevents them from self-association and subsequent aggregation into toxic forms, thereby reducing the risk of T2D (Figure 13.3A and B),[51,52] To test this hypothesis, we crossed hIAPP transgenic (hIAPP-Tg) mice[53] with systemic ZnT8-KO mice, to investigate whether ZnT8 deletion ameliorates hIAPP cytotoxicity (Figure 13.3C). Mice of various genotypes were fed a standard diet (STD) until 8 weeks of age, followed by a high-fat diet (HFD) for 10 weeks starting at 8 weeks of age. All groups demonstrated comparable and progressive weight gain (Figure 13.4A and B). At 8 weeks, glucose tolerance was similar between control (Ctrl) and ZnT8-KO mice, or *hIAPP* and *hIAPP*-ZnT8-KO mice fed the STD diet (Figure 13.4C). However, after 9 weeks of a HFD, glucose tolerance tended to be improved in ZnT8-KO mice (compared with Ctrl mice), and in *hIAPP*-ZnT8-KO mice (compared with *hIAPP* mice) (Figure 13.4D), although insulin levels remained unchanged (Figure 13.4E), suggesting improved insulin sensitivity. To further investigate this point, we conducted an insulin tolerance test, which demonstrated that insulin sensitivity tended to be improved in ZnT8-KO mice (compared with Ctrl mice), and in *hIAPP*-ZnT8-KO mice (compared with *hIAPP* mice) (Figure 13.4F). This suggests that deletion of ZnT8 may improve insulin sensitivity in peripheral tissues, potentially contributing to improved glucose tolerance under metabolic stress (Figure 13.4G). At present, the exact reason for the improved insulin sensitivity and glucose tolerance in ZnT8-KO mice under metabolic stress remains unclear. However, this observation is intriguing, as it resembles the phenotype observed in humans with LOF mutations in ZnT8. Recent reports showed that ZnT8 deficiency protects the liver from acetaminophen toxicity,[54] and therefore, ZnT8 might have a role in inducing protective effects under stress. Further analysis is needed to elucidate the underlying mechanisms of the improvement of insulin sensitivity.

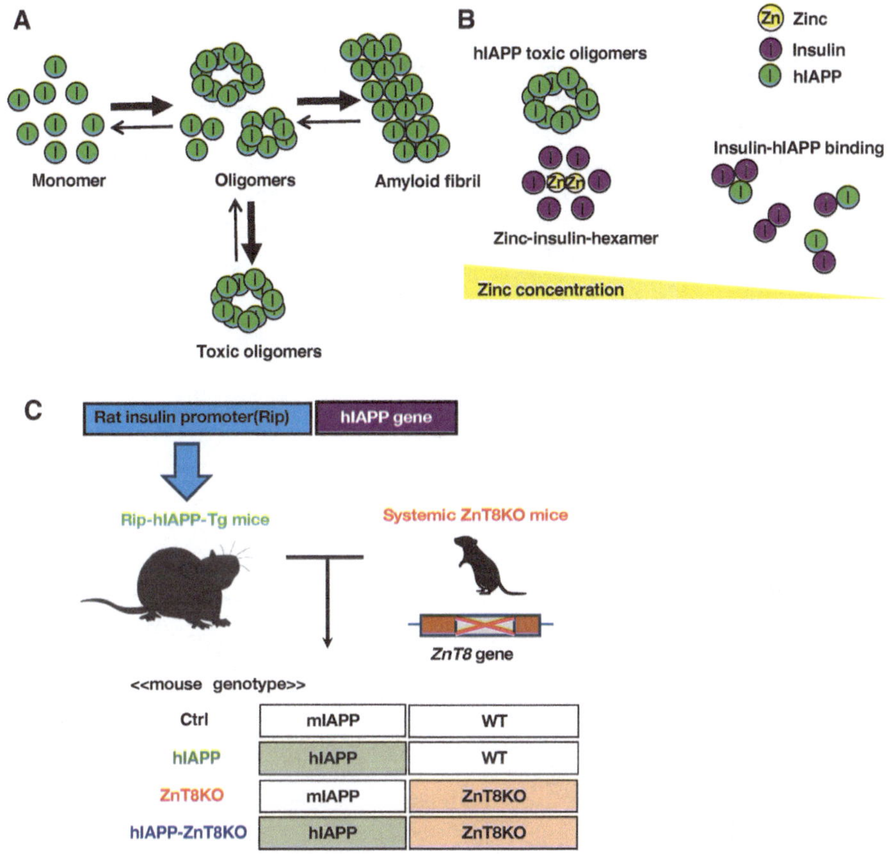

Figure 13.3 Schematic model of the associations among hIAPP, insulin, zinc, and ZnT8. (A) hIAPP can easily form toxic oligomers that induce apoptosis and amyloidogenesis in the β-cells of type 2 diabetes patients. (B) When the zinc concentration in insulin secretory granules is high, zinc is used to form zinc-insulin hexamers, and hIAPP can easily form toxic oligomers. On the other hand, when the zinc concentration is low in insulin secretory granules (such as in insulin secretory granules of loss of function of hZnT8), insulin exists as monomers or dimers, which preferentially bind to hIAPP monomers and prevent hIAPP from self-associating and aggregating. (C) Schematic representation of the experimental mouse models mentioned in this review.

In conclusion, ZnT8 plays a dual role in glucose homeostasis, affecting both insulin secretion in pancreatic β-cells and insulin degradation in hepatocytes. Genetic variations and ZnT8 deficiency may alter ZnT8 activity, potentially reducing the risk of T2D under specific conditions, particularly in response to metabolic stress. Further research is needed to fully understand the molecular mechanisms by which ZnT8 modulates these processes, and to investigate its therapeutic potential for diabetes prevention and treatment.

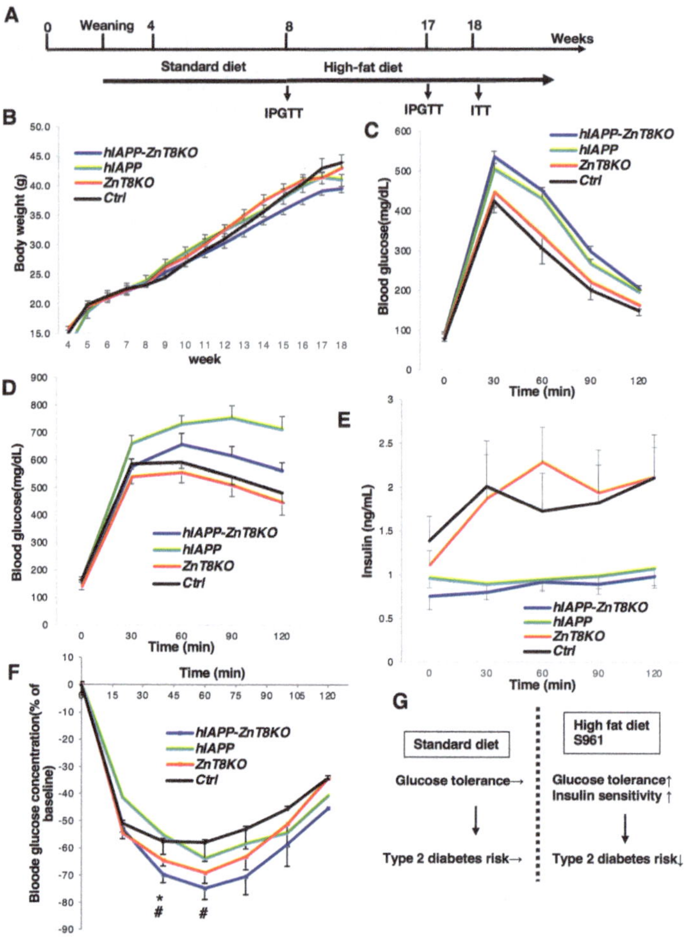

Figure 13.4 Phenotypes of ZnT8-KO mice. (A) Schematic representation of the timecourse of the metabolic experiments related to (B)–(F). (B) Body weights of Ctrl, ZnT8-KO, *hIAPP*, and *hIAPP*-ZnT8-KO mice. Ctrl: $n = 4$–9; ZnT8-KO: $n = 11$–14; *hIAPP*: $n = 5$–7; and *hIAPP*-ZnT8-KO: $n = 6$–9. (C) Glucose tolerance test results of 8-week-old mice. Ctrl: $n = 7$; ZnT8-KO: $n = 8$; *hIAPP*: $n = 9$; and *hIAPP*-ZnT8-KO: $n = 8$. (D) Glucose tolerance test of 17-week-old mice. Ctrl: $n = 7$; ZnT8-KO: $n = 11$; *hIAPP*: $n = 9$; and *hIAPP*-ZnT8-KO: $n = 9$. (E) Insulin levels of the mice in (D). (F) Insulin tolerance test results of 18-week-old mice. Ctrl: $n = 7$; ZnT8-KO: $n = 10$; *hIAPP*: $n = 6$; and *hIAPP*-ZnT8-KO: $n = 6$. (G) Summary of the phenotypes of ZnT8-KO mice. In (D), *hIAPP vs. hIAPP*-ZnT8-KO, $p < 0.05$ by the two-tailed unpaired Student *t*-test. In (F), Ctrl *vs.* ZnT8-KO at 40 min and 60 min, $^{\#}p < 0.05$ by the two-tailed unpaired Student *t*-test; and *hIAPP vs. hIAPP*-ZnT8-KO at 40 min, $^{*}p < 0.05$ by the two-tailed unpaired Student *t*-test.

Abbreviations

T2D	Type 2 diabetes
SNP	Single nucleotide polymorphism
LOF	Loss-of-function
TMDs	Transmembrane domains
HFD	High-fat diet
IPGTT	intraperitoneal glucose tolerance test
ITT	insulin tolerance test
STD	Standard diet
PTP1B	Protein tyrosine phosphatase 1B
ER	Endoplasmic reticulum
UPR	Unfolded protein response
ZnT	Zn transporter

Acknowledgements

We thank Dr H. A. Popiel for helpful suggestions and discussions, and Ms Y. Miyazaki, Ms Y. Tamura, Ms W. Mizutani, and Ms A. Suda for their technical assistance. This work was supported by a Grant-in-Aid for Scientific Research on Innovative Areas "Integrated Bio-metal Science" (MEXT KAKENHI grant number JP22H04801 to A. F.), and grants from the Ministry of Education, Culture, Sports, Science, and Technology, Japan (grant numbers JP22K8618 and JP19K08971 to A. F.), and a grant from Gunma University for the promotion of scientific research (to A. F.).

References

1. D. A. Scott, Crystalline insulin, *Biochem. J.*, 1934, **28**, 1592–1602.
2. D. A. Scott and A. M. Fisher, The insulin and the zinc content of normal and diabetic pancreas, *J. Clin. Invest.*, 1938, **17**, 725–728.
3. E. J. Dodson, G. G. Dodson, D. C. Hodgkin and C. D. Reynolds, Structural relationships in the two-zinc insulin hexamer, *Can. J. Biochem.*, 1979, **57**, 469–479.
4. M. F. Dunn, Zinc-ligand interactions modulate assembly and stability of the insulin hexamer – a review, *Biometals*, 2005, **18**, 295–303.
5. J. C. Hou, L. Min and J. E. Pessin, Insulin granule biogenesis, trafficking and exocytosis, *Vitam. Horm.*, 2009, **80**, 473–506.
6. M. Tamaki, *et al.*, The diabetes-susceptible gene SLC30A8/ZnT8 regulates hepatic insulin clearance, *J. Clin. Invest.*, 2013, **123**, 4513–4524.
7. V. Zoete, M. Meuwly and M. Karplus, Study of the insulin dimerization: binding free energy calculations and per-residue free energy decomposition, *Proteins*, 2005, **61**, 79–93.
8. J. Brange, D. R. Owens, S. Kang and A. Volund, Monomeric insulins and their experimental and clinical implications, *Diabetes Care*, 1990, **13**, 923–954.

9. S. Kang, J. Brange, A. Burch, A. Volund and D. R. Owens, Subcutaneous insulin absorption explained by insulin's physicochemical properties. Evidence from absorption studies of soluble human insulin and insulin analogues in humans, *Diabetes Care*, 1991, **14**, 942–948.
10. N. Havu, G. Lundgren and S. Falkmer, Zinc and manganese contents of micro-dissected pancreatic islets of some rodents. A microchemical study in adult and newborn guinea pigs, rats, Chinese hamsters and spiny mice, *Acta Endocrinol.*, 1977, **86**, 570–577.
11. D. F. Steiner, *et al.*, The role of prohormone convertases in insulin biosynthesis: evidence for inherited defects in their action in man and experimental animals, *Diabetes Metab.*, 1996, **22**, 94–104.
12. S. O. Emdin, G. G. Dodson, J. M. Cutfield and S. M. Cutfield, Role of zinc in insulin biosynthesis. Some possible zinc-insulin interactions in the pancreatic B-cell, *Diabetologia*, 1980, **19**, 174–182.
13. J. C. Hutton, The insulin secretory granule, *Diabetologia*, 1989, **32**, 271–281.
14. A. B. Chausmer, Zinc, insulin and diabetes, *J. Am. Coll. Nutr.*, 1998, **17**, 109–115.
15. M. Hashemipour, *et al.*, Effect of zinc supplementation on insulin resistance and components of the metabolic syndrome in prepubertal obese children, *Hormones*, 2009, **8**, 279–285.
16. W. Maret, Zinc in pancreatic islet biology, insulin sensitivity, and diabetes, *Prev. Nutr. Food Sci.*, 2017, **22**, 1–8.
17. W. Maret, Zinc in cellular regulation: the nature and significance of "zinc signals", *Int. J. Mol. Sci.*, 2017, **18**, 2285.
18. H. Haase and W. Maret, Protein tyrosine phosphatases as targets of the combined insulinomimetic effects of zinc and oxidants, *Biometals*, 2005, **18**, 333–338.
19. A. Fukunaka, *et al.*, Zinc transporter ZIP13 suppresses beige adipocyte biogenesis and energy expenditure by regulating C/EBP-beta expression, *PLoS Genet.*, 2017, **13**, e1006950.
20. T. Hara, *et al.*, Physiological roles of zinc transporters: molecular and genetic importance in zinc homeostasis, *J. Physiol. Sci.*, 2017, **67**, 283–301.
21. T. Kambe, Molecular architecture and function of ZnT transporters, *Curr. Top. Membr.*, 2012, **69**, 199–220.
22. T. J. Nicolson, *et al.*, Insulin storage and glucose homeostasis in mice null for the granule zinc transporter ZnT8 and studies of the type 2 diabetes-associated variants, *Diabetes*, 2009, **58**, 2070–2083.
23. M. Tamaki, *et al.*, Downregulation of ZnT8 expression in pancreatic beta-cells of diabetic mice, *Islets*, 2009, **1**, 124–128.
24. L. D. Pound, *et al.*, Characterization of the human SLC30A8 promoter and intronic enhancer, *J. Mol. Endocrinol.*, 2011, **47**, 251–259.
25. M. El Muayed, *et al.*, Acute cytokine-mediated downregulation of the zinc transporter ZnT8 alters pancreatic beta-cell function, *J. Endocrinol.*, 2010, **206**, 159–169.

26. C. Merriman and D. Fu, Down-regulation of the islet-specific zinc transporter-8 (ZnT8) protects human insulinoma cells against inflammatory stress, *J. Biol. Chem.*, 2019, **294**, 16992–17006.
27. A. Fukunaka, *et al.*, Zinc and iron dynamics in human islet amyloid polypeptide-induced diabetes mouse model, *Sci. Rep.*, 2023, **13**, 3484.
28. K. G. Slepchenko, *et al.*, Synchrotron fluorescence imaging of individual mouse beta-cells reveals changes in zinc, calcium, and iron in a model of low-grade inflammation, *Metallomics*, 2021, **13**, mfab051.
29. B. Lefebvre, *et al.*, Regulation and functional effects of ZNT8 in human pancreatic islets, *J. Endocrinol.*, 2012, **214**, 225–232.
30. Y. Fu, *et al.*, Down-regulation of ZnT8 expression in INS-1 rat pancreatic beta cells reduces insulin content and glucose-inducible insulin secretion, *PLoS One*, 2009, **4**, e5679.
31. E. Ohana, *et al.*, Identification of the Zn^{2+} binding site and mode of operation of a mammalian Zn^{2+} transporter, *J. Biol. Chem.*, 2009, **284**, 17677–17686.
32. R. Saxena, *et al.*, Genome-wide association analysis identifies loci for type 2 diabetes and triglyceride levels, *Science*, 2007, **316**, 1331–1336.
33. R. Sladek, *et al.*, A genome-wide association study identifies novel risk loci for type 2 diabetes, *Nature*, 2007, **445**, 881–885.
34. L. J. Scott, *et al.*, A genome-wide association study of type 2 diabetes in Finns detects multiple susceptibility variants, *Science*, 2007, **316**, 1341–1345.
35. C. Merriman, Q. Huang, G. A. Rutter and D. Fu, Lipid-tuned zinc transport activity of human ZnT8 protein correlates with risk for type-2 diabetes, *J. Biol. Chem.*, 2016, **291**, 26950–26957.
36. L. D. Pound, *et al.*, Deletion of the mouse Slc30a8 gene encoding zinc transporter-8 results in impaired insulin secretion, *Biochem. J.*, 2009, **421**, 371–376.
37. L. D. Pound, *et al.*, The physiological effects of deleting the mouse SLC30A8 gene encoding zinc transporter-8 are influenced by gender and genetic background, *PLoS One*, 2012, **7**, e40972.
38. K. Lemaire, *et al.*, Insulin crystallization depends on zinc transporter ZnT8 expression, but is not required for normal glucose homeostasis in mice, *Proc. Natl. Acad. Sci. U. S. A.*, 2009, **106**, 14872–14877.
39. N. Wijesekara, *et al.*, Beta cell-specific Znt8 deletion in mice causes marked defects in insulin processing, crystallisation and secretion, *Diabetologia*, 2010, **53**, 1656–1668.
40. R. K. Mitchell, *et al.*, Molecular genetic regulation of Slc30a8/ZnT8 reveals a positive association with glucose tolerance, *Mol. Endocrinol.*, 2016, **30**, 77–91.
41. J. Xu, *et al.*, Pancreatic beta cell-selective zinc transporter 8 insufficiency accelerates diabetes associated with islet amyloidosis, *JCI Insight*, 2021, **6**, 143037.
42. T. V. O'Halloran, M. Kebede, S. J. Philips and A. D. Attie, Zinc, insulin, and the liver: a menage a trois, *J. Clin. Invest.*, 2013, **123**, 4136–4139.

43. J. Flannick, *et al.*, Loss-of-function mutations in SLC30A8 protect against type 2 diabetes, *Nat. Genet.*, 2014, **46**, 357–363.
44. O. P. Dwivedi, *et al.*, Loss of ZnT8 function protects against diabetes by enhanced insulin secretion, *Nat. Genet.*, 2019, **51**, 1596–1606.
45. S. Kleiner, *et al.*, Mice harboring the human SLC30A8 R138X loss-of-function mutation have increased insulin secretory capacity, *Proc. Natl. Acad. Sci. U. S. A.*, 2018, **115**, E7642–E7649.
46. Q. Ma, *et al.*, ZnT8 loss-of-function accelerates functional maturation of hESC-derived beta cells and resists metabolic stress in diabetes, *Nat. Commun.*, 2022, **13**, 4142.
47. M. Jaffredo, *et al.*, Electrophysiological characterization of inducible pluripotent stem cell-derived human beta-like cells and an SLC30A8 disease model, *Diabetes*, 2024, **73**, 1255–1265.
48. K. E. Syring, *et al.*, Potential positive and negative consequences of ZnT8 inhibition, *J. Endocrinol.*, 2020, **246**, 189–205.
49. A. Piro, *et al.*, Beta cell specific ZnT8 gene deficiency and resulting loss in zinc content significantly improve insulin secretion, *Mol. Cell. Endocrinol.*, 2024, **594**, 112376.
50. A. Mukherjee, D. Morales-Scheihing, P. C. Butler and C. Soto, Type 2 diabetes as a protein misfolding disease, *Trends Mol. Med.*, 2015, **21**, 439–449.
51. A. Fukunaka and Y. Fujitani, Role of zinc homeostasis in the pathogenesis of diabetes and obesity, *Int. J. Mol. Sci.*, 2018, **19**, 476.
52. P. Nedumpully-Govindan and F. Ding, Inhibition of IAPP aggregation by insulin depends on the insulin oligomeric state regulated by zinc ion concentration, *Sci. Rep.*, 2015, **5**, 8240.
53. J. Janson, *et al.*, Spontaneous diabetes mellitus in transgenic mice expressing human islet amyloid polypeptide, *Proc. Natl. Acad. Sci. U. S. A.*, 1996, **93**, 7283–7288.
54. W. Su, *et al.*, ZnT8 deficiency protects from APAP-induced acute liver injury by reducing oxidative stress through upregulating hepatic zinc and metallothioneins, *Front. Pharmacol.*, 2021, **12**, 721471.

CHAPTER 14

Zinc Transporters and Zinc Signaling in Disease Pathogenesis

TAKAFUMI HARA, EMI YOSHIGAI AND TOSHIYUKI FUKADA*

Molecular and Cellular Physiology, Faculty of Pharmaceutical Sciences, Tokushima Bunri University, 180 Nishihama-Boji, Yamashiro, Tokushima, 770-8514, Japan
*Email: fukada@ph.bunri-u.ac.jp

14.1 Regulation of Cellular Zinc Homeostasis: The Role of Zinc Transporters

Zinc transporters regulate zinc homeostasis tightly to ensure appropriate cellular function and prevent toxic effects of zinc imbalance.[1–3] These transporters, which are expressed on plasma and organelle membranes, belong to two solute carrier (SLC) protein families: SLC30A/zinc transporters (ZnTs) and SLC39A/Zrt/Irt-like proteins (ZIPs), which consist of 10 and 14 members, respectively. The SLC39A/ZIP family increases cytoplasmic zinc levels, whereas the SLC30A/ZnT family decreases them. Each zinc transporter shows different expression patterns in specific organelles, which contributes to their various physiological functions.[2] In particular, zinc ions transported by zinc transporters act as signal molecules, regulating various cellular processes. Over the past decade, the physiological functions of the ZIP family members have been reported in various studies involving human genetic analysis and genetically modified knockout (KO) and conditional knockout (cKO) mice, with the research findings demonstrating a significant

Coordination Chemistry Perspectives No. 2
Zinc in Biology: Molecular Structures, Cellular Processes and Living Systems
Edited by Hiroyuki Yasui and Taiho Kambe

Published by the Royal Society of Chemistry, www.rsc.org

link between these zinc transporters and human diseases. Interestingly, disruption in zinc transporter function has been suggested to cause various diseases, and the molecular mechanisms involved are diverse. Therefore, each zinc transporter possesses specific physiological functions that are indispensable for numerous cellular functions.[1,3] Furthermore, the targeting of zinc transporters offers potential therapeutic benefits, where modulating the activity of specific transporters would restore zinc homeostasis in pathophysiological conditions, providing a novel approach for treating zinc-related disorders. In this chapter, we introduce the functions of the various ZIP family members and summarize recent research findings (including our own) on their relationship to the pathophysiological molecular mechanisms of various diseases and disorders.

14.2 Roles of ZIP7 in Skin and Intestinal Development and the Immune System

ZIP7, which is expressed on the endoplasmic reticulum (ER), transports zinc from the ER lumen to the cytosol.[2] It is involved in various cellular responses, including protein folding in the ER (see Chapter 6), and is important for maintaining skin[4] and intestinal homeostasis[5,6] and regulating the immune system[7] (Figure 14.1). Pathological mutations in the ZIP7-encoding gene *SLC39A7* and their associations to specific diseases have been identified.

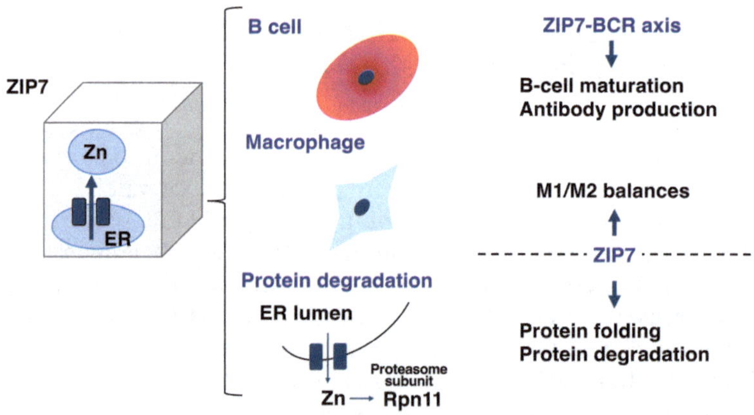

Figure 14.1 Physiological functions controlled by ZIP7 zinc signaling. ZIP7, a zinc transporter expressed in the endoplasmic reticulum (ER), is ubiquitously present throughout the body. In B cells, ZIP7 plays a critical role in immunoglobulin expression and cell maturation. In macrophages, it regulates the expression of M1 and M2 marker molecules, thereby influencing immune responses. Specifically, in controlling ER function, ZIP7 is involved in protein degradation by supplying zinc to 26S proteasome non-ATPase regulatory subunit 14 (Rpn11), an enzyme associated with the proteasome system. This highlights its role in maintaining ER homeostasis and enzyme functionality. BCR, B-cell receptor.

Additionally, ZIP7 inhibitors are being explored for drug discovery. Herein, we discuss the various functions of ZIP7, including recent research findings.

14.2.1 Roles of ZIP7 in Connective Tissue and Intestinal Development

In mice with cKO of the intestine- and skin-specific *Slc39a7* genes, impaired epithelial formation in the intestines[5] and disrupted collagen layer formation in the skin[4] were observed. These phenotypes result from ZIP7 deficiency, causing excessive zinc accumulation in the ER and thus ER stress. Since ER stress is associated with various cellular processes, ZIP7-regulated zinc dynamics and ER functions are likely linked to multiple diseases.

14.2.2 Role of ZIP7 in the Immune System

In the immune system, ZIP7 influences B cell and macrophage activity. The transporter regulates M1 and M2 macrophage marker expression, and its suppression reduces proinflammatory cytokine expression.[8] A single nucleotide polymorphism that results in reduced ZIP7 function has been identified, and patients with this mutation exhibit immunodeficiency and agammaglobulinemia. In patients with *SLC39A7* gene mutations, impaired B-cell receptor (BCR) signaling and abnormal B-cell maturation are observed, indicating the importance of ZIP7 in maintaining B cells.[7] These findings also suggest that ZIP7 contributes to systemic immune responses.

14.2.3 Molecular Mechanism of ZIP7-mediated Protein Degradation

A recent study has indicated that ZIP7 promotes ER-associated degradation by facilitating the deubiquitination and degradation of misfolded proteins *via* the zinc-dependent metalloproteinase Rpn11.[9] In a *Drosophila* model of neurodegeneration, ZIP7 overexpression reversed blindness, indicating the potential of this transporter as a therapeutic target for diseases caused by protein misfolding.

14.2.4 Modulation of ZIP7 Using Specific Compounds

NVS-ZP7-4, a ZIP7 inhibitor, was recently identified through drug discovery screening.[10] This compound inhibits Notch transport by ZIP7, thereby increasing ER stress and inducing cell apoptosis. Therefore, the potential of ZIP7 as a drug target is promising, and further research is expected to expand its therapeutic applications, especially in cancer treatment (see Section 14.7 "Therapeutic potential of the zinc transporters").

14.3 Roles of ZIP8 in Connective Tissue, Muscle, and Cardiovascular Health

ZIP8 (encoded by the *SLC39A8* gene) is involved in the intracellular transport of iron, zinc, and manganese.[2] The metal ion transporter is ubiquitously expressed in cell and lysosomal membranes across various tissues, with notably high expression levels in the kidneys, lungs, and testes. Genetic mutations in *SLC39A8* have been implicated in various diseases, including Crohn's disease, cardiovascular disease, schizophrenia, and Parkinson's disease. Herein, we explore the connections between ZIP8 and various diseases.[11–16]

14.3.1 Pathophysiological Implications of Genetic Mutations in *SLC39A8*

In *Slc39a8*-KO mice, fetal lethality was observed. In contrast, in *Slc39a8*-cKO mice, the manganese levels in the liver were significantly decreased and iron levels were increased, whereas zinc levels remained unaffected.[17] This suggests that ZIP8 plays a role in manganese metabolism. However, other divalent manganese ion (Mn^{2+}) transporters, such as DMT1, ZIP8, and ZIP14, are present in the liver and kidneys, providing alternative pathways for the transport of manganese ions.[18] Therefore, although the precise mechanism through which ZIP8 dysfunction alone causes manganese deficiency remains unclear, this transporter is recognized as having a significant role in manganese metabolism.

14.3.2 Diseases Related to Pathogenic Mutations in ZIP8

ZIP8 is essential for maintaining systemic manganese homeostasis and regulating manganese-dependent enzymes. Mutations in ZIP8 lead to severe disorders of manganese metabolism, which underlie various diseases associated with manganese dysregulation. Therefore, the significance of ZIP8 as a manganese transporter has garnered considerable attention. Several *SLC39A8* gene mutations have been associated with a wide range of systemic pathologies, such as increased risks of schizophrenia, strabismus, reduced blood manganese levels,[11,16] Crohn's disease,[13] idiopathic scoliosis,[14] growth retardation,[19] and obesity.[20,21] In the context of ZIP8-related manganese metabolism, not only is the gastrointestinal absorption pathway critical but so too are the routes for reabsorbing manganese excreted into the bile ducts and renal tubules. These pathways may account for the systemic manganese deficiency observed in patients with *SLC39A8* mutations.

14.4 Roles of ZIP10 in Immune Function and Skin Development

ZIP10 (encoded by the *SLC39A10* gene), a plasma membrane-localized protein, plays a crucial role in B-cell differentiation and maturation,

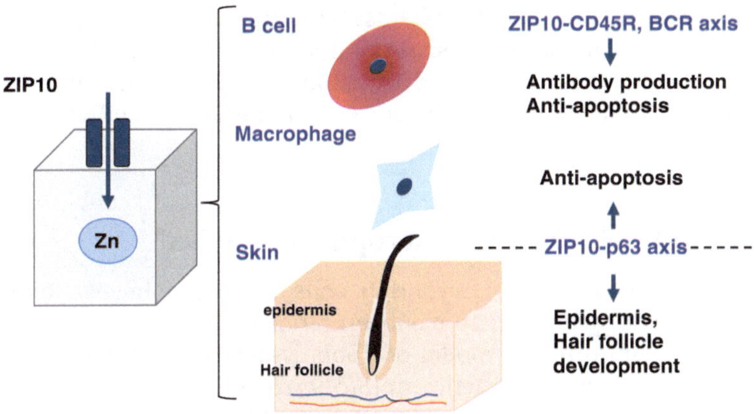

Figure 14.2 Physiological functions controlled by ZIP10 zinc signaling. ZIP10, a zinc transporter expressed on the cell membrane, plays a crucial role in various physiological processes. Its functions are diverse, encompassing skin homeostasis, macrophage survival, and B-cell differentiation and proliferation. In B cells, ZIP10 is vital for antibody production and cell proliferation. It modulates B-cell receptor (BCR) signaling, which is essential for the differentiation and proliferation of B cells, thus playing a critical role in adaptive immunity. In macrophages, ZIP10 also works in conjunction with p63, contributing to cell survival. This role is particularly significant during inflammatory responses, where ZIP10 regulation helps manage cell viability and immune responses. In the skin, ZIP10 is localized in the epithelium and the outer root sheath cells of hair follicles. It collaborates with the transcription factor p63 to maintain the homeostasis of skin and hair follicles, ensuring appropriate skin function and hair growth.

macrophage survival, and skin and hair follicle formation[1,2] (Figure 14.2). Herein, we summarize current knowledge about the involvement of this zinc transporter in immune function and skin homeostasis.

14.4.1 Role of ZIP10 in Immune Function

ZIP10 is highly expressed in the spleen, thymus, and lymph nodes, with particularly high expression levels in B cells at the early stages of development.[22,23] Deletion of the *Slc39a10* gene in pro-B cells significantly reduced the number of B cells and level of plasma immunoglobulins in mice, underscoring the essential role of this zinc transporter in B-cell differentiation. Furthermore, ZIP10 deficiency in B cells impairs their differentiation, increases caspase activity, and induces apoptosis. These effects are also observed when the cells are treated with zinc ion chelators, suggesting a specific regulatory role of ZIP10 in caspase activity and pro-B-cell survival. Additionally, ZIP10 is crucial for mature B-cell function, as B-cell proliferation in response to BCR stimulation is blocked in *Slc39a10*-KO mice. This is likely because ZIP10-mediated zinc signaling regulates the activity of CD45, a receptor protein tyrosine phosphatase essential for BCR signaling. Thus, ZIP10-mediated zinc homeostasis is vital for fine-tuning both early and late stages of B-cell development and function. These findings emphasize the importance of this zinc transporter in antibody-mediated immune responses.

14.4.2 Role of ZIP10 in Macrophage Survival

The role of ZIP10 extends beyond B cells to macrophages. Reduced expression of macrophage ZIP10 has been reported in patients who recovered from sepsis.[24] In a mouse model of lipopolysaccharide-induced sepsis, the macrophage-specific ZIP10-deficient (*Slc39a10*-cKO) mice exhibited a higher survival rate than that of the wild-type mice. This improvement was attributed to decreased cytokine production and reduced macrophage numbers in the KO mice. Mechanistically, apoptosis was significantly induced in the ZIP10-deficient macrophages stimulated with lipopolysaccharide, with p63 contributing to the activation of apoptotic signals in the absence of the metal ion transporter. Therefore, ZIP10 is essential for macrophage survival, as its deficiency induces apoptosis in these cells, reducing the number of immune cells that secrete inflammatory factors such as cytokines, thereby preventing the severe inflammatory response associated with sepsis and enhancing the animal survival rate. These findings suggest that ZIP10 plays a protective role in macrophages during inflammatory responses, and that modulating the transporter function could be a feasible strategy to prevent severe inflammation.

14.4.3 Roles of ZIP10 in Skin and Hair Follicle Development

ZIP10 plays a critical role in the development of skin and hair follicles.[25,26] It is expressed in the epidermis and the outer root sheath of hair follicles. Studies on epidermis-specific *Slc39a10*-cKO mice have revealed epidermal malformation and hair follicle regression, suggesting that ZIP10-regulated zinc homeostasis is integral to the maintenance of skin homeostasis. As a component of ointments used to treat skin diseases, zinc likely exerts its effects through zinc signaling pathways mediated by ZIP10. RNA sequencing analysis of ZIP10-expressing skin cells revealed characteristics of stemness, indicating that the transporter may influence the regulation of stem cells, which are essential for skin differentiation. Furthermore, the maintenance of homeostasis *via* the p63 signaling pathway has been confirmed as a key mechanism of ZIP10. The relationship between ZIP10 and atopic dermatitis has also been reported, where the expression of the transporter is significantly reduced in the epidermis of patients with this common skin disease.[27] These findings underscore the importance of ZIP10 in controlling skin homeostasis and highlight its potential as a therapeutic target for skin diseases.

14.5 Roles of ZIP13 in Connective Tissue, Muscle, and Cardiovascular Health

ZIP13 (encoded by the *SLC39A13* gene), which is expressed on the membrane of the Golgi apparatus, has been found to be integral to connective tissue development,[28] myogenic differentiation[29] and cardiovascular homeostasis[30,31] (Figure 14.3). The metal ion transporter is expressed in mesenchymal cells and is widely expressed in various organs. Furthermore, it is recognized as an important molecule and its gene mutations have been identified as the causes of various diseases.[28,32] Some recent reports on the physiological and pathophysiological functions of ZIP13 are summarized below.

14.5.1 Association of ZIP13 with Ehlers–Danlos Syndrome Spondylodysplastic Type 3

Loss of ZIP13 function is known to be a causative factor of Ehlers–Danlos syndrome spondylodysplastic type 3 (EDSSPD3), the main symptoms of

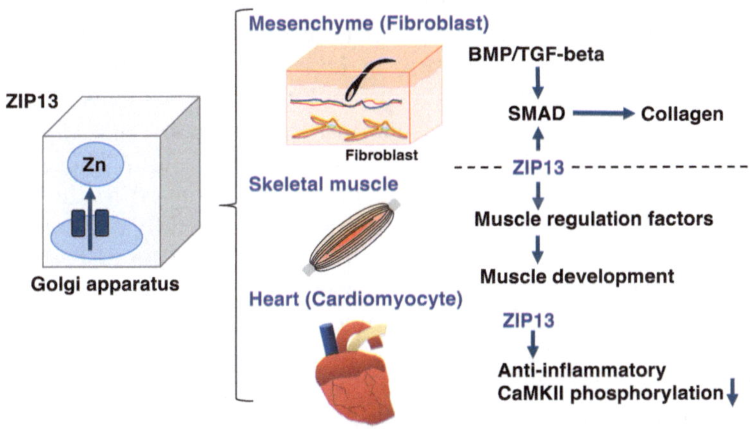

Figure 14.3 Physiological functions controlled by ZIP13 zinc signaling. ZIP13 is a zinc transporter localized in the Golgi apparatus. It is expressed ubiquitously, particularly in mesenchymal cells, playing crucial roles in several physiological processes. For example, ZIP13 contributes to the structural integrity and function of the skin through its involvement in collagen production. In skeletal muscles, ZIP13 is essential for cell differentiation and maturation, highlighting its importance in muscle development and function. ZIP13 zinc signaling also plays a significant role in intracellular signaling pathways. It has been reported to facilitate the nuclear translocation of Smad proteins through phosphorylation by bone morphogenetic protein (BMP)/transforming growth factor-beta (TGF-β), thereby regulating the expression of target molecules involved in various cellular processes. Furthermore, ZIP13 is involved in circulatory homeostasis. Its regulation of cardiac function through the control of calcium/calmodulin-dependent protein kinase (CaMK) activity in the myocardium underscores its role in maintaining overall cardiovascular health.

which are lesions of the hard and connective tissues, such as fragile skin and hypoplasia of bone and teeth. Previous studies by the authors' research group verified that loss of ZIP13 function causes EDSSPD3.[28] Whereas Ehlers–Danlos syndrome is a connective tissue disorder,[33] EDSSPD3 is a disease that causes fragile skin and severely slanted eyelids due to a decrease in collagen, as well as bone and tooth formation abnormalities, short stature, and varicose veins.[28] The pathogenesis of the cases reported by our research group is a mutation of an amino acid due to a single base substitution, which causes the 64th glycine to be mutated to aspartic acid, resulting in changes in the molecular weight and polarity of the ZIP13 protein and its immediate degradation by the ubiquitin–proteasome system after translation and thereby the loss of its function.[32,34] Two other research groups also reported cases of patients with EDSSPD3[35,36] caused by polymorphic *SLC39A13* mutations leading to loss of the transporter functions. This disease is rare, with only nine patients reported among four families worldwide. Polymorphisms in the transporter gene as well as myopathy have also been confirmed in the other patients with EDSSPD3 from other families.[28,35–37] Although transcription and translation proceed normally with these mutations, the produced protein is immediately degraded by the ubiquitin–proteasome system, resulting in loss of ZIP13 function.[32,34] ZIP13 is expressed mainly in mesenchymal cells, and abnormalities caused by the loss of its function also appear in tissues formed by these cells. Characteristic symptoms of EDSSPD3 also include decreased muscle tone (muscular hypotonia)[29] and loss of fat layers (lipoatrophy).[28] With regard to the molecular mechanisms involved in the disease phenotype caused by the loss of ZIP13 function, impaired bone morphogenic protein (BMP) and transforming growth factor-beta (TGF-β) signaling was observed in the corresponding tissues and cells of *Slc39a13*-KO mice.[28] These findings indicate that ZIP13 plays a crucial role in connective tissue development through its involvement in the BMP and TGF-β signaling pathways. Thus, the *Slc39a13*-KO mouse serves as a novel model linking zinc metabolism, BMP/TGF-β signaling, and connective tissue dysfunction.

14.5.2 Role of ZIP13 in Skeletal Muscle Homeostasis

Given that skeletal muscles are differentiated from mesenchymal stem cells, and patients with EDSSPD3 characteristically develop muscle atrophy, ZIP13 might be involved in muscle homeostasis. However, the involvement of ZIP13 in skeletal muscle formation and function has not been demonstrated. In one study using a mouse myoblast cell line (C2C12) and induced pluripotent stem cells derived from patients with EDSSPD3 (EDSSPD3-iPSCs), ZIP13 was found to be upregulated during myogenic stimulation.[29] Knockdown of *SLC39A13* disrupted myotubular differentiation, indicating the crucial role of the metal ion transporter in early myogenic differentiation. Furthermore, myocytes derived from EDSSPD3-iPSCs displayed incomplete myogenic differentiation, which could be corrected by genomic editing of the pathogenic *SLC39A13* mutation. As molecular markers, muscle differentiation-related genes were confirmed to be expressed at a lower level in the absence of ZIP13, indicating that zinc homeostasis mediated by this transporter is crucial for muscle maintenance.[29]

A further challenge lies in elucidating the involvement of ZIP13 and the zinc homeostasis it mediates in muscle satellite cells, which are essential for muscle differentiation. The acquisition of more knowledge about this relationship will provide deeper insights into the mechanisms of muscle differentiation and zinc regulation. In summary, these aforementioned findings suggest that ZIP13 is essential for proper myogenic differentiation and highlight the potential utility of EDSSPD3-iPSCs as a model for studying the molecular basis of clinical features stemming from ZIP13-related diseases.

14.5.3 Role of ZIP13 in Cardiovascular Homeostasis

Cardiovascular disease is one of the leading causes of death worldwide.[38,39] Despite the development of various drugs, current treatments remain insufficient, necessitating the discovery of new therapeutic targets.[40] Recent studies have identified ZIP13 as a regulator of cardiovascular homeostasis and stroke.[30,31,41] *Slc39a13* mRNA expression levels were elevated in primary neonatal mouse cardiomyocytes and mouse cardiac tissue treated with the cardiotoxic drug doxycycline,[30] suggesting that increased ZIP13 expression plays a critical role in the cardiac stress response. Moreover, zinc plays a vital role in cardiomyocytes, as its deficiency affected the morphology and proliferation of cultured primary neonatal mouse cardiomyocytes. Notably, primary neonatal cardiomyocytes from *Slc39a13*-KO mice exhibited arrhythmic beating.[30]

To further investigate ZIP13 activity, we performed RNA sequencing analysis on cardiomyocytes from *Slc39a13*-KO mice, identifying 606 differentially expressed genes, of which those related to inflammation and cell adhesion were significantly upregulated. Additionally, *in vivo* telemetry echocardiography revealed that *Slc39a13*-KO mice develop arrhythmias accompanied by elevated expression levels of collagen type I alpha 1 chain (Col1a1, a cardiac fibrosis marker), endothelial nitric oxide synthase (eNOS, a vascular inflammation-related enzyme), and Golgi matrix protein GM130.[30] Furthermore, heart-specific *Slc39a13*-KO mice showed impaired myocardial ischemia/reperfusion injury that was mediated through the activation of calcium/calmodulin-dependent protein kinase type II alpha chain (CaMKII) due to the mitochondrial dysfunction.[31]

These findings underscore the physiological significance of ZIP13 in maintaining cardiovascular homeostasis by mitigating inflammation and stress responses. Therefore, the optimization of ZIP13 expression or function may represent a novel therapeutic approach for managing cardiovascular diseases.

14.6 Roles of ZIP14 in Metal Transport and Diseases

ZIP14 (encoded by the *SLC39A14* gene), which is located on the cell membrane, is responsible for the transport of several trace metals, including manganese, zinc, iron, and cadmium.[3] Similar to the other ZIP family members, its dysregulation can lead to severe pathologies[42,43] (Figure 14.4), as summarized in this section.

14.6.1 Role of ZIP14 in Manganese Transport

Although manganese is an essential trace element, its transport and regulation remain incompletely understood. Recent studies have identified patients with a novel autosomal recessive manganese transporter defect caused by mutations in *SLC39A14*, which result in manganese overaccumulation.[42] This condition, characterized by hypermagnesemia and progressive parkinsonism-dystonia, is evident through distinctive brain magnetic resonance imaging findings.

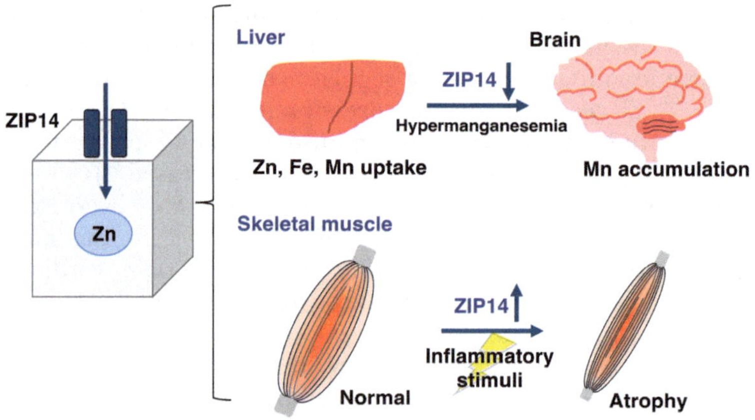

Figure 14.4 Physiological functions controlled by ZIP14 zinc signaling. ZIP14, a zinc transporter located on the cell membrane, is highly expressed in the liver and pancreas. Besides zinc, ZIP14 also transports other divalent metal ions, including those of manganese (Mn), iron (Fe), and cadmium (Cd). In the liver, ZIP14 plays a critical role in iron uptake, thereby regulating serum iron levels. Genetic mutations in *SLC39A14* have been linked to early-onset Parkinson's-like symptoms. This is due to the impaired liver uptake of manganese and its reduced absorption from the blood to the intestinal epithelium *via* ZIP14 during its excretion, resulting in its elevated levels in the blood and subsequent accumulation in the brain. ZIP14 also plays a significant role in cancer cachexia, a debilitating condition characterized by severe muscle wasting commonly seen in patients with advanced cancer. Inflammatory cytokines secreted by cancer cells stimulate an ectopic increase in ZIP14 expression in skeletal muscles. This heightened ZIP14 expression leads to an increased uptake of zinc into skeletal muscles, thereby disrupting muscle differentiation and inducing muscle atrophy.

ZIP14 functions as the primary manganese transporter, and its gene mutations impede uptake of the metal without affecting blood concentrations of zinc, iron, and cadmium. Additionally, intestine-specific ZIP14 deficiency reduces the excretion of manganese into the intestinal tract, leading to overloading of the metal in the brain and associated locomotor defects.[44] Thus, ZIP14 is a critical molecule in manganese metabolism.

Chelation therapy, in which chemicals are used to effectively bind manganese and enhance its urinary excretion, has been shown to be promising in patients with *SLC39A14* mutations.[42] The early administration of chelation therapy following the confirmation of gene mutations may significantly improve symptoms and potentially prevent disease progression.

14.6.2 Role of ZIP14 in Cancer Cachexia

Cancer cachexia, a debilitating syndrome characterized by a significant loss of skeletal muscle mass, is commonly observed in patients with metastatic cancer.[45] Muscle atrophy due to cachexia is particularly evident in the advanced stages of cancer, severely affecting many patients. Despite its prevalence, the underlying mechanisms of cachexia are not fully understood, and no fundamental treatments currently exist.[46] However, a recent study has implicated ZIP14 in the molecular mechanism underlying cachexia-associated muscle atrophy induced by cancer.[43]

In the muscles of cachectic mice and patients with metastatic cancer, ZIP14 expression is upregulated by the inflammatory cytokines tumor necrosis factor-alpha (TNF-α) and TGF-β secreted from the cancerous tissues. The increased expression of ZIP14 facilitates zinc uptake, which in turn suppresses the expression of myoblast determination protein (MyoD) and myocyte-specific enhancer factor 2c (Mef2c), both key muscle differentiation factors. In cancer cachexia models using *Slc39a14*-deficient mice, muscle atrophy was significantly milder than that in wild-type mice, underscoring the critical role of ZIP14 in cancer cachexia-induced muscle atrophy. Therefore, ZIP14 represents a novel and promising therapeutic target for treating cancer cachexia.

14.7 Therapeutic Potential of the Zinc Transporters

As summarized above, various physiological functions of ZIP transporters have been reported, highlighting their critical roles in disease contexts. If the expression and function of these transporters can be appropriately regulated, it could lead to the development of appropriate preventive and therapeutic strategies for related diseases.

ZIP13 (*SLC39A13*) is associated with connective tissue and muscle differentiation, with loss-of-function mutations causing hereditary diseases.[28] Furthermore, because of its involvement in skeletal muscle development[29] and cardiomyocyte function,[30] ZIP13 is considered a significant therapeutic target for cardiovascular diseases. Therefore, if the transplantation of healthy mesenchymal or muscle stem cells and the gene editing of stem cells harboring pathogenic *SLC39A13* mutations become feasible, ZIP13-related diseases could be effectively treated. Additionally, as signaling pathways related to ZIP13 are being elucidated, identifying complementary signaling pathways and applying compounds that enhance those pathways hold promise as strategies for treatment development. Future research should focus on elucidating the detailed mechanisms of ZIP13 function and its broader implications in the molecular mechanisms of human diseases.

ZIP10 is an essential molecule for hematopoiesis,[47] immune function,[22–24] skin development,[25–27] and embryogenesis.[48] Its involvement in regulating B-cell differentiation, macrophage survival, and epidermal formation indicates that it has extensive critical roles in maintaining organismal homeostasis. Notably, the necrosis of hematopoietic stem cells during embryogenesis due to zinc deficiency caused by ZIP10 loss of function suggests that zinc is crucial not only for postnatal growth but also during the developmental stages. Therefore, ZIP10-regulated zinc homeostasis warrants extensive investigation, and the metal ion transporter is considered an important therapeutic target.

Moreover, controlling skin development *via* the modulation of ZIP10 function would not only emphasize the reported benefits of zinc for skin homeostasis but also provide new treatment strategies based on a deeper understanding of the underlying molecular mechanisms of skin diseases. Given the numerous reports suggesting the contribution of zinc to inflammation suppression, inflammatory diseases could very likely be broadly controlled by regulating the function of ZIP10 expressed in macrophages and B cells. Such ZIP10-mediated control of inflammatory responses is expected to contribute to new inflammation control treatments. Future research should focus on not only elucidating the detailed mechanisms of ZIP10-mediated zinc signaling but also identifying small compounds that regulate ZIP10 function.

ZIP14 is a transporter that mediates the transport of various divalent metals and plays a crucial role in maintaining their homeostasis.[42] Notably, the high-level expression of ZIP14 in skeletal muscle, which is considered one of the causes of muscle atrophy in cancer cachexia, suggests that inhibition of the transporter could be a potential therapeutic strategy.[43] Specifically, symptoms were alleviated in cancer cachexia models using *Slc39a14*-KO mice, indicating that ZIP14 plays a significant role in muscle atrophy and thus is a valuable therapeutic target.

However, the potential side effects arising from the inhibition of ZIP14 expressed in the liver and brain are a challenge that needs to be addressed. Since loss of ZIP14 function is associated with iron overload and manganese accumulation, inhibitors that specifically target ZIP14-expressing tissues need to be efficiently delivered (*e.g.*, by using drug delivery systems), and their efficacy must be controlled through the absorption, distribution, metabolism, and excretion (ADME) processes. Give the current lack of effective treatments for muscle atrophy in cancer cachexia and the significant unmet need for the development of therapeutic drugs, further advanced research and development are warranted and expected.

Abbreviations

BMP	Bone morphogenetic protein
cKO	Conditional knockout
EDSSPD3	Ehlers–Danlos syndrome spondylodysplastic type 3
iPSCs	Induced pluripotent stem cells
KO	Knockout
TGF-β	Transforming growth factor-beta
ZIP	Zrt/Irt-like protein
ZnT	Zinc transporter

Acknowledgements

This article was supported by grants from the joint research program of the Institute for Molecular and Cellular Regulation at Gunma University (to EY, TH, and TF), the Japan Society for the Promotion of Science KAKENHI (20H03409), the Princess Takamatsu Cancer Research Fund, SECOM Science and Technology Foundation, Terumo Life Science Foundation, The Uehara Memorial Foundation, the Astellas Foundation for Research on Metabolic Disorders, The Science Research Promotion Fund (to TF), the Japan Society for the Promotion of Science KAKENHI (18K06711, 23K05687), the Koyanagi Foundation, the Naitou Foundation, the Mochida Memorial Foundation for Medical and Pharmaceutical Research, the KOSE Cosmetology Research Foundation (to TH), KAKENHI (21J40212, 22K11871, 22KJ3069), and the Urakami Foods and Food Culture Promotion Foundation (to EY).

References

1. T. Fukada and T. Kambe, Molecular and genetic features of zinc transporters in physiology and pathogenesis, *Metallomics*, 2011, **3**, 662–674.
2. T. Hara, E. Yoshigai, T. Ohashi and T. Fukada, Zinc transporters as potential therapeutic targets: An updated review, *J. Pharmacol. Sci.*, 2022, **148**, 221–228.
3. T. Hara, *et al.*, Physiological roles of zinc transporters: molecular and genetic importance in zinc homeostasis, *J. Physiol. Sci.*, 2017, **67**, 283–301.
4. B.-H. Bin, *et al.*, Requirement of zinc transporter SLC39A7/ZIP7 for dermal development to fine-tune endoplasmic reticulum function by regulating protein disulfide isomerase, *J. Invest. Dermatol.*, 2017, **137**, 1682–1691.
5. W. Ohashi, *et al.*, Zinc transporter SLC39A7/ZIP7 promotes intestinal epithelial self-renewal by resolving ER stress, *PLoS Genet.*, 2016, **12**, e1006349.

6. W. Ohashi, T. Hara, T. Takagishi, K. Hase and T. Fukada, Maintenance of intestinal epithelial homeostasis by zinc transporters, *Dig. Dis. Sci.*, 2019, **64**, 2404–2415.
7. C. Anzilotti, *et al.*, An essential role for the Zn^{2+} transporter ZIP7 in B cell development, *Nat. Immunol.*, 2019, **20**, 350–361.
8. W. Xie, Q. Xue, L. Niu and K.-W. Wong, Zinc transporter SLC39A7 relieves zinc deficiency to suppress alternative macrophage activation and impairment of phagocytosis, *PLoS One*, 2020, **15**, e0235776.
9. X. Guo, *et al.*, The Zn^{2+} transporter ZIP7 enhances endoplasmic-reticulum-associated protein degradation and prevents neurodegeneration in Drosophila, *Dev. Cell*, 2024, **59**, 1655–1667.e6.
10. E. Nolin, *et al.*, Discovery of a ZIP7 inhibitor from a Notch pathway screen, *Nat. Chem. Biol.*, 2019, **15**, 179–188.
11. L. Sunuwar, *et al.*, Pleiotropic ZIP8 A391T implicates abnormal manganese homeostasis in complex human disease, *JCI Insight*, 2020, **5**, e140978.
12. J.-H. Kim, *et al.*, Regulation of the catabolic cascade in osteoarthritis by the zinc-ZIP8-MTF1 axis, *Cell*, 2014, **156**, 730–743.
13. D. Li, *et al.*, A pleiotropic missense variant in SLC39A8 is associated with crohn's disease and human gut microbiome composition, *Gastroenterology*, 2016, **151**, 724–732.
14. G. Haller, *et al.*, A missense variant in SLC39A8 is associated with severe idiopathic scoliosis, *Nat. Commun.*, 2018, **9**, 4171.
15. K. M. Boycott, *et al.*, Autosomal-recessive intellectual disability with cerebellar atrophy syndrome caused by mutation of the manganese and zinc transporter gene SLC39A8, *Am. J. Hum. Genet.*, 2015, **97**, 886–893.
16. J. H. Park, *et al.*, SLC39A8 deficiency: a disorder of manganese transport and glycosylation, *Am. J. Hum. Genet.*, 2015, **97**, 894–903.
17. V. Zhang, *et al.*, A mouse model characterizes the roles of ZIP8 in systemic iron recycling and lung inflammation and infection, *Blood Adv.*, 2023, **7**, 1336–1349.
18. H. Fujishiro, Y. Yano, Y. Takada, M. Tanihara and S. Himeno, Roles of ZIP8, ZIP14, and DMT1 in transport of cadmium and manganese in mouse kidney proximal tubule cells, *Metallomics*, 2012, **4**, 700–708.
19. L. G. Riley, *et al.*, A SLC39A8 variant causes manganese deficiency, and glycosylation and mitochondrial disorders, *J. Inherited Metab. Dis.*, 2017, **40**, 261–269.
20. S. I. Berndt, *et al.*, Genome-wide meta-analysis identifies 11 new loci for anthropometric traits and provides insights into genetic architecture, *Nat. Genet.*, 2013, **45**, 501–512.
21. E. K. Speliotes, *et al.*, Association analyses of 249 796 individuals reveal 18 new loci associated with body mass index, *Nat. Genet.*, 2010, **42**, 937–948.
22. S. Hojyo, *et al.*, Zinc transporter SLC39A10/ZIP10 controls humoral immunity by modulating B-cell receptor signal strength, *Proc. Natl. Acad. Sci. U. S. A.*, 2014, **111**, 11786–11791.

23. T. Miyai, *et al.*, Zinc transporter SLC39A10/ZIP10 facilitates anti-apoptotic signaling during early B-cell development, *Proc. Natl. Acad. Sci. U. S. A.*, 2014, **111**, 11780–11785.
24. H. Gao, *et al.*, Metal transporter Slc39a10 regulates susceptibility to inflammatory stimuli by controlling macrophage survival, *Proc. Natl. Acad. Sci. U. S. A.*, 2017, **114**, 12940–12945.
25. B.-H. Bin, *et al.*, Requirement of zinc transporter ZIP10 for epidermal development: Implication of the ZIP10–p63 axis in epithelial homeostasis, *Proc. Natl. Acad. Sci. U. S. A.*, 2017, **114**, 12243–12248.
26. B.-H. Bin, *et al.*, The epithelial zinc transporter ZIP10 epigenetically regulates human epidermal homeostasis by modulating histone acetyltransferase activity, *Br. J. Dermatol.*, 2019, **180**, 869–880.
27. K. Nakajima, *et al.*, Possible involvement of zinc transporter ZIP10 in atopic dermatitis, *J. Dermatol.*, 2020, **47**, e51–e53.
28. T. Fukada, *et al.*, The zinc transporter SLC39A13/ZIP13 is required for connective tissue development; its involvement in BMP/TGF-β signaling pathways, *PLoS One*, 2008, **3**, e3642.
29. M. Shoji, *et al.*, Possible involvement of zinc transporter ZIP13 in myogenic differentiation, *Sci. Rep.*, 2024, **14**, 8052.
30. T. Hara, *et al.*, Role of Scl39a13/ZIP13 in cardiovascular homeostasis, *PLoS One*, 2022, **17**, e0276452.
31. J. Wang, X. Cheng, H. Zhao, Q. Yang and Z. Xu, Downregulation of the zinc transporter SLC39A13 (ZIP13) is responsible for the activation of CaMKII at reperfusion and leads to myocardial ischemia/reperfusion injury in mouse hearts, *J. Mol. Cell. Cardiol.*, 2021, **152**, 69–79.
32. B.-H. Bin, *et al.*, Biochemical characterization of human ZIP13 protein: a homo-dimerized zinc transporter involved in the spondylocheiro dysplastic Ehlers-Danlos syndrome, *J. Biol. Chem.*, 2011, **286**, 40255–40265.
33. F. Malfait, R. J. Wenstrup and A. De Paepe, Clinical and genetic aspects of Ehlers-Danlos syndrome, classic type, *Genet. Med.*, 2010, **12**, 597–605.
34. B.-H. Bin, *et al.*, Molecular pathogenesis of spondylocheirodysplastic Ehlers-Danlos syndrome caused by mutant ZIP13 proteins, *EMBO Mol. Med.*, 2014, **6**, 1028–1042.
35. C. Giunta, *et al.*, Spondylocheiro dysplastic form of the Ehlers-Danlos syndrome–an autosomal-recessive entity caused by mutations in the zinc transporter gene SLC39A13, *Am. J. Hum. Genet.*, 2008, **82**, 1290–1305.
36. C. Kumps, *et al.*, The connective tissue disorder associated with recessive variants in the SLC39A13 zinc transporter gene (Spondylo-Dysplastic Ehlers–Danlos Syndrome Type 3): Insights from four novel patients and follow-up on two original cases, *Genes*, 2020, **11**, 420.
37. M. Dusanic, *et al.*, Novel nonsense mutation in SLC39A13 initially presenting as myopathy: case report and review of the literature, *MSY*, 2018, **9**, 100–109.
38. K. Mc Namara, H. Alzubaidi and J. K. Jackson, Cardiovascular disease as a leading cause of death: how are pharmacists getting involved?, *Integr. Pharm. Res. Pract.*, 2019, **8**, 1–11.

39. G. A. Roth, *et al.*, Global burden of cardiovascular diseases and risk factors, 1990-2019: update From the GBD 2019 study, *J. Am. Coll. Cardiol.*, 2020, **76**, 2982–3021.
40. S. Choi, X. Liu and Z. Pan, Zinc deficiency and cellular oxidative stress: prognostic implications in cardiovascular diseases, *Acta Pharmacol. Sin.*, 2018, **39**, 1120–1132.
41. M. Traylor, *et al.*, Genetic basis of lacunar stroke: a pooled analysis of individual patient data and genome-wide association studies, *Lancet Neurol.*, 2021, **20**, 351–361.
42. K. Tuschl, *et al.*, Mutations in SLC39A14 disrupt manganese homeostasis and cause childhood-onset parkinsonism–dystonia, *Nat. Commun.*, 2016, **7**, 11601.
43. G. Wang, *et al.*, Metastatic cancers promote cachexia through ZIP14 upregulation in skeletal muscle, *Nat. Med.*, 2018, **24**, 770–781.
44. T. B. Aydemir, *et al.*, Metal transporter Zip14 (Slc39a14) deletion in mice increases manganese deposition and produces neurotoxic signatures and diminished motor activity, *J. Neurosci.*, 2017, **37**, 5996–6006.
45. V. E. Baracos, L. Martin, M. Korc, D. C. Guttridge and K. C. H. Fearon, Cancer-associated cachexia, *Nat. Rev. Dis. Primers*, 2018, **4**, 1–18.
46. Management of Cancer Cachexia: ASCO Guideline, *J. Clin. Oncol.*, 2020, **38**, 2438–2453.
47. X. He, *et al.*, The zinc transporter SLC39A10 plays an essential role in embryonic hematopoiesis, *Adv. Sci.*, 2023, **10**, e2205345.
48. K. M. Taylor, *et al.*, Zinc transporter ZIP10 forms a heteromer with ZIP6 which regulates embryonic development and cell migration, *Biochem. J.*, 2016, **473**, 2531–2544.

CHAPTER 15

Roles of Zinc and Zinc Transporters in Breast Cancer

TOMOKA TAKATANI-NAKASE,*[a] CHIHIRO MATSUI,[a] MANAMI SAKITANI,[a] KODAI NAKAO[a] AND IKUHIKO NAKASE[b]

[a] Department of Pharmaceutics, School of Pharmacy and Pharmaceutical Sciences, Mukogawa Women's University, 11-68, Koshien Kyuban-cho, Nishinomiya, Hyogo 663-8179, Japan; [b] Depatment of Biological Chemistry, Graduate School of Science, Osaka Metropolitan University, 1-1, Gakuen-cho, Naka-ku, Sakai, Osaka 599-8531, Japan
*Email: nakase@mukogawa-u.ac.jp

15.1 Introduction

Breast cancer is the most frequently diagnosed cancer in women and the leading cause of cancer-related deaths among women worldwide.[1,2] Recent studies have clarified the molecular and cellular mechanisms of the pathology of breast cancer and evidence-based theoretical treatments, including pharmacotherapy using molecular target drugs, which are based on the molecular characteristics of breast cancer cells.[3,4] However, breast cancer cells undergo aggressive malignant transformation during treatment, even in the early stages, causing resistance to therapy, metastasis, and recurrence.[5–7] Little is known about the mechanism of malignant progression, limiting treatment.

Recent evidence has shown that zinc (Zn) and Zn transporters are closely related to the malignant processes of breast cancer, making them an appealing approach for novel therapeutic strategies.[8–12] Zn is a trace element that is essential for survival; intracellular Zn levels are spatiotemporally regulated by two classes of transporters: 14 ZRT, IRT-like protein family

Coordination Chemistry Perspectives No. 2
Zinc in Biology: Molecular Structures, Cellular Processes and Living Systems
Edited by Hiroyuki Yasui and Taiho Kambe

Published by the Royal Society of Chemistry, www.rsc.org

members (ZIPs; SLC39) and nine Zn transporter family members (ZnT; SLC30). ZIP family members facilitate the influx of Zn into the cytosol from either extracellular or intracellular stores, such as the endoplasmic reticulum. In comparison, ZnT family members mediate the Zn efflux from the cytosol to other cellular compartments.[13,14] These transporters maintain Zn homeostasis within cells and throughout the body by managing the cellular Zn balance in a tissue-specific manner.[8–14] The Zn carried by the transporter targets molecules selectively as a signaling effector, such that the Zn signaling network spatiotemporally regulates cell functions, and Zn transporter aberrations are related to the onset and progression of various cancers[14] (Table 15.1). The patterns of ZIP dysregulation differ among cancers and reflect the complicated roles of Zn in various cancer types such as breast cancer,[8,15–28] hepatic cancer,[29,30] lung cancer,[31] oral and digestive system cancer,[32–36] otorhinolaryngologic and head & neck cancer,[37–39] ovarian cancer,[40,41] pancreatic cancer,[42–46] prostate cancer[27,47–49] and urologic cancer,[50] as well as in phenotypes associated with tumorigenesis, metastasis, and therapeutic resistance (Table 15.1). In cancer-linked cachexia, ZIP14 is upregulated in skeletal muscle.[51]

Zn levels in the serum and breast tumor tissue of breast cancer patients are lower and higher, respectively, than in healthy individuals.[52–55] The clinical and biological significance of ZIP6 in estrogen receptor (ER)-positive breast cancer,[15–21] ZIP7 in the resistance of ER-positive breast cancer cells to tamoxifen antagonists,[8,22–26] and ZIP10 in breast cancer cell invasion and metastasis have been shown.[20,21,28] To understand the molecular functions of these transporters, we have investigated the ZIP6-centered Zn network and ZIP7 in cell survival within the tumor microenvironment. ZIP10 is also associated with cell death and may have potential therapeutic applications.

This review summarizes our understanding of the roles of ZIP6, ZIP7, and ZIP10, focusing on breast cancer cell behaviors linked to malignant progression and prognosis. We also propose Zn transporters and Zn-targeted molecules as candidate markers for novel treatment strategies.

15.2 ZIP6 Downregulation-switched Zn Network in Breast Cancer Under Hypoxia

Accumulating evidence suggests that ZIP6 is a key molecule determining the fate of ER-positive breast cancer cells, and that its expression is controlled by the surrounding environment and malignant progression.[7–11,15–21] Inside solid tumors, such as breast cancer, there are areas of low oxygen pressure where the oxygen supply from blood vessels is insufficient. These areas not only facilitate the survival of breast cancer cells but are also deeply involved in the malignancy of the cancer, such as by showing resistance to chemotherapy and radiotherapy.[56–58] In cell culture systems, when human ER-positive MCF-7 breast cancer cells are cultured in three dimensions, they form colonies that look like spherical moss balls, and the interior of these

Table 15.1 Dysregulation of Zn transporter ZIP (SLC39a) in cancer.

Cancer type	Protein name (gene name)	Expression	Ref.
Breast cancer			
ER + breast	ZIP6 (SLC39a6)	Downregulated	15–18
Metastatistic ER + breast	ZIP6 (SLC39a6)	Upregulated	19–21
Anti-hormone therapy resistance	ZIP7 (SLC39a7)	Upregulated	22–26
Breast cancer stem-like cell	ZIP7 (SLC39a7)	Upregulated	8
Breast cancer cell	ZIP9 (SLC39a9)	Apoptosis induced by ZIP9 activation	27
Metastatic breast (TNBC)	ZIP10 (SLC39a10)	Upregulated	20, 21 and 28
Hepatic cancer			
Hepatocellular	ZIP4 (SLC39a4)	Upregulated	29
Hepatocellular	ZIP6 (SLC39a6)	Upregulated	30
Lung cancer			
Lung	ZIP4 (SLC39a4)	Upregulated	31
Oral and digestive system cancer			
Oral squamous	ZIP4 (SLC39a4)	Upregulated	32
Esophageal	ZIP5 (SLC39a5)	Upregulated	33
Esophageal squamous-cell	ZIP6 (SLC39a6)	Upregulated	34
Gastric	ZIP6 (SLC39a6)	Upregulated	35
Colorectal	ZIP6 (SLC39a6)	Upregulated	36
Otorhinolaryngologic and head and neck cancer			
Nasopharyngeal	ZIP4 (SLC39a4)	Upregulated	37
Glioma	ZIP4 (SLC39a4)	Upregulated	38
Neuroblastoma	ZIP8 (SLC39a8)	Proliferation inhibited by ZIP8 knockdown	39
Ovarian cancer			
Ovarian	ZIP4 (SLC39a4)	Upregulated	40
Ovarian	ZIP13 (SLC39a13)	Upregulated	41
Pancreatic cancer			
Pancreatic	ZIP3 (SLC39a3)	Downregulated	42
Pancreatic	ZIP4 (SLC39a4)	Upregulated	43
Pancreatic	ZIP4 (SLC39a4)	Upregulated in cancer cell-derived exosomes	44
Pancreatic	ZIP6 (SLC39a6)	Upregulated	45
Pancreatic	ZIP11 (SLC39a11)	Upregulated	46
Prostate cancer			
Prostate	ZIP1 (SLC39a1)	Downregulated	47 and 48
Prostate	ZIP2 (SLC39a2)	Downregulated	47 and 48
Prostate	ZIP3 (SLC39a3)	Downregulated	47 and 48

Table 15.1 (*Continued*)

Cancer type	Protein name (gene name)	Expression	Ref.
Prostate	ZIP4 (SLC39a4)	Downregulated	47 and 48
Prostate	ZIP6 (SLC39a6)	Upregulated	47 and 49
Prostate	ZIP9 (SLC39a9)	Apoptosis induced by ZIP9 activation	27 and 47
Prostate	ZIP14 (SLC39a14)	Downregulated	47
Urologic cancer			
Renal cell	ZIP11 (SLC39a11)	ZIP11 variant as risk factor	50
Bladder	ZIP11 (SLC39a11)	ZIP11 variant as risk factor	50
Cancer cachexia			
Cachexia in metastatic cancer	ZIP14 (SLC39a14)	Upregulated in skeletal muscle	51

colonies is hypoxic, as confirmed by the hypoxia sensing probe LOX-1.[8–10] The formation of these colonies disappeared after treatment with a cell-permeable Zn-specific chelating agent (*N,N,N′,N′*-tetrakis(2-pyridylmethyl)-ethylenediamine; TPEN), and was restored by supplying Zn using a Zn ionophore. Furthermore, when MCF-7 cells were cultured under an *in vitro hypoxia stimulated model* of a low-oxygen, low-nutrient environment, they had a high survival rate that depended on activation of the hypoxia-inducible factor HIF-1. Intracellular Zn levels are correlated with the HIF-1 activity and survival rate under low-oxygen conditions, suggesting that Zn contributes to resistance under hypoxic environments *via* the activation of HIF-1. We analyzed the HIF-1 signaling-transcription pathway, and identified metastasis-associated protein 1 (MTA-1) as a candidate zinc-targeting molecule (Figure 15.1). ZIP6 has been identified in MCF-7 cells,[59] and the ZIP6-centered Zn network was found to play an important role in ER-positive breast cancer.[8–10,15–21] The specific stable knockdown of ZIP6 in MCF-7 cells disrupted intracellular Zn homeostasis, and resulted in a high survival rate under hypoxic conditions *via* the suppression of apoptosis and epithelial-mesenchymal transition.[15] In tumor samples from breast cancer patients, ZIP6 expression decreases as the breast cancer stage progresses, and breast cancer patients with low ZIP6 expression have significantly lower recurrence-free survival rates and poorer prognoses than breast cancer patients with high ZIP6 expression.[8–10,17,18] We also found that a high expression ratio of ZIP7 to ZIP6 is necessary for tolerance of hypoxic environments, and treatment with a ZIP7-specific inhibitor significantly suppressed the survival in a hypoxic environment. Based on these findings, the Zn network originating from ZIP6 is thought to control the determination of cancer malignancy in the primary breast cancer lesion, and the level of ZIP6 expression may be a prognostic factor. Further detailed analysis of the ZIP6-mediated Zn

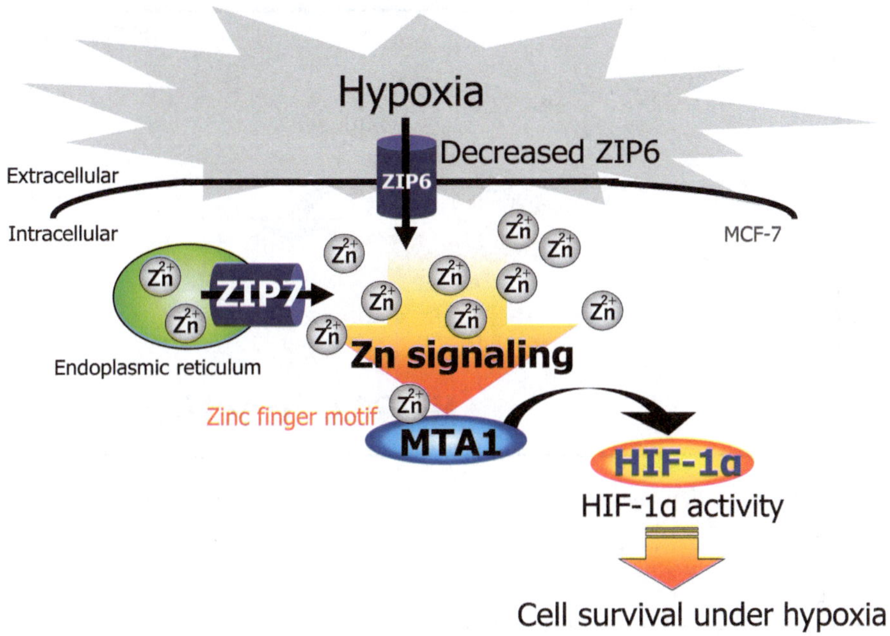

Figure 15.1 ZIP6/ZIP7 contribute to cell survival under hypoxic stress. ZIP6/ZIP7-mediated Zn is suggested to regulate cell survival through HIF-1α activity. ZIP6 downregulation-switched Zn network is required for adaptation to hypoxic stress.

network, such as the interaction between Zn transporters and the involvement of Zn levels inside and outside the cell, will provide important directions for establishing breast cancer diagnosis and treatment methods.

15.3 ZIP7 and Malignant Breast Cancer

15.3.1 Impact of Zn and ZIP7 on BCSCs

Breast cancer stem-like cells (BCSCs) can initiate and maintain tumor growth and are believed to be linked to the malignant progression of breast cancer, such as tumor chemoresistance, recurrence, and metastasis.[60–63] BCSCs self-renew by generating unlimited copies and give rise to mature non-stem cell progeny *via* differentiation, producing phenotypically different cells.[60,63] However, the molecular mechanisms that regulate BCSCs remain largely unknown, and should be examined to improve the outcomes of patients with breast cancer.

The spheroid culture method is a powerful three-dimensional *in vitro* cell culture model that maintains the characteristics of BCSCs, such as marker expression, cell self-renewal potential, and high tumorigenicity *in vivo*.[63–65] With an assay using MCF-7, we demonstrated that treatment with TPEN significantly suppressed the sphere formation efficiency and the nuclear

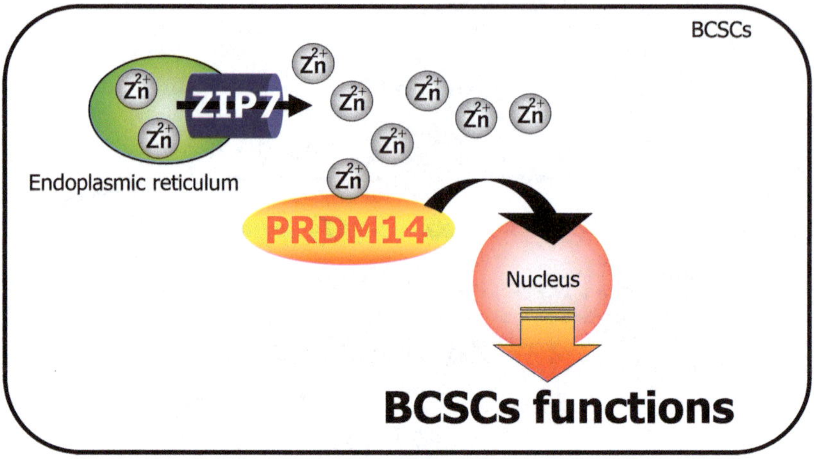

Figure 15.2 ZIP7 plays an important regulatory role for maintaining BCSCs. ZIP7 plays a key role in BCSC maintenance through PRDM14 activity. This is the first report of a biological role for Zn transporters in BCSCs.

expression of PR domain zinc finger protein 14 (PRDM14), a cancer stem cell transcription factor essential for maintaining BCSCs[66] (Figure 15.2). This suppression was neutralized by the addition of Zn using Zn ionophores, suggesting that intracellular Zn-dependent PRDM14 activity is essential for the ability to form spheres. We also found that the sphere formation efficiency was blocked by inhibiting the ability of ZIP7 to carry Zn from the endoplasmic reticulum to the cytoplasm. ZIP7 directly drives pro-survival pathways, such as the AKT, MAPK, and mTOR pathways for tumorigenesis,[22] and is increased in endocrine-resistant breast cancer.[26] As a result, ZIP7 expression is strongly associated with poor outcomes in breast cancer patients.[23] The levels of ZIP7 and phosphorylated ZIP7, which transports Zn, are promising markers of malignant breast cancer.[22,23] Our data show that ZIP7 mediates the maintenance of BCSCs *via* Zn-dependent PRDM14 activity. MCF-7 cells, in which ZIP6, which is important in ER-positive breast cancer, was knocked down, had high sphere-formation efficiency, and suppressing the ZIP7 activity suppressed sphere formation. We are currently investigating the relationship between ZIP6 and ZIP7 to clarify Zn signaling in BCSCs. PRDM14 expression is high in BCSCs as an essential transcription factor, whereas PRDM14 is not found in normal tissue stem cells.[66] Stage II or III breast cancer patients with high PRDM14 expression had poorer prognoses than those with low or undetectable PRDM14 expression. These findings suggest that PRDM14 could be used to develop definitive therapeutic strategies. Nucleic acid therapy targeting PRDM14 is currently being developed. However, it is difficult to target the PRDM14 molecule using small-molecule therapeutics, including nucleic acid drugs, because PRDM14 is expressed and activated in the nucleus.[67] Further understanding of the mechanism underlying Zn-mediated BCSCs *via* PRDM14 could accelerate progress toward the development of novel molecular targeted therapies for BCSCs.

15.3.2 The Possible Involvement of ZIP7 in Breast Cancer Dormancy

Breast cancer cell dormancy is believed to be involved in breast cancer recurrence.[68–70] After 1 month of tamoxifen treatment, MCF-7 cells could not survive, but tamoxifen-resistant cells isolated from MCF-7 cells survived and had unusual shapes. The expression of Ki67, a proliferation marker, was reduced to below the detection limit in tamoxifen-resistant cells, indicating that their proliferative ability had stopped. However, the cells began to proliferate when the tamoxifen was removed, suggesting that tamoxifen-resistant cells enter a dormant state under long-term tamoxifen treatment. Using this model, the effect of the ZIP7 inhibitor NVS-ZP7-4 was examined in tamoxifen-resistant cells. NVS-ZP7-4 induced cell death and blocked the process of entering a dormant state during tamoxifen treatment. Little is known about the mechanism involving zinc and zinc transporters in breast cancer dormancy and further study is needed. Our results based on zinc open possibilities for suppressing recurrence after tamoxifen resistance.

15.4 ZIP10 and Pharmacotherapy in Breast Cancer

15.4.1 Development of More Effective Tamoxifen Treatment Using Zn in ER-positive Breast Cancer

Approximately 70% of breast cancer cases are ER-positive, and endocrine therapy such as the anti-estrogenic drug tamoxifen is the treatment of choice, with relatively high efficacy.[1–3] However, endocrine therapy is a long treatment that frequently causes side effects and leads to drug resistance. Moreover, tamoxifen is ineffective in some ER-positive patients.[71–73] Furthermore, recurrence and death can occur after more than 5 years of tamoxifen treatment.[71–73] To improve the effectiveness and safety of tamoxifen therapy, it is necessary develop new treatment strategies based on the mechanism underlying the sensitivity to and efficacy of tamoxifen.

The effect of tamoxifen is regulated not only by the presence or absence of estrogen receptors but also by receptor binding-related molecules and the activity of cell survival and cell death signaling pathways in breast cancer cells. However, the key mediator remains unknown.[71] To develop more effective endocrine therapy using tamoxifen, we investigated whether the anticancer activity of tamoxifen could be controlled by the addition of Zn.[8] MCF-7 cells treated with tamoxifen plus Zn died in a Zn concentration-dependent manner, as compared to with tamoxifen alone. Moreover, the combination of tamoxifen and Zn suppressed tumor growth in mice *in vivo*, without serious side effects. This effect of the combination of tamoxifen and Zn was caused by the supply of Zn to the cytoplasm *via* ZIP10, followed by the

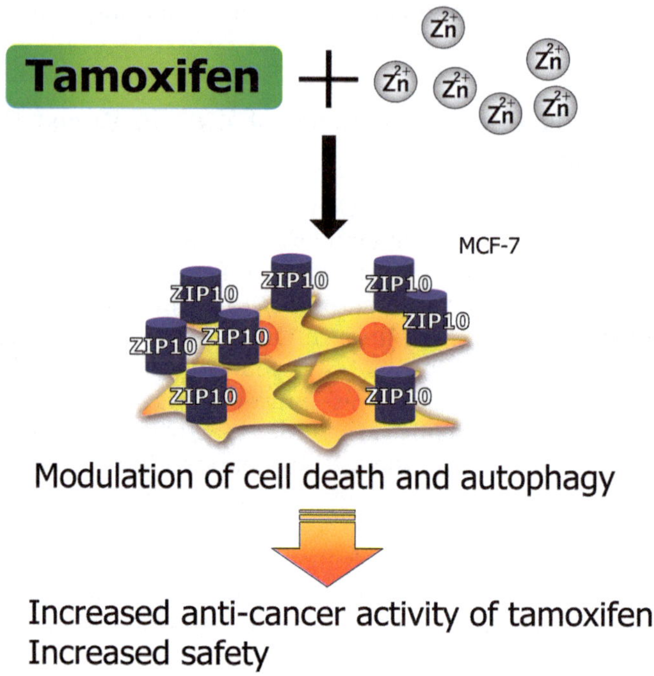

Figure 15.3 Zn and ZIP10 mediate the anti-cancer activity of tamoxifen. The anticancer activity of tamoxifen is successfully enhanced by the addition of Zn, and ZIP10 is a key molecule for this function. Zn is a powerful tool for optimizing breast cancer treatment using tamoxifen.

induction of non-apoptotic cell death accompanied by autophagy (Figure 15.3). Tamoxifen treatment suppresses estrogen signaling, and the expression of ZIP6, one of the genes activated by estrogen signaling, was downregulated. The use of Zn in combination with tamoxifen induced cell death and reduced the IC50 by approximately 1/2 to 1/100 that of tamoxifen alone, even in tamoxifen-resistant MCF-7 cells. Therefore, the combination of tamoxifen and Zn restored the effect of tamoxifen, even in tamoxifen-resistant breast cancer cells. These data provide evidence that Zn will be useful for optimizing breast cancer treatment using tamoxifen.

In colorectal cancer, the anticancer drug 5-fluorouracil increases the intracellular Zn concentration, which suppresses tumor growth by activating the tumor suppressor gene p53 and inhibiting NF-κB.[74] Furthermore, the combination of 5-fluorouracil and Zn has significantly higher anti-cancer activity than treatment with 5-fluorouracil alone, and can also restore the sensitivity of 5-fluorouracil-resistant cells.[74] The combination of Zn and the anticancer drug Adriamycin has also successfully treated colorectal cancer without recurrence, even when the Adriamycin dose was reduced in an *in vivo* xenograft mouse model.[75] These results for cancer do not clarify the Zn network or its involvement with zinc transporters, and they do not show the autophagic cell death-increasing effect of the combination of Zn and tamoxifen that we observed in our breast cancer research. The mechanism of the increased effect of anticancer drugs in combination with Zn may differ depending on the types of cancer and anticancer drug, even if the result is the same. Further study and development of dosing formulations according to cancer type would be useful.

15.4.2 Development of a TNBC Therapeutic Strategy Using Zn

Triple-negative breast cancer (TNBC) accounts for approximately 20% of all breast cancers.[1,2] TNBC does not express the estrogen receptor (ER), progesterone receptor, or human epidermal growth factor receptor type 2 (HER2), and it is impossible to use hormone therapy or molecular-targeted drugs.[76,77] The response rate to chemotherapy is also low. Therefore, the prognosis is extremely poor. There is an urgent need to develop a safe effective targeted therapeutic strategy for TNBC.[77,78] We are studying drug repositioning, which involves discovering new effects of existing drugs and applying them to other diseases. We have reported on the possibility of controlling the anticancer activity of combination drugs in TNBC by using the autophagy-inducing function, which is an off-target effect of the anti-estrogen drug tamoxifen that is independent of the ER.[78] Furthermore, we have shown that Zn, which is important for the functional regulation of breast cancer cells, can selectively control target molecules depending on the type of zinc transporter; on that basis, ZIP10 is strongly related to the promotion of cell death in breast cancer cells. We focused on TNBC cells with high ZIP10 expression and low ZIP6 expression, and investigated whether tamoxifen in combination with Zn has anticancer activity in TNBC. The viability of MCF-7 cells treated with tamoxifen alone was suppressed in a concentration-dependent manner, whereas TNBC cells were less likely to die and had a very high IC50. The combination of tamoxifen and zinc significantly induced cell death, as compared to either treatment alone. Furthermore, tamoxifen in combination with Zn acetate significantly reduced the cell viability. Polaprezinc, a Zn-containing gastric ulcer treatment, also had anticancer activity in combination with tamoxifen. These results demonstrate that the combination of tamoxifen and Zn has high activity against TNBC, suggesting that these existing drugs combined with Zn are promising candidates for drug repositioning.

15.5 Conclusion

The Zn network in breast cancer is becoming clear at cellular and molecular levels. Zn transporters change their expression levels and molecular species in response to the environment surrounding breast cancer cells. The suppression of ZIP6 expression in ZIP6 knockdown breast cancer cells and the cell-culture conditions caused the induction and activation of ZIP7 or ZIP10, determining the phenotype and function of malignant breast cancer cells and the effect of anticancer therapy. A better understanding of the Zn network and Zn transporters should lead to elucidation of the mechanism underlying breast cancer malignancy and the development of new strategies using Zn-related molecules. We will continue breast cancer research in pioneering areas using Zn biology.

Acknowledgements

This work was supported in part by JSPS KAKENHI grants 24790099, 15K07955, 18K06676, 21K06567 and 24K09827, and The Naito Foundation (for T. T.-N.) and Takeda Science Foundation grant J20ZZ00356 (for I. N.). We thank Dr Haruki Torii and Ms Chinami Ikushima for their kind advice and technical assistance.

References

1. F. Bray, M. Laversanne, H. Sung, J. Ferlay, R. L. Siegel, I. Soerjomataram and A. Jemal, *Ca-Cancer J. Clin.*, 2024, **74**(3), 229.
2. https://www.who.int/news-room/fact-sheets/detail/breast-cancer (last accessed September 2024).
3. T. Shimoi, S. E. Nagai, T. Yoshinami, M. Takahashi, H. Arioka, M. Ishihara, Y. Kikawa, K. Koizumi, N. Kondo, Y. Sagara, M. Takada, T. Takano, J. Tsurutani, Y. Naito, R. Nakamura, M. Hattori, F. Hara, N. Hayashi, T. Mizuno, M. Miyashita, N. Yamashita, T. Yamanaka, S. Saji, H. Iwata and T. Toyama, *Breast Cancer*, 2020, **27**(3), 322.
4. H. J. Burstein, *N. Engl. J. Med.*, 2020, **383**(26), 2557.
5. Early Breast Cancer Trialists' Collaborative Group (EBCTCG), *Lancet*, 2005, **365**(9472), 1687.
6. H. Dillekås, M. S. Rogers and O. Straume, *Cancer Med.*, 2019, **8**(12), 5574.
7. A. I. Riggio, K. E. Varley and A. L. Welm, *Br. J. Cancer*, 2021, **124**(1), 13.
8. T. Takatani-Nakase, C. Matsui, M. Sakitani and I. Nakase, *Metallomics Res.*, 2022, **2**(1), 29.
9. T. Takatani-Nakase, *Biol. Pharm. Bull.*, 2018, **41**(10), 1517.
10. T. Takatani-Nakase, C. Matsui and K. Takahashi, *Biomed. Res. Clin. Prac.*, 2016, **1**(3), 71.
11. T. Hara, T. A. Takeda, T. Takagishi, K. Fukue, T. Kambe and T. Fukada, *J. Physiol. Sci.*, 2017, **67**(2), 283.
12. T. Hara, E. Yoshigai, T. Ohashi and T. Fukada, *J. Pharmacol. Sci.*, 2022, **148**(2), 221.
13. T. Kambe, K. M. Taylor and D. Fu, *J. Biol. Chem.*, 2021, **296**, 100320.
14. M. Bendellaa, P. Lelièvre, J. L. Coll, L. Sancey, A. Deniaud and B. Busser, *Int. J. Cancer*, 2024, **154**(1), 7.
15. C. Matsui, T. Takatani-Nakase, Y. Hatano, S. Kawahara, I. Nakase and K. Takahashi, *FEBS Lett.*, 2017, **591**(20), 3348.
16. V. Lopez and S. L. Kelleher, *Exp. Cell Res.*, 2010, **316**(3), 366.

17. G. Kasper, A. A. Weiser, A. Rump, K. Sparbier, E. Dahl, P. Wild, U. Schwidetzky, E. Castanos-Velez and K. Lehmann, *Int. J. Cancer*, 2005, **117**(6), 961.
18. K. M. Taylor, H. E. Morgan, K. Smart, N. M. Zahari, S. Pumford, I. O. Ellis, J. F. R. Robertson and R. I. Nicholson, *Mol. Med.*, 2007, **13**(7–8)), 396.
19. C. Hogstrand, P. Kille, M. L. Ackland, S. Hiscox and K. M. Taylor, *Biochem. J.*, 2013, **455**(2), 229.
20. T. Takatani-Nakase, C. Matsui, S. Maeda, S. Kawahara and K. Takahashi, *PLoS One*, 2014, **9**(2), e90136.
21. T. Takatani-Nakase, *Yakugaku Zasshi*, 2013, **133**(11), 1195.
22. T. Nimmanon, S. Ziliotto, S. Morris, L. Flanagan and K. M. Taylor, *Metallomics*, 2017, **9**(5), 471–481.
23. S. Ziliotto, J. M. W. Gee, I. O. Ellis, A. R. Green, P. Finlay, A. Gobbato and K. M. Taylor, *Metallomics*, 2019, **11**(9), 1579.
24. K. M. Taylor, S. Hiscox, R. Nicholson, C. Hogstrand and P. Kille, *Sci. Signaling*, 2012, 5(210), 1.
25. C. Hogstrand, P. Kille, R. I. Nicholson and K. M. Taylor, *Trends Mol. Med.*, 2009, **15**(3), 101.
26. K. M. Taylor, P. Vichova, N. Jordan, S. Hiscox, R. Hendley and R. I. Nicholson, *Endocrinology*, 2008, **149**(10), 4912.
27. P. Thomas and J. Dong, *J. Steroid Biochem. Mol. Biol.*, 2021, **211**, 105906.
28. N. Kagara, N. Tanaka, S. Noguchi and T. Hirano, *Cancer Sci.*, 2007, **98**(5), 692.
29. B. P. Weaver, Y. Zhang, S. Hiscox, G. L. Guo, U. Apte, K. M. Taylor, C. T. Sheline, L. Wang and G. K. Andrews, *PLoS One*, 2010, **5**(10), e13158.
30. R. Shen, F. Xie, H. Shen, Q. Liu, T. Zheng, X. Kou, D. Wang and J. Yang, *PLoS One*, 2013, **8**(2), e56542.
31. C. Huang, X. Cui, X. Sun, J. Yang and M. Li, *Oncotarget*, 2016, **7**(41), 66935.
32. S. Ishida, A. Kasamatsu, Y. Endo-Sakamoto, D. Nakashima, N. Koide, T. Takahara, T. Shimizu, M. Iyoda, M. Shiiba, H. Tanzawa and K. Uzawa, *Biochem. Biophys. Res. Commun.*, 2017, **483**(1), 339.
33. Q. Li, J. Jin, J. Liu, L. Wang and Y. He, *Oncol. Res.*, 2016, **24**(3), 205.
34. C. Wu, D. Li, W. Jia, Z. Hu, Y. Zhou, D. Yu, T. Tong, M. Wang, D. Lin, Y. Qiao, Y. Zhou, J. Chang, K. Zhai, M. Wang, L. Wei, W. Tan, H. Shen, Y. Zeng and D. Lin, *Nat. Genet.*, 2013, **45**(6), 632.
35. J. Gao, W. Ren, C. Xiao, L. Wang, Q. Huang, Z. Zhang, Y. Dang, P. Weng, H. Wang, X. Fang, M. Zhuang, L. Lin and S. Chen, *BMC Cancer*, 2019, **19**(1), 1069.
36. V. Barresi, G. Valenti, G. Spampinato, N. Musso, S. Castorina, E. Rizzarelli and D. F. Condorelli, *J. Cell Biochem.*, 2018, **119**(12), 9707.
37. Q. Zeng, Y. M. Liu, J. Liu, J. Han, J. X. Guo, S. Lu, X. M. Huang, P. Yi, J. Y. Lang, P. Zhang and C. T. Wang, *Cell Death Dis.*, 2019, **10**(8), 588.
38. Y. Lin, Y. Chen, Y. Wang, J. Yang, V. F. Zhu, Y. Liu, X. Cui, L. Chen, W. Yan, T. Jiang, G. W. Hergenroeder, S. A. Fletcher, J. M. Levine, D. H. Kim, N. Tandon, J. J. Zhu and M. Li, *Neuro-Oncol.*, 2013, **15**(8), 1008.

39. Z. Mei, P. Yan, Y. Wang, S. Liu and F. He, *Mol. Med. Rep.*, 2018, **18**(1), 477–485.
40. Q. Fan, Q. Cai, P. Li, W. Wang, J. Wang, E. Gerry, T. L. Wang, I. M. Shih, K. P. Nephew and Y. Xu, *Oncotarget*, 2017, **8**(52), 90090.
41. X. Cheng, J. Wang, C. Liu, T. Jiang, N. Yang, D. Liu, H. Zhao and Z. Xu, *J. Exp. Clin. Cancer Res.*, 2021, **40**(1), 199.
42. L. C. Costello and R. B. Franklin, *Pancreat. Disord. Ther.*, 2013, **4**, 002.
43. X. Cui, Y. Zhang, J. Yang, X. Sun, J. P. Hagan, S. Guha and M. Li, *Cell Cycle*, 2014, **13**(7), 1180.
44. H. Jin, P. Liu, Y. Wu, X. Meng, M. Wu, J. Han and X. Tan, *Cancer Sci.*, 2018, **109**(9), 2946.
45. J. Unno, K. Satoh, M. Hirota, A. Kanno, S. Hamada, H. Ito, A. Masamune, N. Tsukamoto, F. Motoi, S. Egawa, M. Unno, A. Horii and T. Shimosegawa, *Int. J. Oncol.*, 2009, **35**(4), 813.
46. B. Zhu, R. Huo, Q. Zhi, M. Zhan, X. Chen and Z. C. Hua, *J. Trace Elem. Med. Biol.*, 2021, **65**, 126734.
47. S. Acevedo, M. F. Segovia and E. de la Fuente-Ortega, *Nutrients*, 2024, **16**(13), 2026.
48. V. Kolenko, E. Teper, A. Kutikov and R. Uzzo, *Nat. Rev. Urol.*, 2013, **10**(4), 219.
49. H. W. Lue, X. Yang, R. Wang, W. Qian, R. Z. Xu, R. Lyles, A. O. Osunkoya, B. P. Zhou, R. L. Vessella, M. Zayzafoon, Z. R. Liu, H. E. Zhau and L. W. Chung, *PLoS One*, 2011, **6**(11), e27720.
50. L. Wu, K. G. Chaffee, A. S. Parker, H. Sicotte and G. M. Petersen, *Tumour Biol.*, 2015, **36**(10), 7431.
51. G. Wang, A. K. Biswas, W. Ma, M. Kandpal, C. Coker, P. M. Grandgenett, M. A. Hollingsworth, R. Jain, K. Tanji, S. López-Pintado, A. Borczuk, D. Hebert, S. Jenkitkasemwong, S. Hojyo, R. V. Davuluri, M. D. Knutson, T. Fukada and S. Acharyya, *Nat. Med.*, 2018, **24**(6), 770.
52. P. Rusch, A. V. Hirner, O. Schmitz, R. Kimmig, O. Hoffmann and M. Diel, *Arch. Gynecol. Obstet.*, 2021, **303**(1), 195.
53. L. Jouybari, F. Kiani, A. Akbari, A. Sanagoo, F. Sayehmiri, J. Aaseth, M. S. Chartrand, K. Sayehmiri, S. Chirumbolo and G. Bjørklund, *J. Trace Elem. Med. Biol.*, 2019, **56**, 90.
54. Y. Feng, J. W. Zeng, Q. Ma, S. Zhang, J. Tang and J. F. Feng, *J. Trace Elem. Med. Biol.*, 2020, **62**, 126629.
55. J. Gumulec, M. Masarik, S. Krizkova, V. Adam, J. Hubalek, J. Hrabeta, T. Eckschlager, M. Stiborova and R. Kizek, *Curr. Med. Chem.*, 2011, **18**(33), 5041.
56. J. Araos, J. P. Sleeman and B. K. Garvalov, *Clin. Exp. Metastasis*, 2018, **35**(7), 563.
57. G. L. Semenza, *Biochim. Biophys. Acta*, 2016, **1863**(3), 382–391.
58. A. L. Harris, *Nat. Rev. Cancer*, 2002, **2**(1), 38–47.
59. D. L. Manning, R. J. Daly, P. G. Lord, K. F. Kelly and C. D. Green, *Mol. Cell. Endocrinol.*, 1988, **59**(3), 205.
60. S. Liu and M. S. Wicha, *J. Clin. Oncol.*, 2010, **28**(25), 4006.

61. B. Beck and C. Blanpain, *Nat. Rev. Cancer*, 2013, **13**(10), 727.
62. A. Kreso and J. E. Dick, *Cell Stem Cell*, 2014, **14**(3), 275.
63. P. Dey, M. Rathod and A. De, *Breast Cancer*, 2019, **11**, 115.
64. T. Tang, Z. Yang, Q. Zhu, Y. Wu, K. Sun, M. Alahdal, Y. Zhang, Y. Xing, Y. Shen, T. Xia, T. Xi, Y. Pan and L. Jin, *FASEB J.*, 2018, **32**, 6965.
65. A. Kreso and J. E. Dick, *Cell Stem Cell*, 2014, **14**(3), 275.
66. H. Taniguchi, D. Hoshino, C. Moriya, H. Zembutsu, N. Nishiyama, H. Yamamoto, K. Kataoka and K. Imai, *Oncotarget*, 2017, **8**(29), 46856.
67. H. Taniguchi and K. Imai, *Methods Mol. Biol.*, 2019, **1974**, 233.
68. T. G. Phan and P. I. Croucher, *Nat. Rev. Cancer*, 2020, **20**(7), 398.
69. M. H. Manjili, *Cancer Res.*, 2017, 77(10), 2564.
70. A. Recasens and L. Munoz, *Trends Pharmacol. Sci.*, 2019, **40**(2), 128.
71. E. A. Musgrove and R. L. Sutherland, *Nat. Rev. Cancer*, 2009, **9**(9), 631.
72. Early Breast Cancer Trialists' Collaborative Group (EBCTCG), C. Davies, J. Godwin, R. Gray, M. Clarke, D. Cutter, S. Darby, P. McGale, H. C. Pan, C. Taylor, Y. C. Wang, M. Dowsett, J. Ingle and R. Peto, *Lancet*, 2011, **378**(9793), 771.
73. J. Cuzick, I. Sestak, S. Cawthorn, H. Hamed, K. Holli, A. Howell and J. F. Forbes, IBIS-I Investigators, *Lancet Oncol.*, 2015, **16**(1), 67.
74. J. M. Arriaga, A. Greco, J. Mordoh and M. Bianchini, *Mol. Cancer Ther.*, 2014, **13**(5), 1369.
75. A. Garufi, V. Ubertini, F. Mancini, V. D'Orazi, S. Baldari, F. Moretti, G. Bossi and G. D'Orazi, *J. Exp. Clin. Cancer Res.*, 2015, **34**(1), 87.
76. K. A. Won and C. Spruck, *Int. J. Oncol.*, 2020, **57**(6), 1245.
77. C. Denkert, C. Liedtke, A. Tutt and G. von Minckwitz, *Lancet*, 2017, **389**(10087), 2430.
78. T. Takatani-Nakase, C. Ikushima, M. Sakitani and I. Nakase, *Life Sci.*, 2024, **339**, 122411.

CHAPTER 16

Role of Zinc in the Pathogenesis of Inflammatory Bowel Disease: From Trace Elements to Zinc Signaling

YUJI NAITO,*[a] YASUKI HIGASHIMURA[b] AND TOMOHISA TAKAGI[c,d]

[a] Human Immunology and Nutrition Science, Graduate School of Medical Science, Kyoto Prefectural University of Medicine, Japan; [b] Department of Food Science, Ishikawa Prefectural University, Japan; [c] Molecular Gastroenterology and Hepatology, Graduate School of Medical Science, Kyoto Prefectural University of Medicine, Japan; [d] Department for Medical Innovation Translational Medical Science, Graduate School of Medical Science, Kyoto Prefectural University of Medicine, Japan
*Email: ynaito@koto.kpu-m.ac.jp

16.1 Hypozincemia in Patients with IBD

The pathogenesis of IBD involves a complex interaction of factors, including impaired intestinal barrier function, gut microbiome dysbiosis, abnormal immune and inflammatory responses, and genetic predispositions. However, the exact etiology remains uncertain. Reduced serum zinc levels have been observed in patients with IBD, likely due to impaired zinc absorption related to intestinal inflammation or a history of bowel resection. However, the causes of hypozincemia are diverse, including the use of chelating agents, increased urinary zinc excretion, dietary limitations, and reliance on fast food. Special

Coordination Chemistry Perspectives No. 2
Zinc in Biology: Molecular Structures, Cellular Processes and Living Systems
Edited by Hiroyuki Yasui and Taiho Kambe

Published by the Royal Society of Chemistry, www.rsc.org

attention should be given to the effects of medications such as diuretics (*e.g.*, spironolactone and furosemide), systemic antifungals, oral iron supplements, and thyroid hormones.

Studies comparing serum zinc levels in patients with IBD have shown that 42.2% of patients with Crohn's disease (CD) and 38.6% of patients with ulcerative colitis (UC) were diagnosed with hypozincemia (serum zinc levels below 66 $\mu g\,dL^{-1}$).[1] Zinc deficiency is more prevalent in patients with CD, likely due to impaired absorption associated with inflammation. A meta-analysis revealed that zinc deficiency is more common in CD, likely due to the malabsorptive nature of the disease and the fact that zinc absorption primarily occurs in the proximal intestine. An intriguing epidemiological study in Japan, which matched the age and sex of newly diagnosed patients with UC, found that a higher zinc intake one year before UC onset was associated with a significantly lower risk of developing the disease (OR 0.39, 95% CI 0.17–0.81).[2] From a preventive medicine perspective, maintaining sufficient zinc intake through diet may help reduce the risk of UC.

According to Japanese standards, serum zinc levels below 60 $\mu g\,dL^{-1}$ indicate zinc deficiency, while levels between 60 $\mu g\,dL^{-1}$ and 80 $\mu g\,dL^{-1}$ indicate marginal zinc deficiency.[3] Higashimura *et al.*[4] reported that 75% of patients with CD had serum zinc levels below 80 $\mu g\,dL^{-1}$, with a strong negative correlation between zinc levels and disease activity. Omatsu *et al.*[5] found that 8.1% of patients with CD had zinc deficiency, while 45.9% had marginal zinc deficiency. In patients with UC, 7.8% had zinc deficiency and 29.7% had marginal zinc deficiency. Notably, significantly lower serum zinc levels were observed in patients with severe CD (as assessed by SES-CD) and severe UC (as assessed by Mayo-ES), based on endoscopic findings. Ishihara *et al.*[6] found that zinc deficiency ($<70\ \mu g\,dL^{-1}$) was observed in 60.2% of pediatric patients with CD, 51.7% of pediatric patients with UC, and 37.2% of healthy controls.

Regular monitoring of blood zinc levels in patients with IBD is critical in clinical practice because zinc deficiency is challenging to estimate based solely on subjective and objective symptoms. Increased intestinal permeability, which may persist even in patients with clinically remitted IBD, can contribute to clinical relapse. Molecular mechanisms underlying increased permeability, such as the downregulation of adhesion molecules like occludin and ZO-1, have been reported, with their expression further reduced under zinc deficiency.[7] Prospective studies evaluating serum zinc levels and IBD-related risks have shown that hospitalization, surgery, and complications are significantly higher in patients with hypozincemia, particularly those with CD, compared to patients with UC.[1]

In Japan, oral zinc preparations are approved for treating zinc deficiency (Figure 16.1). Monitoring and therapeutic interventions are essential for improving long-term outcomes in patients with IBD. Sakurai *et al.*[8] retrospectively evaluated the efficacy of zinc acetate hydrate in patients with IBD. Their analysis, primarily of patients with CD ($n = 32$), demonstrated that after four weeks of treatment, the average serum zinc concentration

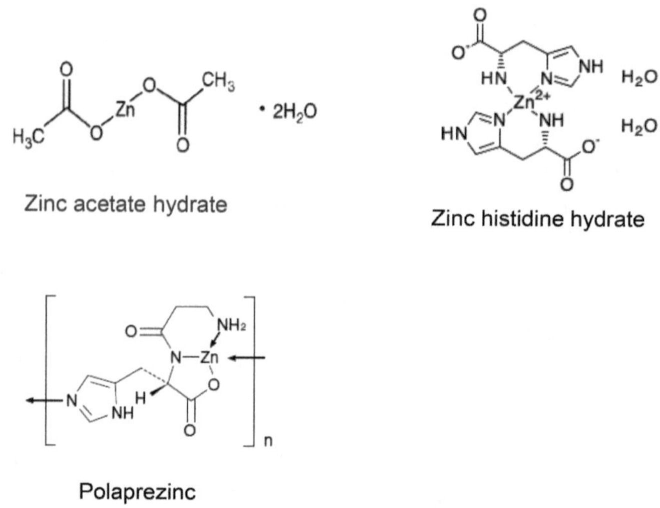

Figure 16.1 Zinc formulations used for oral administration in Japan.

significantly increased from 55.5 μg dL^{-1} to 91.0 μg dL^{-1}, with 71.9% of patients achieving normal levels (≥80 μg dL^{-1}). Additionally, the Crohn's Disease Activity Index (CDAI), a measure of disease activity, significantly decreased from 171.5 to 129. The study also noted a wide range of zinc acetate hydrate dosages (25–150 mg per day) required for individual patients, emphasizing the importance of adjusting dosages based on monitored serum zinc levels. To prevent copper deficiency due to excessive zinc administration, monitoring should include both copper and zinc levels.

16.2 Role of Metallothionein in Zinc Signaling

The discovery of zinc transporters has revealed that intracellular zinc concentrations are regulated by transporters that import and export zinc across the cell membrane. Moreover, changes in intracellular zinc levels affect signaling pathways and, consequently, cellular functions, leading to the emergence of zinc biology as a distinct academic field. Metallothionein (MT), localized in the cytoplasm, plays a vital role in maintaining intracellular zinc homeostasis. MT deficiency can result in reduced intracellular zinc levels and, consequently, abnormal cellular function. To assess the role of MT in intestinal inflammation, we used MTI/II-knockout mice to induce dextran sulfate sodium (DSS)-induced colitis.[9] We found that DSS-induced colitis was exacerbated macroscopically and histologically in MTI/II knockout mice compared to wild-type mice, with increased expression of inflammatory cytokines such as tumor necrosis factor (TNF)-α, interferon (IFN)-γ, and interleukin (IL)-17. Further investigations involved isolating peritoneal macrophages from each mouse and examining their response to lipopolysaccharide (LPS). It was observed that in MTI/II-deficient macrophages, the response to LPS was significantly enhanced, with elevated IFN-γ and IL-17 responses.

16.3 Dynamics of Intestinal Inflammatory Cytokines and Zinc

Various inflammatory cytokines are overexpressed in the intestinal mucosa of patients with IBD, and experimental colitis mouse models corroborate this observation. To investigate the role of zinc in colitis, we examined DSS-induced colitis in Zip14-knockout mice and 2,4,6-trinitrobenzene sulfonic acid (TNBS)-induced colitis in mice treated with the zinc chelator, *N*,*N*,*N*′,*N*′-tetrakis(2-pyridylmethyl)ethylenediamine (TPEN). We first confirmed that Zip14 is strongly expressed in the colon by examining the expression of various ZIP transporters. Immunohistological studies by Guthrie *et al.*[10] revealed that Zip14 is expressed on the basolateral side of intestinal epithelial cells and in endosomes. Zip14 deficiency has been shown to trap zinc in endosomes, reduce intracellular zinc concentration, weaken intracellular zinc signaling, and decrease the expression of tight junction proteins such as claudin and occludin, leading to impaired barrier function (Figure 16.2).

Using a mouse model of TNBS-induced colitis (commonly used as a model for Crohn's disease), we evaluated the effects of the zinc chelator TPEN.[4]

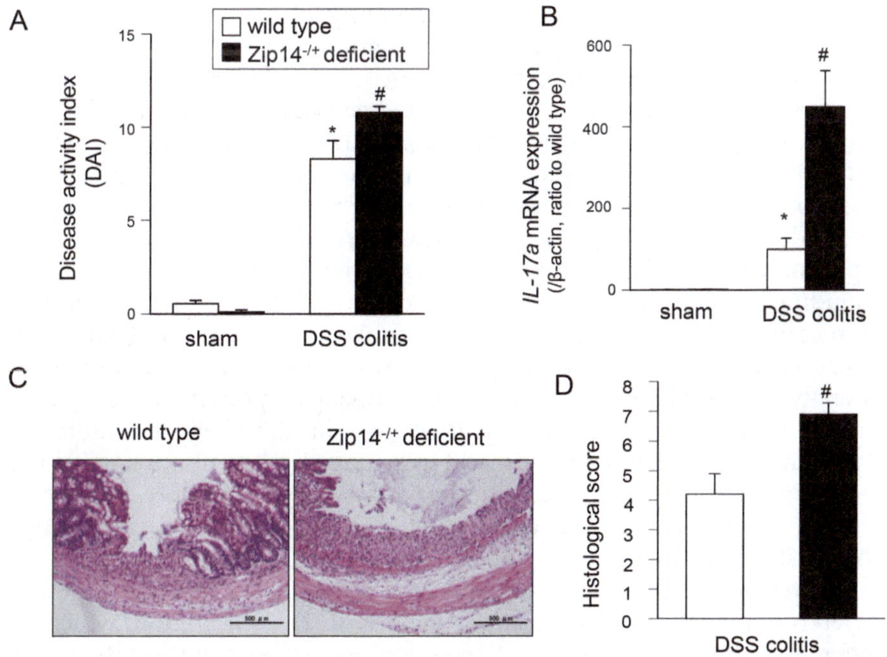

Figure 16.2 Role of zinc transporter Zip14 in the pathogenesis of dextran sodium sulfate (DSS)-induced colitis in mice. $^{*}p<0.05$ *vs.* sham group, $^{\#}p<0.05$ *vs.* wild type mice treated with DSS.

Four days after TNBS enema administration, ulcerative lesions and transmural inflammation in the colon were observed macroscopically and histologically, and these lesions were markedly exacerbated following TPEN administration. Additionally, the proportion of IL-17A$^+$ CD4$^+$ T cells in the mucosa and the tissue content of IL-17A protein were significantly increased after TPEN treatment. Serum zinc levels were also significantly reduced following TPEN administration. These findings suggest that zinc influences the differentiation of T cells into Th17 cells. However, *in vitro* Th17 differentiation assays using naïve T cells showed no direct effect of TPEN or zinc on the differentiation of naïve T cells into Th17 cells, indicating that zinc or zinc signaling may not directly regulate this differentiation process.

16.4 Macrophage Phenotype and Zinc Signaling

T-cell differentiation into Th1, Th2, and Th17 cells is influenced by cytokines secreted by macrophages. In this study, we first analyzed the macrophage phenotype at lesion sites using a TNBS colitis model. We fractionated CD45$^+$CD11b$^+$ macrophages and analyzed them using Ly6C as an M1 macrophage marker and CD206 as an M2 marker. The results indicated an increase in the M1 macrophage fraction in TNBS colitis, which was significantly augmented by TPEN administration.[4] Conversely, M2 macrophages were reduced in TNBS colitis, and this decrease was further enhanced by TPEN. These findings suggest zinc signaling is involved in macrophage phenotype differentiation during inflammation.

To elucidate the detailed mechanism of the involvement of zinc in macrophage differentiation, we conducted cellular experiments using bone marrow-derived macrophages (BMDMs). BMDMs differentiate into M1 and M2 cells in response to LPS/IFNγ stimulation. When cytoplasmic free zinc was labeled with Newport Green, it was observed that cytoplasmic free zinc levels decreased during M1 phenotype differentiation. Furthermore, the expression levels of zinc transporters involved in intracellular zinc influx, including Zip6, Zip8, and Zip10, declined, along with a significant reduction in metallothionein-bound zinc. Additionally, treatment of BMDMs with TPEN led to a depletion of cytoplasmic zinc and the induction of the M1 phenotype, with high concentrations of IL-23p19 protein detected in the culture supernatant. Secretion of IL-23 was inhibited by the addition of $ZnSO_4$. As a confirmatory experiment, we induced differentiation using naïve T cells isolated from the spleen, and IL-17A was secreted in response to the BMDM culture supernatant. This induction capability was blocked using an anti-IL-23p19 antibody.

The results thus far indicate that zinc signaling is involved in the M1 differentiation of macrophages, particularly in IL-23 secretion, which subsequently influences T cell IL-17A secretion downstream of IL-23. IL-23, a key molecule in gastrointestinal inflammation, acts upstream of IL-17. It exists as a dimer composed of p19 and p40 subunits and plays a major role in the inflammatory signaling pathway *via* the IL-23 receptor. Although clinical

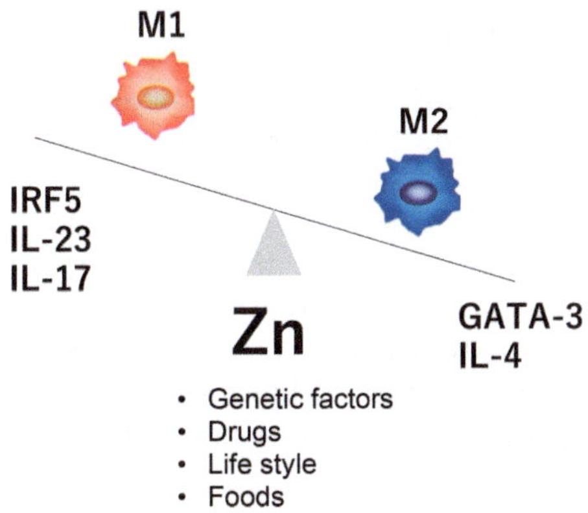

Figure 16.3 Zinc acts as an intracellular signal and controls the phenotypes of macrophages in a transcription factor-dependent manner.

evidence on the role of zinc signaling in the effectiveness of IL-23 antibodies (such as mirikizumab targeting p19 or ustekinumab targeting p40) in IBD is lacking, many patients with IBD present with decreased serum zinc levels. Maintaining adequate blood zinc levels through supplementation could enhance the efficacy of IL-23 antibody therapies, allow for effective results at lower doses, and improve remission maintenance.

As shown by the macrophage phenotype analysis in TNBS colitis, M2 phenotype macrophages were reduced in colitis pathology, with a further decrease observed following TPEN treatment. Zinc signaling has also been reported to play a role in M2 differentiation. Kido *et al.*[11] demonstrated that zinc signaling induces IL-4 secretion *via* the transcription factor GATA-3, contributing to its anti-inflammatory and anti-allergic effects. These findings suggest that intracellular zinc signaling regulates the activation of various transcription factors, which subsequently control macrophage differentiation and modulate Th1 and Th2 cytokine responses (Figure 19.3).

16.5 Conclusion

In this review, we discuss the role of zinc not only as a trace metal nutrient essential for vital functions but also as a regulator of intestinal inflammation by modulating intracellular zinc levels in the context of IBD. However, monitoring of serum zinc levels in clinical practice for IBD remains insufficient. Given the involvement of zinc in IL-23 signaling and the critical role of IL-23 in IBD pathology, it is crucial to monitor serum zinc levels before and during IBD treatment. Oral zinc supplementation should therefore be considered for addressing zinc deficiency. Maintaining adequate serum zinc levels may improve the efficacy of IL-23 antibody therapies or reduce the required dose, potentially enhancing IBD management outcomes.

Conflicts of Interest

YN received scholarship funds from Taiyo Kagaku Co. Ltd, Morinaga Co. Ltd, Miyarisan Pharma Co. Ltd, Morishita-Jintan Co. Ltd, Fujikko Co. Ltd, Mizkan Co. Ltd, a collaboration research fund from Taiyo Kagaku Co., Ltd; and received lecture fees by Takeda Pharma. Co. Ltd, Mochida Pharma. Co. Ltd, Biofermin. Co. Ltd, Otsuka Pharma. Co. Ltd, and Miyarisan Pharma. Co. Ltd. TT received a collaboration research fund from PreMedica Inc. The present research was partly supported by these funds. Neither the funding agency nor any outside organization has participated in the study design or have any competing interests. These companies have approved the final version of the manuscript.

Acknowledgements

This work was partly supported by the MAFF Commissioned project study on "Project for the realization of foods and dietary habits to extend healthy life expectancy" allotted to YN (Grant Number JPJ009842). The authors would like to thank all the participants of the Iwaki Health Promotion Project (COI-NEXT, JST Grant Number JPMJPF2210). We would like to thank Editage (www.editage.jp) for the English language editing.

References

1. S. Siva, D. T. Rubin, G. Gulotta, K. Wroblewski and J. Pekow, *Inflammatory Bowel Dis.*, 2017, **23**, 152.
2. Y. Kobayashi, S. Ohfuji, K. Kondo, W. Fukushima, S. Sasaki and N. Kamata, *et al.*, *J. Gastroenterol. Hepatol.*, 2019, **34**, 1703.
3. H. Kodama, M. Tanaka, Y. Naito, K. Katayama and M. Moriyama, *Int. J. Mol. Sci.*, 2020, **21**, 2941.
4. Y. Higashimura, T. Takagi, Y. Naito, K. Uchiyama, K. Mizushima and M. Tanaka, *et al.*, *J. Crohns Colitis*, 2020, **14**, 856.
5. T. Omatsu, T. Takaga, Y. Nakahata, S. Hayashi and R. Mukai, *et al.*, *J. Clin. Biochem. Nutr.*, 2025, **76**, 50.
6. J. Ishihara, K. Arai, T. Kudo, R. Nambu, H. Tajiri and T. Aomatsu, *et al.*, *Dig. Dis. Sci.*, 2022, **67**, 2485.
7. A. Finamore, M. Massimi, L. Conti Devirgiliis and E. Mengheri, *J. Nutr.*, 2008, **138**, 1664.
8. K. Sakurai, S. Furukawa, T. Katsurada, S. Otagiri, K. Yamanashi and K. Nagashima, *et al.*, *Intest. Res.*, 2022, **20**, 78.
9. T. Tsuji, Y. Naito, T. Takagi, M. Kugai, H. Yoriki and R. Horie, *et al.*, *Int. J. Mol. Med.*, 2013, **31**, 1037.
10. G. J. Guthrie, T. B. Aydemir, C. Troche, A. B. Martin, S. M. Chang and R. J. Cousins, *Am. J. Physiol.: Gastrointest. Liver Physiol.*, 2015, **308**, G171.
11. T. Kido, K. Ishiwata, M. Suka and H. Yanagisawa, *Immunology*, 2019, **156**, 356.

Subject Index